Stem Cell Biology and Regenerative Medicine

Series Editor
Kursad Turksen

More information about this series at http://www.springer.com/series/7896

Alice Pébay • Raymond C.B. Wong
Editors

Lipidomics of Stem Cells

Humana Press

Editors
Alice Pébay
The University of Melbourne
& Centre for Eye Research Australia
Melbourne, VIC, Australia

Raymond C.B. Wong
The University of Melbourne
& Centre for Eye Research Australia
Melbourne, VIC, Australia

ISSN 2196-8985 ISSN 2196-8993 (electronic)
Stem Cell Biology and Regenerative Medicine
ISBN 978-3-319-49342-8 ISBN 978-3-319-49343-5 (eBook)
DOI 10.1007/978-3-319-49343-5

Library of Congress Control Number: 2017932291

Printed on acid-free paper

This Humana Press imprint is published by Springer Nature
The registered company is Springer International Publishing AG
The registered company address is: Gewerbestrasse 11, 6330 Cham, Switzerland

Preface

This volume of *Stem Cell Biology and Regenerative Medicine* aims at covering the current knowledge on the role of lipids in stem cell pluripotency and differentiation. We would like to thank all the authors to this volume who have shared their expertise.

We also wish to thank Dr. Kursad Turksen for his support during the process of compiling this book. Finally, a special thank you goes to Michael Koy for his help during the preparation of the volume.

Melbourne, VIC, Australia Alice Pébay
Melbourne, VIC, Australia Raymond C.B. Wong

Contents

1 **Lysophosphatidic Acid and Sphingosine-1-Phosphate in Pluripotent Stem Cells** ... 1
Grace E. Lidgerwood and Alice Pébay

2 **Morphogenetic Sphingolipids in Stem Cell Differentiation and Embryo Development** ... 11
Guanghu Wang and Erhard Bieberich

3 **Autotaxin in Stem Cell Biology and Neurodevelopment** ... 41
Babette Fuss

4 **Lysophosphatidic Acid (LPA) Signaling in Neurogenesis** ... 65
Whitney S. McDonald and Jerold Chun

5 **Fate Through Fat: Neutral Lipids as Regulators of Neural Stem Cells** ... 87
Laura K. Hamilton and Karl J.L. Fernandes

6 **Cannabinoids as Regulators of Neural Development and Adult Neurogenesis** ... 117
Alline C. Campos, Juan Paraíso-Luna, Manoela V. Fogaça, Francisco S. Guimarães, and Ismael Galve-Roperh

7 **Ceramide-1-Phosphate and Its Role in Trafficking of Normal Stem Cells and Cancer Metastasis** ... 137
Gabriela Schneider and Mariusz Z. Ratajczak

8 **The Emerging Role of Sphingolipids in Cancer Stem Cell Biology** ... 151
Alexander C. Lewis, Jason A. Powell, and Stuart M. Pitson

9 **Lysophosphatidic Acid Signalling Enhances Glioma Stem Cell Properties** ... 171
Wayne Ng

10 New Developments in Free Fatty Acids and Lysophospholipids: Decoding the Role of Phospholipases in Exocytosis 191
Vinod K. Narayana, David Kvaskoff, and Frederic A. Meunier

Index 207

Contributors

Erhard Bieberich Department of Neuroscience and Regenerative Medicine, Medical College of Georgia, Augusta University, Augusta, GA, USA

Alline C. Campos Department of Pharmacology, Medical School of Ribeirão Preto, University of São Paulo, Ribeirão Preto, SP, Brazil

Jerold Chun Sanford Burnham Prebys Medical Discovery Institute, La Jolla, CA, USA

Karl J.L. Fernandes Research Center of the University of Montreal Hospital (CRCHUM), Tour Viger, Montreal, QC, Canada

Department of Neurosciences, Faculty of Medicine, Université de Montréal, Montréal, QC, Canada

Manoela V. Fogaça Department of Pharmacology, Medical School of Ribeirão Preto, University of São Paulo, Ribeirão Preto, SP, Brazil

Babette Fuss Department of Anatomy and Neurobiology, Virginia Commonwealth University School of Medicine, Richmond, VA, USA

Ismael Galve-Roperh Department of Biochemistry and Molecular Biology I, School of Biology, Complutense University, and Neurochemistry Universitary Research Institute, Madrid, Spain

CIBERNED, Center for Networked Biomedical Research in Neurodegenerative Diseases, Madrid, Spain

Francisco S. Guimarães Department of Pharmacology, Medical School of Ribeirão Preto, University of São Paulo, Ribeirão Preto, SP, Brazil

Laura K. Hamilton Research Center of the University of Montreal Hospital (CRCHUM), Tour Viger, Montreal, QC, Canada

Department of Neurosciences, Faculty of Medicine, Université de Montréal, Montréal, QC, Canada

David Kvaskoff Clem Jones Centre for Ageing Dementia Research, Queensland Brain Institute, The University of Queensland, Brisbane, QLD, Australia

Heidelberg University Biochemistry Centre, Heidelberg, Germany

Alexander C. Lewis Centre for Cancer Biology, University of South Australia and SA Pathology, Adelaide, SA, Australia

Grace E. Lidgerwood Centre for Eye Research Australia, Royal Victorian Eye and Ear Hospital, The University of Melbourne, Melbourne, VIC, Australia

Ophthalmology, Department of Surgery, The University of Melbourne, Melbourne, Australia

Whitney S. Mcdonald Sanford Burnham Prebys Medical Discovery Institute, La Jolla, CA, USA

Frederic A. Meunier Clem Jones Centre for Ageing Dementia Research, Queensland Brain Institute, The University of Queensland, Brisbane, QLD, Australia

Vinod K. Narayana Clem Jones Centre for Ageing Dementia Research, Queensland Brain Institute, The University of Queensland, Brisbane, QLD, Australia

Wayne Ng University of Melbourne, Parkville, VIC, Australia

Department of Surgery, Centre for Medical Research, Royal Melbourne Hospital, Parkville, VIC, Australia

Melbourne Brain Centre at Royal Melbourne Hospital, Parkville, VIC, Australia

Juan Paraíso-Luna Department of Biochemistry and Molecular Biology I, School of Biology, Complutense University, and Neurochemistry Universitary Research Institute, Madrid, Spain

CIBERNED, Center for Networked Biomedical Research in Neurodegenerative Diseases, Madrid, Spain

Alice Pébay Centre for Eye Research Australia, Royal Victorian Eye and Ear Hospital, The University of Melbourne, Melbourne, Australia

Ophthalmology, Department of Surgery, The University of Melbourne, Melbourne, Australia

Stuart M. Pitson Centre for Cancer Biology, University of South Australia and SA Pathology, Adelaide, SA, Australia

Jason A. Powell Centre for Cancer Biology, University of South Australia and SA Pathology, Adelaide, SA, Australia

Mariusz Z. Ratajczak Stem Cell Institute at the James Graham Brown Cancer Center, University of Louisville, Louisville, KY, USA

Department of Regenerative Medicine, Warsaw Medical University, Warsaw, Poland

Gabriela Schneider Stem Cell Institute at the James Graham Brown Cancer Center, University of Louisville, Louisville, KY, USA

Guanghu Wang Department of Neuroscience and Regenerative Medicine, Medical College of Georgia, Augusta University, Augusta, GA, USA

About the Editors

Alice Pébay, Ph.D., is a Principal Research Fellow at the University of Melbourne and the principal investigator of the Neuroregeneration Unit at the Centre for Eye Research Australia. She holds a Ph.D. in neuroscience and has extensive expertise in cell biology, having published more than 50 peer-reviewed articles and chapters in the field of stem cell biology and lysophospholipid biology. Assoc. Prof. Pébay's current research focuses on the role of lysophospholipids in human pluripotent stem cell pluripotency and differentiation towards retinal lineages, and on modeling human diseases using patient specific induced pluripotent stem cells. Assoc. Prof. Pébay also has a key interest in the role of lysophopshatidic acid in neurotrauma and in the cellular mechanisms involved in the genetic disease, Friedreich's Ataxia.

Raymond C.B. Wong, Ph.D., is a Senior Research Fellow at the University of Melbourne and the principal investigator of the Cellular Reprogramming Unit at the Centre for Eye Research Australia. He is a stem cell biologist specialising in human pluripotent stem cells and reprogramming. He completed his Ph.D. in stem cell biology (Monash University) and overseas postdoctoral training in the University of California, Irvine, and subsequently National Institutes of Health. His previous research in the past 13 years has led to improvement in methods of growing and generating human pluripotent stem cells. Dr. Wong's current research focuses on developing methods to turn human pluripotent stem cells into retinal cells, as well as utilizing human stem cells to model and study pathological progression of various retinal diseases to improve treatment options.

Chapter 1
Lysophosphatidic Acid and Sphingosine-1-Phosphate in Pluripotent Stem Cells

Grace E. Lidgerwood and Alice Pébay

Abbreviations

ABC	ATP-binding cassette
ATX	Autotaxin
ENNP2	Ectonucleotide pyrophosphatase phosphodiesterase 2
ERK	Extracellular signal-regulated kinase
HDAC	Histone deacetylase
hESC	Human embryonic stem cell
iPSC	Induced pluripotent stem cell
JNK	c-jun N-terminal kinase
LPA	Lysphosphatidic acid
MAP	Mitogen-activated protein
mESC	Mouse embryonic stem cell
PDGF	Platelet-derived growth factor
PI3K	Phosphoinositide 3-kinase
PPAR	Peroxisome proliferator-activated receptor
S1P	Sphingosine-1-phosphate
SPhK	Sphingosine kinase
TRAF2	TNF receptor-associated factor 2
VEGF	Vascular endothelial growth factor

G.E. Lidgerwood • A. Pébay (✉)
Centre for Eye Research Australia, Royal Victorian Eye and Ear Hospital, The University of Melbourne, Melbourne, Australia

Ophthalmology, Department of Surgery, The University of Melbourne, Melbourne, Australia
e-mail: Lidgerwood.g@unimelb.edu.au; apebay@unimelb.edu.au

A. Pébay, R.C.B. Wong (eds.), *Lipidomics of Stem Cells*, Stem Cell Biology and Regenerative Medicine, DOI 10.1007/978-3-319-49343-5_1

1.1 Introduction

Lipidomics refers to the analysis of lipids in cells, tissues, or organisms. Lipids are one of the main classes of biomolecules necessary to life, yet are probably the least understood and studied biomolecules. It is estimated that there are between 9,000 and 100,000 different lipid species [1, 2]. This massive range reflects how little is known about this class of biomolecules. Few techniques are currently available to the study of lipids, and it is very difficult to isolate and analyze lipids, explaining why lipidomics somehow lags behind the study of other biomolecules. Lipids are the major compounds of the biological membranes that serve as the physical barrier, giving structural integrity to the cell and its components. They also play an important metabolic function in terms of energy storage. Lipids are also integral to membrane trafficking and can be found in vesicles such as exosomes. Lipids with cell signaling functions are often referred to as bioactive lipids, as opposed to lipids that form the structural composition of cell membranes or those used for energy, and have an array of biological functions, including mediating inflammation; regulating cell growth and polarity; and determining cell fate decisions. This essential signaling feature of bioactive lipids occurs in a variety of different pathways; lipids can engage with specific receptors to activate a cascade of downstream signaling pathways, or through indirect means, i.e., via membrane trafficking or as lipid rafts.

1.2 Lipid Homeostasis in Stem Cell Biology

A stem cell is a cell that is capable of self-renewing by undergoing indefinite symmetrical cell divisions, giving rise to daughter cells that are genetically identical to the original parent cell. Under the right conditions, stem cells can also differentiate into specialized cells that have specific functions in the body. Adult stem cells are generally of multipotent potential, meaning they are capable of differentiating into restricted lineages. Pluripotent stem cells, on the other hand, are capable of giving rise to all cell types of the body. There are two main sources of pluripotent stem cells: embryonic stem cells (ESCs), which are derived from the inner cell mass of a preimplantation blastocyst; and induced pluripotent stem cells (iPSCs), somatic cells that have been reprogrammed into a pluripotent state, and exhibit functional similarities to ESCs. Pluripotency is maintained by the expression of particular genes, which is intricately controlled by the homeostasis of a range of regulatory signaling molecules and epigenetic factors. Subtle changes in cellular conditions ultimately determine the fate of pluripotent stem cells. Historically, scientists have focused on the role of signaling proteins and genetic factors in the maintenance of pluripotency; however, more recently, signaling lipids have surfaced as potential regulators of stem cell maintenance and differentiation.

Lipid homeostasis is fundamental to development and cellular homeostasis, and lipid dysregulations can lead to developmental abnormalities as well as neurodegeneration [3–5]. It is likely that changes in the lipidomic signature of a cell from pluripotency to differentiation will reflect a change in substrate availability during these events and may also give rise to a predictive model of differentiation and maturity. For instance, there is evidence that specific lipids play fundamental roles in neural development [6–8] but less is known about the general profile of lipids in pluripotency and upon differentiation. There are in fact a limited number of large lipidomic studies that have been performed within the stem cell field. Nonetheless, there is some suggestion that depending on their level of pluripotency or differentiation, cells will show a different distribution of heterogenous lipids [9]. Further, the lipidome is also modified upon mouse ESC differentiation [10]. Interestingly, Wang et al. [11] demonstrated in a landmark publication that in *C. elegans*, germ line stem cell longevity is regulated by an active control of lipid metabolism [11]. Further, the lipidomic profiling of mouse retinal stem cells identified a distinct glycerophospholipid signature, which when altered, participates in the regulation of proliferation or differentiation [12]. Similarly, the peroxisome proliferator-activated receptor (PPAR) pathway acts as a metabolic switch to control hematopoietic stem cell maintenance or differentiation, by regulating the oxidation of fatty acids [13], thus suggesting a direct impact of lipids on cell fate. Human iPSCs are composed of less saturated fatty acids than human (h)ESCs, which may indicate metabolic differences in these two cell types [14]. This exemplifies how lipid homeostasis is most likely fundamental to pluripotency and differentiation.

1.3 LPA and S1P Synthesis and Degradation

Lysophospholipids are bioactive lipids consisting of one *O*-acyl chain, generated by the hydrolytic cleavage of fatty acids from glycerophospholipids by phospholipases. Two main categories of lipids form lysophospholipids: those derived from glycerol, glycerophospholipids (including LPA) and those with a sphingomyelin backbone, sphingolipids (including S1P). Both these classes of lipids play an integral role in cell fate, including in regulating pluripotency and differentiation of various types of stem cells. LPA and S1P are the most characterized lysophospholipids in terms of effects in pluripotent stem cells, and will thus be the major focus of this chapter.

1.3.1 LPA

LPA can be synthesized and degraded through a variety of pathways [8, 15]. Autotaxin/ectonucleotide pyrophosphatase phosphodiesterase 2 (ENNP2) is responsible for most of the production of extracellular LPA. This secreted enzyme has a lysophospholipase D domain able to cleave lysophospholipids, in particular lysophosphatidylcholine, into LPA. Other enzymes can also generate extracellular LPA: secreted phospholipases A1 and A2, which can deacylate phosphatidic acid. Intracellular LPA, on the other hand, can be generated by other enzymatic pathways that include activities of intracellular phospholipases A1 and A2; glycerol 3-phosphate acyltransferase, which acylates glycerol 3-phosphate; or monoacylglycerol kinase, which phosphorylates monoacylglycerol. LPA degradation is then mediated by lipid phosphate phosphatases 1–3, which dephosphorylates LPA to monoacylglycerol.

1.3.2 S1P

Sphingolipids are acyl (fatty acid) derivatives of the amino alcohol, sphingosine, and encompass a range of bioactive lipids, including S1P. In contrast to LPA synthesis, S1P can only be generated by one pathway, involving the phosphorylation of sphingosine by sphingosine kinases (SphK) 1 and 2. S1P can then be degraded by S1P lyase, or dephosphorylated into sphingosine by S1P phosphatases and nonspecific lipid phosphatases, or converted to ceramide by ceramide synthase [15, 16]. S1P is synthesized intracellularly and thus needs to be excreted in order to act as an extracellular ligand. This export is likely to occur through ATP-binding cassette (ABC) transporters [16]. S1P is also present in the nucleus and in the mitochondria, where it is synthesized by SphK2 [17, 18].

1.4 LPA and S1P Signaling

LPA and S1P act extracellularly mainly through the binding to their specific G protein-coupled receptors. There are currently six confirmed LPA receptors (LPA_{1-6}) and five S1P receptors ($S1P_{1-5}$) [19]. Other extracellular receptors have been implicated as LPA receptors, including the purinergic receptors P_2Y_5 and P_2Y_{10}, GPR87 and the TRPV1 channel [8]. LPA and S1P receptors are known to act through G_q and $G_{12/13}$, G_i and potentially G_s, to modulate multiple signaling pathways including: stimulation of phospholipase C/protein kinase C and modification in intracellular calcium concentration; stimulation of the phosphoinositide 3-kinase (PI3K)/AKT pathway; stimulation of Ras/mitogen-activated protein (MAP) kinase pathways including of extracellular signal-regulated kinases (ERK) 1/2; inhibition and potential stimulation of adenylate cyclase pathways; activation of small G proteins and subsequent stimulation of the Rho/ROCK pathway; and activation of phospholipases A_2 and D [19].

Both LPA and S1P can thus act as extracellular mediators by binding their cellular membrane receptors, but they can also act as intracellular receptors. Some research indeed suggests that the nuclear receptor PPARγ can also bind LPA [8]. As for S1P, it is now clearly demonstrated that it is an intracellular nuclear mediator, with direct interaction with key molecules that are not S1P receptors [20].

Intracellularly, S1P is known to counteract the pro-apoptotic effects of ceramide, contributing to the S1P-ceramide rheostat [20]. Intracellular S1P has also been shown to modulate NF-κB signaling by interacting with protein kinase Cδ and TNF receptor-associated factor 2 (TRAF2) [20]. It can also directly interact with events controlling mitochondrial respiration [18]. Finally, within the nucleus, S1P has been shown to bind and inhibit histone deacetylases (HDACs) 1/2, which most likely has consequences on gene regulation and epigenetics [17]. This could be particularly relevant to pluripotency.

Given the complexity of LPA and S1P signaling, it is not surprising that these molecules induce pleiotropic biological effects in different cells, including stem cells [21, 22].

1.5 Role of LPA and S1P in Pluripotent Stem Cells

LPA and S1P have been implicated in events regulating survival, autophagy, apoptosis, proliferation, differentiation, cytoskeleton rearrangements, polarity, and migration. Lysophospholipids also control events of pluripotency and differentiation in both adult and embryonic stem cells and in various species (as reviewed in [8, 15, 23, 24]). Both mouse and human pluripotent stem cells express LPA and S1P receptors, with some variations. Mouse ESCs express $LPA_{1,2,3}$ [25] and $S1P_{1-5}$ [26–29] although $S1P_4$ expression depends on the mESC lines [30]. Human ESCs and iPSCs express LPA_{1-5} and $S1P_{1-5}$ [23, 31–33] with some expression variation depending in cell lines, as observed with mouse ESCs. These differences could be artifacts of cell culture methods. Although unlikely - given the redundancy in signaling pathways modulated by the various receptors - the difference in receptor expression between pluripotent stem cell lines might indicate some potential variation in these bioactive lipids' cellular effects.

Both LPA [25] and S1P [29, 30] stimulate proliferation of mESCs. LPA's effect is dependent on the activation of the phospholipase C pathway, leading to modifications of intracellular calcium concentration, itself inducing expression of the early gene *c-myc* and subsequent proliferation [25]. LPA also induces Erk phosphorylation and downstream c-fos activation in the pluripotent stem cells [34]. Given the role of c-myc in pluripotency and reprogramming of somatic cells into iPSCs [35], it is interesting to note that LPA is able to induce its expression in ESCs. Likewise, S1P stimulates mESC proliferation, at least through its receptor-mediated activation of the Erk pathway [29, 30]. Other pathways might intervene. In particular, Ryu et al. [29] suggest that S1P promotes mESC proliferation by the $S1P_{1/3}$-induced transactivation of the vascular endothelial growth factor (VEGF) receptor, Flk-1, and subsequent phosphorylation of Jnk and Erk [29]. Together with the demonstration that S1P induces VEGF expression in mESCs [29], this data suggests an important interaction between S1P and VEGF in mESC pluripotency. Finally, the knocking down of S1P lyase in mESCs is accompanied by a large increase in S1P levels, increased proliferation and expression of the mouse pluripotency markers sse4 and oct4, as well as an increase in stat3 signaling, all suggestive that endogenous S1P metabolism is highly regulated in mESCs and is key to pluripotency [28].

In hESCs, we reported that we did not observed an effect of LPA alone (up to 10 μM) on their maintenance [31], which was similarly reported by others using a different culture medium [36]. LPA has however been described as blocking Wnt pro-differentiation effects in hESCs [36]. Of note, it was also described that low concentrations of LPA (up to 100 nM) slightly increases the number of pluripotent cells in conditions favoring differentiation (mTeSR without basic fibroblast growth factor), while 1 μM induces death of hESCs [32]. This data is at odds with the previous reports, which could be partially explained by the fact that LPA was reconstituted and prepared in water in place of solvents (generally chloroform or ethanol/water) necessary for LPA solubilization. Together, these data suggest that LPA may be important for the maintenance of pluripotency, most likely as a "counter actor," an anti-differentiation agent, rather than a direct pro-pluripotency factor.

Recently, LPA was shown to modulate the Hippo pathway in both hESCs and human iPSCs, by activating YAP/TAZ [37, 38]. This is interesting in terms of pluripotency and differentiation, as the Hippo pathway is fundamental to development and is key to stem cell pluripotency and differentiation (for review of the pathway, see [39]). Indeed, when active, the YAP/TAZ transcriptions factors would be involved in self-renewal of hESCs and iPSCs, while inactivation of the pathway was shown to be linked to differentiation [37]. Interestingly, the activation of YAP by LPA results in the stimulation of a naïve state in hESCs and human iPSCs [38], allowing the generation of transgene-free human naïve pluripotent stem cells, clearly indicative of a fundamental role of LPA in human pluripotency.

On the other hand, S1P, in combination with platelet-derived growth factor (PDGF), was shown to maintain hESCs undifferentiated, in G_i-, ERK-, and SphK-dependent mechanisms [31]. This maintenance of pluripotency was observed with cells cultivated on feeder and feeder-free, and in the absence of serum, thus demonstrating a direct effect of S1P on hESCs. Interestingly, S1P alone was not able to maintain hESCs undifferentiated, and PDGF was shown to stimulate SphK, thus allowing the generation of intracellular S1P [31]. It is thus feasible that the presence of both extracellular S1P- and intracellular S1P-mediated effects contribute to the maintenance of pluripotency and further work to clarify this point would be interesting. S1P was also shown to be anti-apoptotic in hESCs, through the phosphorylation of ERK 1/2, but independent of the PI3K pathway [40]. S1P can also induce the phosphorylation of p38 and to a lesser extent of c-jun N-terminal kinases (JNK) in hESCs, but the significance of these activated pathways remains to be established [23]. Finally, S1P does not induce intracellular calcium modification, suggesting that the phospholipase C pathway is not essential to hESC pluripotency and survival [40]. This pro-survival effect of S1P was also observed by an increased expression of anti-apoptotic genes and cell cycle-related genes, and a down-regulation of pro-apoptotic genes [41].

Little is known on the basal levels of LPA and S1P in pluripotent stem cells. High performance liquid chromatography—mass spectrometry revealed that many sphingolipid intermediates are present in hESCs, in particular ceramide and low levels of intracellular S1P [42]. It was recently demonstrated that hESCs and human iPSCs express cilia that are regulated by the ceramide/sphingomyelinase pathway [43]. Given the close relationship between ceramide and S1P, it is possible that intracellular S1P might also be involved in ciliogenesis, a fundamental process of developing cells.

1.6 Discussion and Conclusion

Little is known of the role of lipids, their interactions, catabolism, metabolism and how these modulate many diverse biological processes, including in stem cells. The world of lipids is complex, in terms of functions, diversity, and numbers, and is probably the least understood "-ome" of biology. With today's technology and given the extremely large numbers of lipids per cell, it is still not possible to assess the entire lipidome of a cell. However, lipidomics is now emerging because tools and strategies used for genomics and proteomics are being applied to the study of lipids. For instance, high performance liquid chromatography, electrospray ionization mass spectrometry, coupled with bioinformatic analysis will allow for large-scale system-level analysis of lipids and pathways involved [44]. These techniques might help answer important questions, such as: are there modifications in the lipidome of cells upon cellular fate? If so, are these a consequence of the cellular transition or are they a driving force behind change?

In terms of signaling lipids, it is clear that these play fundamental role in stem cell biology. In particular, LPA and S1P modulate various effects in various stem cells, both pluripotent and multipotent (as reviewed in [15]). In pluripotent stem cells, there seems to be some difference in effects of LPA and S1P between species, but it is clear that these molecules positively influence pluripotency and survival. A further understanding of the role played by intracellular S1P in pluripotency, epigenetics, and on the Hippo pathway would most likely be very informative. Likewise, a clearer picture of the interactions between LPA and Wnt signaling in pluripotent stem cells and upon differentiation would provide new knowledge in our understanding of the complexity of lysolipid signaling in pluripotency.

Acknowledgements This work was supported by an Australian Postgraduate Award Scholarship (GL), an Australian Research Council (ARC) Future Fellowship (AP, FT140100047), the University of Melbourne and Operational Infrastructure Support from the Victorian Government.

References

1. van Meer G (2005) Cellular lipidomics. EMBO J 24(18):3159–3165
2. Yetukuri L, Ekroos K, Vidal-Puig A, Oresic M (2008) Informatics and computational strategies for the study of lipids. Mol Biosyst 4(2):121–127
3. Fyrst H, Saba JD (2010) An update on sphingosine-1-phosphate and other sphingolipid mediators. Nat Chem Biol 6(7):489–497

4. Piomelli D, Astarita G, Rapaka R (2007) A neuroscientist's guide to lipidomics. Nat Rev Neurosci 8(10):743–754
5. Wymann MP, Schneiter R (2008) Lipid signalling in disease. Nat Rev Mol Cell Biol 9(2):162–176
6. Kraut R (2011) Roles of sphingolipids in Drosophila development and disease. J Neurochem 116(5):764–778
7. Rotstein NP, Miranda GE, Abrahan CE, German OL (2010) Regulating survival and development in the retina: key roles for simple sphingolipids. J Lipid Res 51(6):1247–1262
8. Frisca F, Sabbadini RA, Goldshmit Y, Pebay A (2012) Biological effects of lysophosphatidic acid in the nervous system. Int Rev Cell Mol Biol 296:273–322
9. Martin MC et al (2013) 3D spectral imaging with synchrotron Fourier transform infrared spectro-microtomography. Nat Methods 10(9):861–864
10. Park H et al (2010) Transcript profiling and lipidomic analysis of ceramide subspecies in mouse embryonic stem cells and embryoid bodies. J Lipid Res 51(3):480–489
11. Wang MC, O'Rourke EJ, Ruvkun G (2008) Fat metabolism links germline stem cells and longevity in *C. elegans*. Science 322(5903):957–960
12. Li J, Cui Z, Zhao S, Sidman RL (2007) Unique glycerophospholipid signature in retinal stem cells correlates with enzymatic functions of diverse long-chain acyl-CoA synthetases. Stem Cells 25(11):2864–2873
13. Ito K et al (2012) A PML-PPAR-delta pathway for fatty acid oxidation regulates hematopoietic stem cell maintenance. Nat Med 18(9):1350–1358
14. Panopoulos AD et al (2012) The metabolome of induced pluripotent stem cells reveals metabolic changes occurring in somatic cell reprogramming. Cell Res 22(1):168–177
15. Pebay A, Bonder CS, Pitson SM (2007) Stem cell regulation by lysophospholipids. Prostaglandins Other Lipid Mediat 84(3–4):83–97
16. Olivera A, Allende ML, Proia RL (2013) Shaping the landscape: metabolic regulation of S1P gradients. Biochim Biophys Acta 1831(1):193–202
17. Hait NC et al (2009) Regulation of histone acetylation in the nucleus by sphingosine-1-phosphate. Science 325(5945):1254–1257
18. Strub GM et al (2011) Sphingosine-1-phosphate produced by sphingosine kinase 2 in mitochondria interacts with prohibitin 2 to regulate complex IV assembly and respiration. FASEB J 25(2):600–612
19. Kihara Y, Maceyka M, Spiegel S, Chun J (2014) Lysophospholipid receptor nomenclature review: IUPHAR Review 8. Br J Pharmacol 171(15):3575–3594
20. Maceyka M, Harikumar KB, Milstien S, Spiegel S (2012) Sphingosine-1-phosphate signaling and its role in disease. Trends Cell Biol 22(1):50–60
21. Bieberich E (2012) It's a lipid's world: bioactive lipid metabolism and signaling in neural stem cell differentiation. Neurochem Res 37(6):1208–1229
22. Ghasemi R, Dargahi L, Ahmadiani A (2016) Integrated sphingosine-1 phosphate signaling in the central nervous system: from physiological equilibrium to pathological damage. Pharmacol Res 104:156–164
23. Pitson SM, Pebay A (2009) Regulation of stem cell pluripotency and neural differentiation by lysophospholipids. Neurosignals 17(4):242–254
24. Kobayashi NR, Hawes SM, Crook JM, Pebay A (2010) G-protein coupled receptors in stem cell self-renewal and differentiation. Stem Cell Rev 6(3):351–366
25. Todorova MG, Fuentes E, Soria B, Nadal A, Quesada I (2009) Lysophosphatidic acid induces Ca2+ mobilization and c-Myc expression in mouse embryonic stem cells via the phospholipase C pathway. Cell Signal 21(4):523–528
26. Kleger A et al (2007) The bioactive lipid sphingosylphosphorylcholine induces differentiation of mouse embryonic stem cells and human promyelocytic leukaemia cells. Cell Signal 19(2):367–377
27. Lee CW, Rivera R, Gardell S, Dubin AE, Chun J (2006) GPR92 as a new G12/13- and Gq-coupled lysophosphatidic acid receptor that increases cAMP, LPA5. J Biol Chem 281(33):23589–23597

28. Smith GS, Kumar A, Saba JD (2013) Sphingosine phosphate lyase regulates murine embryonic stem cell proliferation and pluripotency through an S1P(2)/STAT3 signaling pathway. Biomolecules 3(3):351–368
29. Ryu JM et al (2014) Sphingosine-1-phosphate-induced Flk-1 transactivation stimulates mouse embryonic stem cell proliferation through S1P1/S1P3-dependent β-arrestin/c-Src pathways. Stem Cell Res 12(1):69–85
30. Rodgers A et al (2009) Sphingosine 1-phosphate regulation of extracellular signal-regulated kinase-1/2 in embryonic stem cells. Stem Cells Dev 18(9):1319–1330
31. Pebay A et al (2005) Essential roles of sphingosine-1-phosphate and platelet-derived growth factor in the maintenance of human embryonic stem cells. Stem Cells 23(10):1541–1548
32. Ermakov A et al (2012) A role for intracellular calcium downstream of G-protein signaling in undifferentiated human embryonic stem cell culture. Stem Cell Res 9(3):171–184
33. Kleger A, Liebau S, Lin Q, von Wichert G, Seufferlein T (2011) The impact of bioactive lipids on cardiovascular development. Stem Cells Int 2011:916180
34. Schuck S, Soloaga A, Schratt G, Arthur JS, Nordheim A (2003) The kinase MSK1 is required for induction of c-fos by lysophosphatidic acid in mouse embryonic stem cells. BMC Mol Biol 4:6
35. Takahashi K et al (2007) Induction of pluripotent stem cells from adult human fibroblasts by defined factors. Cell 131(5):861–872
36. Blauwkamp TA, Nigam S, Ardehali R, Weissman IL, Nusse R (2012) Endogenous Wnt signalling in human embryonic stem cells generates an equilibrium of distinct lineage-specified progenitors. Nat Commun 3:1070
37. Hsiao C et al (2016) Human pluripotent stem cell culture density modulates YAP signaling. Biotechnol J 11((5)):662–675
38. Qin H et al (2016) YAP induces human naive pluripotency. Cell Rep 14(10):2301–2312
39. Yu F-X, Guan K-L (2013) The Hippo pathway: regulators and regulations. Genes Dev 27(4):355–371
40. Wong RC, Tellis I, Jamshidi P, Pera M, Pebay A (2007) Anti-apoptotic effect of sphingosine-1-phosphate and platelet-derived growth factor in human embryonic stem cells. Stem Cells Dev 16(6):989–1001
41. Avery K, Avery S, Shepherd J, Heath PR, Moore H (2008) Sphingosine-1-phosphate mediates transcriptional regulation of key targets associated with survival, proliferation, and pluripotency in human embryonic stem cells. Stem Cells Dev 17(6):1195–1205
42. Brimble SN et al (2007) The cell surface glycosphingolipids SSEA-3 and SSEA-4 are not essential for human ESC pluripotency. Stem Cells 25(1):54–62
43. He Q et al (2014) Primary cilia in stem cells and neural progenitors are regulated by neutral sphingomyelinase 2 and ceramide. Mol Biol Cell 25(11):1715–1729
44. Wenk MR (2010) Lipidomics: new tools and applications. Cell 143(6):888–895

Chapter 2
Morphogenetic Sphingolipids in Stem Cell Differentiation and Embryo Development

Guanghu Wang and Erhard Bieberich

Abbreviations

Akt	AK strain transforming (Akt kinase)
aPKC	Atypical PKC
C1P	Ceramide-1-phosphate
CECs	Ceramide-enriched compartments
EGF	Endothelial growth factor
ERK	Extracellular regulated kinase
ES cell	Embryonic stem cell
EV	Extracellular vesicle
FB1	Fumonisin B1
FGF-2	Fibroblast growth factor 2
FTY720	Fingolimod
GPCR	G protein-coupled receptor
Grp94	Glucose-regulated protein 94
GSK3	Glycogen synthase kinase 3
GSLs	Glycosphingolipids
HDAC	Histone deacetylase
hESC	Human ES cell
HSP90	Heat shock protein 90
iPSC	Induced pluripotent stem cell
Jak	Janus kinase
LIF	Leukemia inhibitory factor
MAPK	Mitogen-activated protein kinase
mESC	Mouse (murine) ES cell

G. Wang • E. Bieberich, Ph.D. (✉)
Department of Neuroscience and Regenerative Medicine, Medical College of Georgia, Augusta University, 1120 15th Street Room CA4012, Augusta, GA 30912, USA
e-mail: ebieberich@augusta.edu

A. Pébay, R.C.B. Wong (eds.), *Lipidomics of Stem Cells*, Stem Cell Biology and Regenerative Medicine, DOI 10.1007/978-3-319-49343-5_2

NPC	Neural precursor cell
nSMase	Neutral sphingomyelinase
OPC	Oligodendrocyte precursor cells
PAR-4	Prostate apoptosis response 4
PDGF	Platelet-derived growth factor
PDMP	*N*-[2-hydroxy-1-(4-morpholinylmethyl)-2-phenylethyl]-decanamide
PHB2	Prohibitin 2
PI3K	Phosphatidyl inositol 3 kinase
PIP	Phosphatidyl inositol phosphate
PKC	Protein kinase C
PLC	Phospholipase C
PP2a	Protein phosphatase 2a
S18	*N*-oleoyl serinol
S1P	Sphingosine-1-phosphate
Shh	Sonic hedgehog
SphK	Sphingosine kinase
SPL	S1P lyase
Spns2	Spinster homolog 2
Stat3	Signal transducer and activator of transcription 3
Wnt	Wingless type MMTV

2.1 Ceramide and Its Derivatives

In this section, we will focus on the function of ceramide and derivatives known to regulate stem cell differentiation, namely, sphingosine-1-phosphate (S1P), ceramide-1-phosphate (C1P), and glycosphingolipids (GSLs) (Fig. 2.1). We will not discuss sphingolipid metabolism or the function of sphingolipids in general cell-signaling pathways. There are excellent reviews and the reader is encouraged to attend to these resources [1, 2]. Instead, we will highlight most recent studies showing the function of sphingolipids in cell-signaling pathways critical for regulation of cell polarity and morphogenesis as part of the stem cell differentiation program.

2.1.1 Ceramide and Ceramide-Enriched Compartments

A morphogenetic lipid will induce a specific stem cell differentiation program and regulate embryo development and morphogenesis. We have proposed that ceramide is such a morphogenetic lipid based on the observation that it is critical for the apicobasal patterning of the primitive ectoderm in embryonic stem (ES) cell-derived embryoid bodies and for promoting neural differentiation [2–6]. Compartmentalization into ceramide-enriched compartments, CECs, allows for localized metabolic release

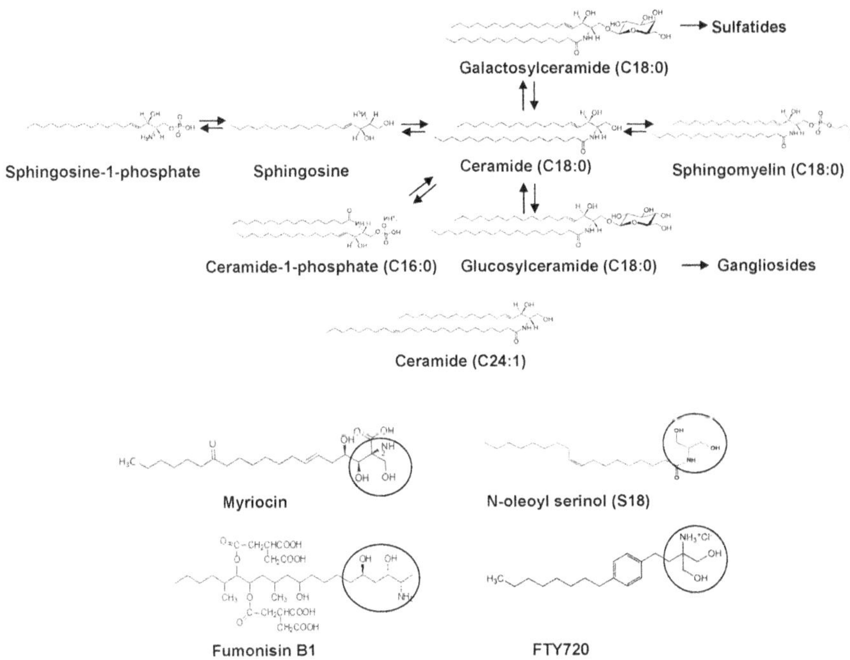

Fig. 2.1. Structure and metabolism of morphogenetic sphingolipids and effectors/analogs. Ceramide is a metabolic hub for the generation of morphogenetic sphingolipids. Myriocin is a serine palmitoyltransferase (SPT) inhibitor. Note the structural difference between C18:0 ceramide (*N*-oleoyl sphingosine) and C24:1 ceramide (*N*-nervonoyl sphingosine). Fumonisin B1 (FB1) is a ceramide synthase inhibitor. FTY720 (fingolimod) is an S1P pro-drug analog. *N*-oleoyl serinol (S18) is a soluble ceramide analog developed in our laboratory. The two β-hydroxy methyl groups (*circled*) of the polar, serine-derived head group are a common structural motif of all ceramide analogs and many other effectors of sphingolipid metabolism

of ceramide derivatives such as ceramide-1-phosphate (C1P, Fig. 2.1) or sphingosine-1-phosphate (S1P, Fig. 2.1), and formation of local sphingolipid-protein complexes that regulate cell polarity. Several years ago, we have termed these hypothetical complexes "sphingolipid-induced protein scaffolds" or SLIPs and proposed their critical function for remodeling of the cytoskeleton and distribution of cell polarity proteins [7]. Recent studies in our and other laboratories support this hypothesis and open the possibility to engineer morphogenesis by changing the composition and compartmentalization of sphingolipids in stem cells.

Our studies and those from other laboratories have demonstrated that sphingolipids including ceramide are organized in lipid microdomains or rafts and CECs [2, 8–18]. In addition, various lipids are distributed in a gradient with cholesterol and sphingomyelin enriched in the cell membrane, while ceramide appears to be enriched in the endosomal compartment [19–21]. Based on these observations, we hypothesize that the lateral anisotropy of sphingolipids leads to raft formation (*X*-axis in Fig. 2.2), which is integrated with a lipid gradient orthogonal to the mem-

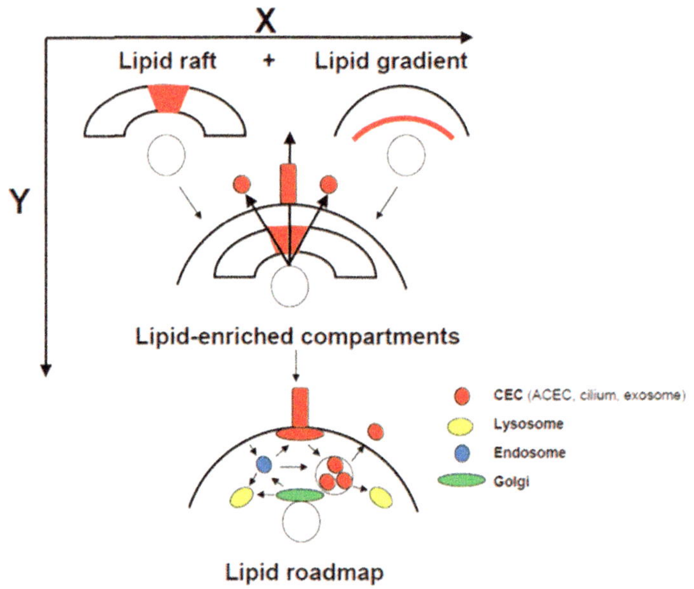

Fig. 2.2. Lipid road map in vesicle trafficking and compartment identity. Integration of lateral membrane anisotropy (lipid rafts or microdomains, here along *X*-axis) with orthogonal lipid gradients (anterograde and retrograde trafficking pathways, here along *Y*-axis) generates a map of vesicles and compartments with distinct lipid composition critical for cell polarity and morphogenesis

brane (*Y*-axis in Fig. 2.2). This integration leads to compartmentalization that regulates intracellular vesicle traffic and polarity similar to a road map directing car traffic (Fig. 2.2, bottom panel). Previous studies noted that sphingolipids are sorted into specific vesicle populations and enriched along distinct trafficking pathways [22–27]. The vesicular identity is even preserved during mitosis when many compartments such as the Golgi apparatus and the nuclear envelope are disintegrated into a myriad of vesicles and yet reassemble in the daughter cells to their original organelles. While only little is known about the sorting mechanisms that direct distinct sphingolipid trafficking pathways toward specific lipid-enriched compartments (including CECs) when exported from the Golgi apparatus/trans-Golgi network or internalized by endocytosis [25, 28–30], one may speculate that they are intimately connected to our model of a lipid road map guiding establishment of cell polarity and ultimately, asymmetric division of progenitor cells and embryo morphogenesis. Our group has shown that two distinct compartments, cilia and exosomes, are enriched with ceramide and directly linked to cell polarity in differentiating stem cells and secretion of growth factors. Formation of these CECs is stimulated by exogenously added ceramide or compromised by inhibitors of enzymes that generate ceramide. Ceramide is enriched at the base and in the membrane of cilia, a cell compartment with sensory and motility functions [4, 11, 31]. It is also enriched in exosomes, lipid vesicles generated in the endosomal compartment and then secreted to transfer cell signaling and growth factors between cells [32].

2.1.1.1 Ceramide and Cilia

Primary cilia are important for stem cell differentiation because they are endowed with growth factor receptors controlling sonic hedgehog, Wnt, FGF, and PDGF cell-signaling pathways [33–53]. Sonic hedgehog binding to its receptor Patched releases the co-receptor Smoothened that is then transported into the cilium and activates the transcription factor Gli, a cilium-controlled process that has been termed "Gli shuttle" [54]. In the neural tube, this mechanism is critical for ventral pattering of the neuroepithelium [33]. In adult neural stem cells and oligodendrocyte precursor cells (OPCs), this mechanism induces the differentiation to neurons and oligodendrocytes, respectively [44, 51, 52]. Factors that regulate ciliogenesis or cilium function are likely to affect and edit these cell-signaling pathways (readers interested in the developmental function of cilia and cilia disorders (ciliopathies) in brain, bone, kidney, and heart are prompted to the following excellent reviews on these topics: [37, 49, 55–61]). While most of research focused on proteins in the regulation of cilia, only very little is known about the function of lipids in ciliogenesis and cilium-induced cell-signaling pathways for stem cell differentiation.

Ceramide is critical for primary cilium formation in mouse and human ES cell-derived neural progenitors [4]. When undifferentiated ES cells were incubated with the ceramide synthase inhibitor Fumonisin B1 (FB1, Fig. 2.1) or the neutral sphingomyelinase (nSMase) inhibitor GW4869, the number and length of primary cilia in neural progenitors were reduced. However, levels of Sox2 and Pax6, two transcription factors expressed in neural progenitors, were not affected. Despite undergoing neural differentiation, progenitors were not able to form rosettes, indicating that loss of ceramide disrupts morphogenesis of the neural tube and ventricular zone during embryonic brain development. Indeed, the *fro/fro* mouse carrying a deletion in *nSMase* shows reduced number and length of ependymal cell motile cilia [31]. Using various inhibitors for ceramide generation including myriocin (Fig. 2.1), FB1 (Fig. 2.1), and GW4869, our group has found that ceramide is not only critical for ciliogenesis, but it is also involved in establishing apicobasal polarity of primitive ectoderm cells and neural progenitors [3, 6].

One of the questions currently investigated in our group is how ceramide regulates the cell-signaling pathways for apicobasal polarity and ciliogenesis. Our working hypothesis is that ceramide enriched in CECs interacts with polarity proteins and the cytoskeleton. Candidate proteins are atypical protein kinase Cζ and ι/λ (aPKC) and glycogen synthase kinase 3β (GSK3), two protein kinases we have shown to bind to ceramide and to regulate acetylation of tubulin in neural cell cilia [3, 10, 31, 62–64]. aPKC as well as GSK3 are also critical for maintaining pluripotency and editing lineage commitment [65–72]. Ceramide binding to these two kinases may very well regulate differentiation of stem cells of various origins. Since ceramide distribution is anisotropic within cellular membranes and even polarized in neural progenitor cells, modulation of aPKC and GSK3 may act through sequestration to CECs and modulation of kinase activity. We have found that the addition of exogenous ceramide, in particular very long chain fatty acid (C24:1) ceramide (Fig. 2.1), increases tubulin acetylation and rescues cilia in neural progenitors with inhibited ceramide biosynthesis [4]. Intriguingly, acetylated tubulin-labeled processes in ES cell-derived neurons were elongated far beyond 500 μm, indicating that ceramide drives neural differentiation and process formation.

Another ceramide target is protein phosphatase 2A (PP2A). Protein phosphatases were among the first enzymes shown to be activated by ceramide [73–76]. Recent research suggests that ceramide functions to sequester and inactivate the PP2A inhibitor protein I2PP2A in the holoenzyme complex [77]. The significance of the endogenous ceramide–PP2A interaction for stem cell differentiation has not been investigated yet. However, inhibition of PP2A has been reported to sustain self-renewal of stem cells and activation of PP2A by exogenous C2 ceramide has been shown to promote neural differentiation [78, 79]. These observations suggest that activation of PP2A by endogenous ceramide promotes stem cell differentiation toward neural cell fate. PP2A has also been found to increase dephosphorylation of aPKC and GSK3 in *Drosophila* neuroblasts and mammalian cells [79, 80], indicating a synergistic effect with direct binding of these two kinases to ceramide by inactivating (sequestering) aPKC and activating GSK3. In addition to direct effects by binding to PP2A, ceramide can upregulate GSK3 activity by inhibiting the phosphatidylinositol 3 kinase (PI3K)-to-Akt pathway, a major GSK3-inactivating cell-signaling pathway known to sustain self-renewal of stem cells [80, 81]. Taken together, regulation of GSK3 by ceramide involves a variety of cell-signaling networks including aPKC (inactivates GSK3 unless sequestered by ceramide), PI3K/Akt (inactivates GSK3 unless inhibited by ceramide), PP2a (activates GSK3 when activated by ceramide), suggesting that ceramide is a bona fide drug target for enhancing neural differentiation in regenerative medicine.

On a separate note, ceramide appears to be important for both, neuronal and glial differentiation of ES cells, since studies in our laboratory have shown that the combination of exogenously added ceramide (or the ceramide analog *N*-oleoyl serinol, S18, Fig. 2.1) and S1P (or the S1P pro-analog FTY720, Fig. 2.1) directs neural cell fate toward oligodendroglial lineage [82] (for more information on S1P, see following section). In addition, ceramide is critical for primary and motile ciliogenesis in astrocytes and ependymal cells, respectively [4, 31]. In summary, these results suggest that ceramide regulates neural cell fate by a common mechanism that involves ciliogenesis and cell-signaling pathways activated by cilia. Therefore, sonic hedgehog and PDGF are likely candidates to be regulated by ceramide.

2.1.1.2 Ceramide and Exosomes

Exosomes belong to the population of extracellular vesicles (EVs), lipid vesicles that are secreted as intercellular carriers by transporting and transferring proteins, lipids, and RNAs (including microRNAs). In addition to exosomes that are generated in multivesicular endosomes, microvesicles or ectosomes blebbing off the cell membrane constitute another portion of EVs. Ceramide has been shown to be required for the formation and secretion of a particular population of exosomes (ESCORT-independent exosomes) although it is not clear whether there is a specific function of ceramide-dependent exosomes vs. other EV fractions [32, 83–85]. Our laboratory has shown that exosomes enriched with ceramide, particularly C18:0 ceramide (Fig. 2.1) play important functions in the etiology of Alzheimer's disease [32, 86]. It is not known if stem cells are involved in this process. Cancer stem cells have been shown to secrete exosomes or shed microvesicles to reprogram the host tissue and accommodate metastases [83, 87–90]. This is mainly achieved by the transfer of mRNAs, microRNAs, and enzymes breaking down the extracellular matrix such as matrix metalloproteases.

In principle, stem or progenitor cells could adopt a similar mechanism to either reprogram the tissue in which they differentiate or to receive instructions for differentiation into a particular tissue. In tissue damage and subsequently tissue regeneration, EVs were found to activate stem cells and induce tissue repair [91–95]. In addition, "instructive" exosomes can be custom-made for the use of stem cells in regenerative medicine [96]. In this case, ceramide may primarily be used for boosting instructive exosome formation. It should be noted that the "ciliogenic" C24:1 ceramide (Fig. 2.1) is structurally different from the "exosomogenic" C18:0 ceramide (Fig. 2.1) and that biophysical studies using synthetic lipid vesicles generated with these two ceramide species showed remarkable differences in shaping membranes. While C18:0 ceramide induces spherical shapes, C24:1 ceramide triggers formation of tubules [97, 98]. In astrocyte-derived exosomes, the major ceramides were C18:0 ceramide (ca. 60%) and C24:1 ceramide (ca. 30%) [32]. Therefore, by being enriched in the exosomal membrane, ceramide (especially neuronal process-inducing C24:1 ceramide) may also participate in induction of stem cell differentiation, particularly toward neural lineage as described in the previous section. It should be noted that exosomes are exquisite lipid carriers comparable to liposomes because of their higher surface (membrane)-to-volume ratios, which is dictated by geometry. Currently, the most promising examples for therapeutic use of (stem cell-derived) EVs are cardiovascular wound repair and protection against ischemia-reperfusion injury in heart and kidney [91, 95, 99–102].

2.1.2 Sphingosine-1-Phosphate

Sphingosine-1-phosphate (S1P) is a metabolic derivative of ceramide and another morphogenetic sphingolipid that has a widespread range of biological effects, including regulation of pluripotency and differentiation, survival and proliferation, migration, and homing. S1P regulates the pertinent cell-signaling pathways in various stem cell types, such as pluripotent stem cells, neural stem cells, mesenchymal stem cells, hematopoietic stem cells, endothelial stem cells, and cardiac precursor cells [2, 103–107].

S1P has a short half-life and its tissue levels are maintained by numerous enzymes and factors [103–105]. S1P is mainly generated intracellularly by two enzymes, sphingosine kinase 1 (SphK1) and 2 (SphK2); irreversibly degraded by S1P lyase (SPL); and hydrolyzed by lipid phosphate phosphatases and S1P-specific phosphatases. It is also exported out of cells by transporter proteins, such as ABC transporters and Spns2 [106, 108–112]. S1P exportation from red blood cells, activated platelets, and endothelial cells comprises most of the extracellular S1P pool, which is usually found at a several-fold higher concentration than that of tissues [112]. SphK1 can also be secreted out and generate S1P outside of cells [112].

Extracellular S1P exerts its function through five cell surface G protein-coupled receptors (GPCRs) S1P1–S1P5 [113] (Fig. 2.3). It stimulates different signal transduction pathways in different cell types depending on the receptors expressed. For example, S1P receptor 1 (S1P1) is coupled exclusively via Gi protein to activate Ras, mitogen-activated protein kinase (MAPK), PI3K/Akt, and phospholipase C pathways [113] (Fig. 2.3). Extracellular S1P has been used to derive or maintain mESCs and hESCs in experimental settings [114–117], demonstrating stimulation of stem cell self-renewal and pluripotency by extracellular S1P. In mESCs, the main pathway allowing maintenance of pluripotency appears to be through the activation of the JAK/STAT3 pathway [117–119]. This notion is supported by studies showing that silencing of the S1P-degrading enzyme, SPL, leads to an increased S1P level concomitant with increased proliferation, and elevated expression of pluripotency markers Ssea1 and Oct-4 in mESCs [120]. The S1P2/Stat3 signaling has been identified to be the major pathway in SPL knockdown-mediated pluripotency. Besides pluripotency maintenance, extracellular S1P plays other crucial roles in stem cells, including proliferation, migration, and homing of various types of progenitor cells (see reviews by [109, 121–124]), and it is critical for vascular development ([109, 125, 126] and reviews by [123, 124]). Extracelluar S1P signaling is important for tumorigenesis and holds great potential as target for disease treatment [105]. S1P promotes cancer stem cell generation and expansion, which contributes greatly to drug resistance, metastasis, and relapse in multiple cancer types [127, 128].

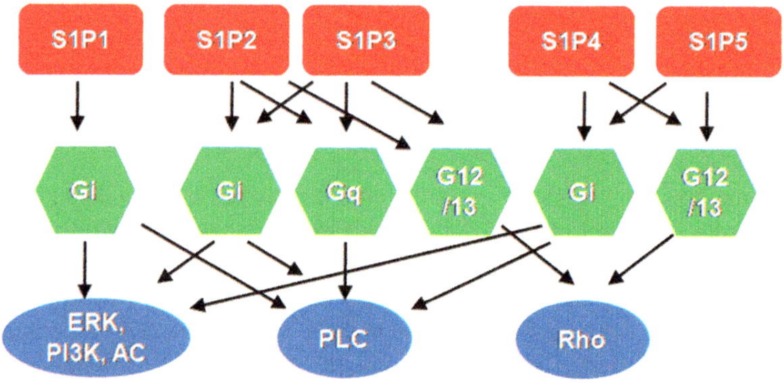

Fig. 2.3. Signaling pathways regulated by extracellular S1P. Extracellular S1P is a ligand for five specific G protein-coupled receptors $S1P_1$–$S1P_5$. Each S1P receptor is coupled to different G proteins; G_i,G_q, G_{12-13}, which regulates stem cell pluripotency, self-renewal, and differentiation through various kinases such as ERK (extracellular signal-regulated kinases), PI3K (phosphatidylinositol-4,5-bisphosphate 3-kinase), AC (adenylyl cyclase), PLC (phospholipase C), and Rho GTPase

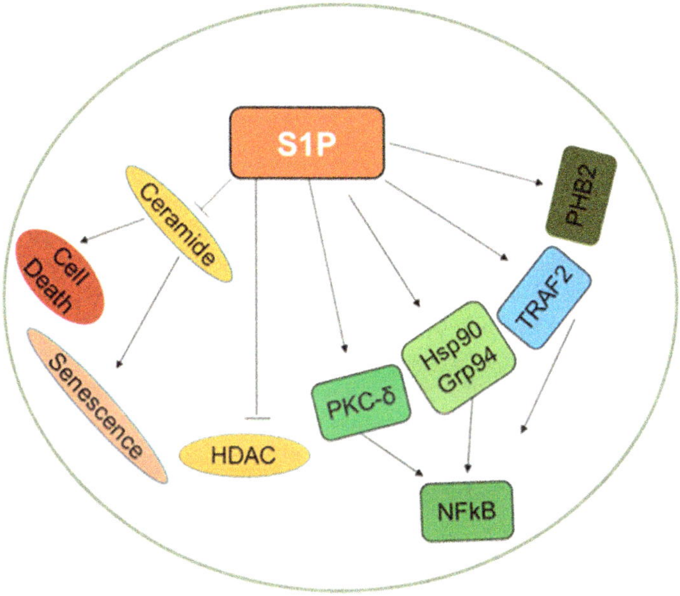

Fig. 2.4. Signaling pathways regulated by intracellular S1P. Intracellular S1P regulates stem cell fate through intracellular targets ceramide, HDAC (histone deacetylases, nuclear), Hsp90 (heat shock protein 90, cytosolic), Grp94 (glucose-regulated protein 94, ER), PHB2 (prohibitin 2, mitochondria), PKCδ (protein kinase C δ, cytosolic), and potentially TRAF2 (TNF receptor associated factor 2, cytosolic)

On the other hand, S1P-primed human mesenchymal stem cells enhance therapeutic potential for pulmonary artery hypertension [129].

Intracellular S1P carries out its function in a receptor-independent manner [104], by either mediating calcium release from the endoplasmic reticulum, or by interacting with its intracellular targets, such as PKCδ, histone deacetylases (HDACs), prohibitin 2 (PHB2), Grp94, and Hsp90α [130, 131] (Fig. 2.4). The intracellular S1P target, PKCδ, is essential for stem cell maintenance and differentiation. Activation of PKCδ mediates cardiac differentiation from ESCs and hematopoietic stem cells [132, 133]. Further, PKCδ activity is required for Jagged-1 induced osteoblast differentiation in hESCs together with canonical Notch signaling [134]. With respect to the function of PKCδ in stem cell pluripotency, it has been found that treatment with PKCδ inhibitors, GF 109203X and rottlerin, prevents early differentiation of mESCs undergoing hypoxia by increasing levels of leukemia inhibitory factor (LIF) receptor and phosphorylated Stat3 [135]. These studies were validated in human pluripotent stem cells by a kinase inhibitor library screening, which identifies PKC inhibitors capable of enhancing pluripotency [136]. Another intracellular target of S1P is histone deacetylase (HDAC). It is known that epigenetic landscapes determine stem cell fate (see reviews [137, 138]). HDACs form the core catalytic component of co-repressor complexes that epigenetically regulate gene expression. Deletion of HDAC1 and

HDAC2 in ES cells caused cell death specifically in undifferentiated cells, concomitant with drastic reduction of pluripotency factors Oct-4, Nanog, Esrrb, and Rex1, indicating that HDAC1 and HDAC2 are essential for pluripotency and renewal of embryonic stem cells [139]. During stem cell differentiation, HDAC inhibition increases expression of neuroectodermal markers and enhances the neuroectodermal specification once neural differentiation is initiated, thereby leading to more neural progenitor cell generation.

In addition to HDACs, other intracellular target proteins of S1P have been identified. S1P activates Prohibitin 2 (PHB2). PHB2 is a pleiotropic factor mainly localized in mitochondria. PHB2 is highly expressed in pluripotent mESCs and decreased during differentiation. Knockdown of PHB2 leads to significant apoptosis, whereas its overexpression results in enhanced proliferation. These results suggest that PHB2 is a crucial regulatory factor for homeostasis and differentiation in mES cells [140]. Similarly, in flat worms (planarians), silencing of PHB2 greatly reduced the number of proliferating neoblasts, which severely impairs tissue regeneration [141]. The Hsp90 family members Hsp90α and Grp94 are newly identified intracellular S1P target proteins [131]. S1P specifically interacts with the N-terminal domain of heat shock proteins during ER stress [131]. Both Hsp90 and Grp94 are essential regulators of stem cell fate. Pharmacological inhibition and genetic knockdown of Hsp90 leads to pluripotency loss in mESCs, which is rescued by Hsp90 re-expression [118]. Hsp90 associates with Oct-4 and Nanog and protects them from degradation by the ubiquitin proteasome system [118]. Hsp90 inhibition predominantly leads to mesoderm differentiation. Because of these effects, Hsp90 inhibitors have been used to specifically eliminate cancer stem cells in a wide range of cancer types [142, 143]. On the other hand, Grp94 deletion leads to defects in mesoderm formation in mice as well as mESCs [144]. Liver-specific deletion of GRP94 leads to hyperproliferation of progenitor cells and acceleration of tumor development in a PTEN-dependent manner, including both hepatocellular carcinoma and cholangiocarcinoma, suggestive of progenitor cell origin [145]. In summary, both intra- and extracellular S1P play profound roles in stem cell biology, which in turn contributes significantly to normal development, morphogenesis, and disease initiation and treatment.

2.1.3 Ceramide-1-Phosphate

Ceramide-1-phosphate (C1P) is synthesized from ceramide by ceramide kinase (Fig. 2.1). It has been shown to induce migration of mesenchymal and hematopoietic stem cells although studies on embryonic stem cells or embryo development are not yet available [146–148]. Its potential as sphingolipid being important for stem cell differentiation (and potentially, morphogenesis) may emerge from its ability to activate phospholipase A_2, an enzyme generating lysophosphatidic acid (LPA) and arachidonic acid, the precursor of eicosanoids [149–151]. Both LPA and eicosanoids involved in stem cell differentiation will be discussed in other chapters of this book.

2.1.4 Glycosphingolipids

Glycosphingolipids (GSLs) are a major class of ceramide derivatives important for differentiation of stem and progenitor cells. Their biosynthesis starts with glycosylation of the C1 hydroxyl group of ceramide using activated glucose or galactose, which can then be followed by the addition of other sugar residues that are either neutral (neutral GSLs) or modified by acidic groups (sulfatides and complex GSLs) (Fig. 2.1). Galactosylceramide is the main (neutral) GSL in brain and comprises about 23% of the total mass of myelin lipids [152]. Galactosylceramide is also known as O1 epitope, a marker for immature oligodendrocytes and the metabolic precursor for galactosulfatide (O4 epitope), a marker for OPCs [153–155]. Determination or isolation of OPCs and oligodendrocytes is achieved by detecting and separating cells with O4(+)/O1(−) and O4(+)/O1(+) epitopes, respectively. Interestingly, the O4 (but not O1) antibody can block terminal differentiation of oligodendrocytes, indicating a functional role of galactosulfatide in differentiation [156, 157].

Galactosulfatide has been suggested to mediate axon-glial contact at the node of Ranvier, a site were the myelin sheath attaches to the axon and leaves a gap for saltatory conduction of the electrical current along the nerve fiber [158–160]. The role of galactosulfatide in OPC differentiation is unclear, while the function of its precursor galactosylceramide is better characterized. It has been reported that galactosylceramides form lipid microdomains or rafts with two other lipids, cholesterol and sphingomyelin in the membrane of the endoplasmic reticulum of OPCs and other cells [161–163]. These lipid rafts interact with sigma receptors important for OPC differentiation. It is not known if galactosulfatide forms lipid rafts as well [161].

In contrast to galactosulfatide, the function of other GSLs, particularly globosides and gangliosides in the regulation of growth factor receptors by lipid rafts is well investigated. Globosides and gangliosides are synthesized from glucosylceramide by first adding galactose (forms lactosylceramide) and then other sugar residues with modification, particularly N-acetyl residues (Fig. 2.1). A rather simple ganglioside termed GD3 has been found to be highly enriched in neural stem cells and to activate EGF receptors in lipid rafts of the plasma membrane [164–169]. Another more complex ganglioside, GM1, has been shown to activate calcium influx into nuclei, which is likely to involve lipid rafts and interaction of Na/Ca exchangers with GM1 in the nuclear membrane [170–174]. While GD3 promotes self-renewal of neural stem and progenitor cells, GM1-induced calcium influx triggers neural differentiation and sustains function of mature neurons. Consistent with consecutive stages of neural differentiation, ganglioside biosynthesis switches from simpler to more complex gangliosides at gestational day E14.5 (mouse), a time point when neural progenitor cells start to divide asymmetrically and give rise to one self-renewing daughter stem cell and one intermediate progenitor eventually undergoing terminal differentiation [175, 176]. We have found that at this time point in brain development, ceramide is also upregulated, suggesting integration of sphingolipid metabolism with neural differentiation [177].

Consistent with the importance of sphingolipid metabolism for neural differentiation, knockout mice for enzymes in ceramide or ganglioside biosynthesis show defects in brain development or function [16, 178–183]. Due to metabolic and functional redundancy (several enzymes can generate the same lipid or different lipids have similar functions), the phenotypes of these knockout mice are not always as severe as predicted by functions determined in vitro. In fact, it appears that the severity of ceramide synthase and glycosyltransferase knockout mice in ceramide and ganglioside biosynthesis is more visible during adult neural differentiation and function than in embryo development. The knockout mice described for deletion of ceramide synthase 1 and 2, glucosylceramide synthase, and alkaline ceramidase 3 are deficient in cerebellar function, particularly due to Purkinje neuron defects or loss [184–189]. The phenotype of the ceramidase synthase 1-deficienct mouse resembles that of the alkaline ceramidase 3 knockout, suggesting that ceramide imbalance is detrimental for adult neural differentiation and function [184, 188]. However, in the ceramide synthase knockout mice, deficiency of a particular ceramide species is accompanied by accumulation of the immediate metabolic ceramide precursors, the long chain bases sphingosine and dihydrosphingosine [188, 190]. Most recently, it was shown that expressing ceramide synthase 2 in the background of ceramide synthase 1 knockout leads to normalization of the long chain bases sphingosine and dihydrosphingosine, while total ceramide levels were not affected [190]. This observation suggests that the phenotype of ceramide synthase knockouts is rather caused by accumulation of long chain bases than lack of ceramide. Interestingly, neurotoxicity of long chain bases has already been described decades ago when the fungus toxin fumonisin B1 (FB1) was found in *Fusarium*-contaminated corn or food for kettle and horses [191–194]. FB1 is a specific inhibitor of ceramide synthases, which leads to reduction of total ceramide and increase of long chain base concentration. In rural areas of South America, eating tortillas contaminated with *Fusarium* leads to a high rate of birth defects, particularly neural tube closure defects and *spina bifida* [195]. This phenotype resembles genetic deficiencies in the Shh pathway, which we already discussed to be activated by primary cilia, and potentially ceramide as regulator for ciliogenesis [196–198]. Currently, it is not known why increased levels of long chain bases or decreased ceramide levels affect neural development, but the phenotypes of the respective knockout mice and effects of inhibitors in ceramide biosynthesis clearly indicate that regulation of sphingolipid metabolism is critical for neural differentiation and function.

2.1.5 Sphingolipids in Stem Cell Therapy and Regenerative Medicine

The plethora of developmental processes regulated by sphingolipids suggests that they are useful in regenerative medicine, particularly for the controlled differentiation of stem cells. Currently, there are three potential avenues tested or hypothetically useful for the application of sphingolipids in stem cell differentiation and regenerative medicine: (1) direct administration of sphingolipids or analogs; (2) generation and administration of sphingolipid-enriched exosomes; and (3) administration of effectors for enzymes in sphingolipid metabolism. Sphingolipids/analogs, exosomes, and enzyme effectors can be added to stem cells in vitro prior to grafting or in vivo, directly into the recipient organism prior to, after, or without stem cell transplantation. Research in our laboratory has focused on in vitro treatment of pluripotent stem cells with ceramide and S1P analogs prior to transplantation into brain. In many ES cell-derived progenitor cell preparations, residual pluripotent stem cells pose the risk of teratoma or other tumor formation after transplantation [199]. We discovered that escaping from apoptosis is one of the reasons why residual pluripotent or progenitor cells (termed "Zombie cells") continue to proliferate [200]. Once apoptosis is reactivated by incubation of progenitors with ceramide analogs, particularly *N*-oleoyl serinol or S18 (Fig. 2.1), the risk of teratoma formation is dramatically reduced. In follow-up studies, we observed that incubation of S18-treated stem cells with the S1P pro-analog FTY720 (Fig. 2.1) directs neural differentiation toward oligodendroglial lineage [5, 82]. Our results suggest that the expression level of prostate apoptosis response-4 (PAR-4), a sensitizer toward ceramide-induced apoptosis, is critical for this specificity. In contrast to residual pluripotent cells with higher PAR-4 expression levels, neural progenitors express only little of PAR-4, while they express the S1P and FTY720 receptor S1P1 (Edg-1), which promotes oligodendrocyte differentiation [5].

The use of FTY720 in improving oligodendrocyte differentiation or function has been hypothesized to be in part responsible for the beneficial effect of fingolimod, the medical preparation of FTY720, in treating multiple sclerosis (MS). The main effect of FTY720 is induction of endocytosis and proteolytic degradation of S1P1 in peripheral T-cells that account for the autoimmune response destroying myelin in MS patients [201]. However, recent research suggests that FTY720 has additional effects on the central nervous system due to its ability to penetrate the blood–brain barrier. For one, it has been found to downregulate S1P1 in reactive astrocytes, which suppresses neuroinflammation aggravating MS. [202, 203] Secondly, it has been shown to protect NPCs and OPCs due to its activating effect on S1P1 [5, 204, 205]. Most likely, the outcome of FTY720 depends on the effective dose and duration of incubation. At low nanomolar concentration and short incubation time, it will activate S1P1 and protect and promote differentiation of OPCs, while at higher concentration and longer incubation time, it will induce S1P1 receptor degradation and prevent neuroinflammation. More recently, several additional molecular targets of FTY720 have been identified, including ceramide synthase (inhibited by FTY720) and PP2A (activated by FTY720), turning this drug into a promising "magic bullet" for treatment of several CNS diseases and cancer [206–210].

While direct administration of sphingolipid analogs to stem cells or in vivo is one potential application, the use of exosomes is another one that rapidly gains interest in regenerative medicine. So far, two avenues have been tested: (1) administration of exosomes to stem cells prior to grafting, and (2) direct injection of exosomes into the blood stream. Exosomes can be stem cell-derived ("stem cell therapy without stem cells") or they can be custom-made and produced by any other appropriate cell type [91–95, 211]. Of the >100 papers currently published on the topic of exosomes in regenerative medicine, the majority focuses on designing exosomes carrying specific microRNAs to reprogram stem cells in vitro and in vivo. Only little is known on the use of sphingolipids in exosome therapy.

Last not least, effectors of sphingolipid metabolism can be directly used in stem cells to "metabolically reprogram" their identity, enhance safety, or boost differentiation toward a particular lineage. While promising in theory, this approach has not yet found significant practical application. The reason maybe twofold: (1) most known effectors of sphingolipid metabolism are enzyme inhibitors that prevent biosynthesis of sphingolipids useful for stem cell differentiation such as ceramide, S1P, and gangliosides; and (2) once biosynthesis of a particular sphingolipid is inhibited, a wealth of important metabolic derivatives of this sphinoglipid are also depleted. Enzyme inhibitors have not found widespread use to manipulate sphingolipid metabolism in stem cells. However, there are anecdotal reports that may change this. D-PDMP, a specific inhibitor of glucosyltransferase, the enzyme that converts ceramide to glucosylceramide, has been applied to neural progenitor cells, but without significant effect on neural differentiation [212]. The non-inhibitor stereoisomer L-PDMP, however, was shown to stimulate neural progenitor proliferation in vitro and in vivo [213–215]. It has been suggested that in contrast to D-PDMP, L-PDMP stimulates glucosylceramide and ganglioside biosynthesis, but it is not known if this compound can be used to enhance stem cells for therapy. In principle, a combination of enzyme inhibitors and sphingolipid analogs can be used to tailor the sphingolipid composition in stem cells and control differentiation. Future studies are needed to determine if this approach is beneficial in stem cell therapy and regenerative medicine.

2.2 Other Lipids

Apart from sphingolipids, many other lipids are known to regulate stem cell differentiation and embryo morphogenesis. These lipids can be post-translational modifications of cell-signaling proteins (e.g., palmitoylation), receptor ligands (e.g., eicosanoids), or cell-signaling lipids to activate or inhibit cell-signaling pathways (e.g., phosphatidylinositol phosphates or PIPs) that sustain self-renewal or promote differentiation of stem and progenitor cells [2]. These lipids often form lipid microdomains or rafts together with sphingolipids due to membrane anisotropy. Therefore, they can cooperate with sphingolipids in editing cell-signaling pathways for stem cell differentiation and morphogenesis. Among lipid modifications of cell-signaling proteins, palmitoylation and cholesterylation of Shh is probably the most prominent example [216, 217]. Cholesterol derivatives such as steroids, as well as eicosanoids and retinoic acid almost exclusively act through receptors. PIPs activate protein kinases in the stem cell survival pathway and promote differentiation toward specific lineages [218, 219]. Similar to ceramide, PIPs are not only cell signaling but also polarity lipids in that their asymmetric distribution recruits and locally activates kinases in the regulation of cell polarity and migration. The integration of cell differentiation and polarity is vital for germ layer formation and embryo morphogenesis. Similar to sphingolipids, generation and localization of other lipids, including cholesterol, eicosanoids, and PIPs is controlled by enzymes in the respective lipid metabolism, which allows for metabolic integration of stem cell metabolism and differentiation.

2.3 Concluding Remarks

The effect of sphingolipids on stem cell differentiation is far more diverse than one could do justice in just one single review or book chapter. However, in order to define an overarching function for lipids in differentiation and development one should let go of discussing these effects for individual lipid classes. We believe that after finishing this chapter, one conclusion can be safely drawn: unlike many proteins with narrowly defined functions, lipids often have overlapping functions and can complement or substitute for each other, regardless of being sphingolipids or other lipid classes. So, what is the "bigger picture" in the role of lipids for stem cell differentiation and development? Why do different lipids have similar effects and can complement or even substitute for each other? And how is this overarching function useful in regenerative medicine to improve stem cells?

In contrast to most proteins, the biosynthesis of which is initiated outside of the membrane, lipids are intrinsic constituents of cellular membranes. Many lipids do not have to be made and then inserted, they are of membrane origin. To change lipid composition, membranes are fused or membrane-resident lipids converted by enzymes. Therefore, lipids are the root cause for determining membrane fluidity and anisotropy, even if regulated by localized enzyme activation or spatially directed vesicle transport. This membrane anisotropy can show itself by localized clustering as in lipid rafts or even asymmetry as in apicobasal polarity or localized membrane protrusions such as cilia and neuronal processes. Membrane anisotropy may rely on lipids in self-organized domains or rafts, involve cytoskeletal and motor proteins that move rafts and vesicles, or endow proteins with lipid moieties to attach to rafts and form spatial gradients and locally defined cell-signaling platforms. Based on these few considerations, one may conclude that the main contributions of lipids to stem cell differentiation and embryo morphogenesis is to endow stem and progenitor cells with polarity, a spatial cue that gives cells orientation in a bigger complex made of constantly morphing layers and tissues during development. Therefore, the term "morphogenetic lipids" is about the function of lipids in the integration of stem cell differentiation and embryo morphogenesis.

How can this function of lipids be utilized in designing differentiation protocols that improve stem cell therapy for regenerative medicine? The linchpin of lipid-regulated stem cell differentiation and its integration with morphogenesis is the association of membrane anisotropy with regulation of the cytoskeleton and cell polarity. Membrane anisotropy is initiated by the formation of lipid microdomains or rafts. Lipid rafts can be self-organized by the biophysical properties of lipids; this has been shown by a plethora of experiments using synthetic vesicles made of pure lipid compositions [15, 98, 220–224]. However, the way rafts morph, move, and interact with other membrane components needs the participation of proteins in a mutually regulating process.

Interestingly, the consequence of this rather inclusive view is that "next generation design" of stem cells in regenerative medicine will rely on reagent cocktails that include effectors for lipid metabolism as well as the associated protein signaling. In a somewhat surprising way, this has already been done from the very beginning of stem cell research. Colchicine, a microtubule-destabilizing drug, has been used to prevent neural differentiation of P19 teratocarcinoma and other types of undifferentiated stem cells [225–227]. Once commitment to neural progenitors is initiated by incubation with retinoic acid, cells become resistant due to acetylation- and detyrosination-induced stabilization of microtubules and incorporation of neurofilaments and microtubule-associated proteins [225, 227, 228]. Retinoic acid induces a several-fold increase in the levels of ceramide in teratocarcinoma cells, which has previously been considered a pro-apoptotic signal [229]. However, we have discovered that very long chain C24:1 ceramide is upregulated during neural differentiation of human ES and iPS cells and promotes acetylation of microtubules due to downregulation or inhibition of HDAC6 [4] (see also above for discussion of ceramide in ciliogenesis). Hence, ceramide may act through a dual effect on promoting neuronal differentiation and concurrent stabilization of microtubules by inhibiting deacetylation. Likewise, another ceramide target recently discovered, GSK3, may promote differentiation through the canonical Wnt/β-catenin cell-signaling pathway as well as increased outgrowth of neuronal processes through the non-canonical pathway and tubulin acetylation through inhibition of HDAC6, respectively.

The GPCR-to-PI3K/Akt-to-GSK3 cell-signaling pathway is one of the major signaling hubs interfacing induction of stem cell differentiation by growth factors with sphingolipid metabolism. Recent studies from our and other laboratories show that this pathway is a node for integrating sphingolipid (S1P and ceramide) and LPA with PIP signaling since S1P and LPA act on GPCRs and inactivate GSK3 through activation of Akt by PIP3 (Fig. 2.5). S1P or LPA counteract ceramide-mediated inhibition of Akt by GPCR-mediated activation of PI3K/Akt. Based on these observations, we conclude that Akt and GSK3-regulated differentiation of stem cells and embryo morphogenesis is balanced by S1P (leads to activation of Akt, inactivation of GSK3, and self-renewal) and ceramide (leads to inactivation of Akt, activation of GSK3, and differentiation). Pharmacological inhibition of Akt with LY294002 and GSK3 with bio/indirubin monoxime has been shown to promote differentiation and pluripotency, respectively [69, 81]. It should be noted, however, that the effect of Akt and GSK3 inhibitors is differential and has opposite effects depending on the duration of incubation or developmental stage. Long-term inhibitor incubation or inhibition of Akt and GSK3 at more committed progenitor stages will prevent differentiation and self-renewal, respectively [65, 230–232].

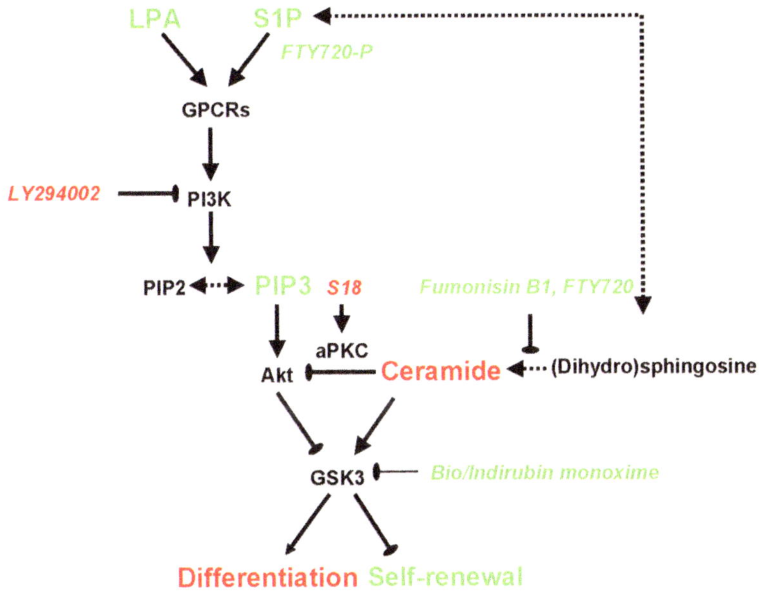

Fig. 2.5. Lipid-regulated GPCR-to-PI3K/Akt-to-GSK3 cell-signaling pathways modulate cell fate decisions in stem cells and morphogenesis. The balance between ceramide and S1P regulates cell fate decision between self-renewal and differentiation in stem cells through different signaling nodes in the GPCR-to-PI3K/Akt-to-GSK3 cells-signaling pathway

The outcome of the GPCR-to-PI3K/Akt-to-GSK3 cell-signaling node is mostly modulated by two growth factors, LIF and fibroblast growth factor-2 (FGF-2), and the pertinent downstream activation of additional cell-signaling pathways, particularly the JAK/STAT3 (via LIF) and ERK (via FGF-2) pathways [66, 81, 233]. Because mouse and human stem cells differ in their response to these growth factors, it is difficult to predict and requires empirical testing to determine which combination of growth factor and modulator of lipid cell-signaling pathways will direct stem cell fate to a desired cell type.

Our research has shown that ceramide may bind and activate GSK3 and in turn, promote acetylation of microtubules and neuronal process formation [4, 31]. On the other hand, we have also found that during differentiation of neural stem cells to OPCs, S1P and ceramide or its analog *N*-oleoyl serinol (S18, Fig. 2.1) may act synergistically once progenitors are committed to glial cell fate [2, 5, 82] (Fig. 2.5). Since S1P can be metabolically derived from ceramide (and vice versa) (Figs. 2.1 and 2.5), sphingolipid metabolism will play an important role in the regulation of stem cell differentiation. The metabolic balance between S1P and ceramide, once predominantly linked to the decision between cell survival and death, has gained a far more subtle and novel function in stem cell differentiation and embryo morphogenesis. Therefore, sphingolipids, particularly S1P and ceramide are morphogenetic lipids and potential drug targets for regenerative medicine.

Acknowledgments This study was supported by grants NIH R01AG034389, R01NS095215, and NSF1121579 to E.B. and American Lung Association RG-351596 to G.W. We are also grateful to institutional support by the Department of Neuroscience and Regenerative Medicine (chair Dr. Lin Mei), Medical College of Georgia at Augusta University.

References

1. Bartke N, Hannun YA (2009) Bioactive sphingolipids: metabolism and function. J Lipid Res 50(Suppl):S91–S96
2. Bieberich E (2012) It's a lipid's world: bioactive lipid metabolism and signaling in neural stem cell differentiation. Neurochem Res 37(6):1208–1229. doi:10.1007/s11064-011-0698-5
3. Krishnamurthy K, Wang G, Silva J, Condie BG, Bieberich E (2007) Ceramide regulates atypical PKC{zeta}/{lambda}-mediated cell polarity in primitive ectoderm cells: a novel function of sphingolipids in morphogenesis. J Biol Chem 282(5):3379–3390
4. He Q, Wang G, Wakade S, Dasgupta S, Dinkins M, Kong JN, Spassieva SD, Bieberich E (2014) Primary cilia in stem cells and neural progenitors are regulated by neutral sphingomyelinase 2 and ceramide. Mol Biol Cell 25(11):1715–1729. doi:10.1091/mbc.E13-12-0730
5. Bieberich E (2010) There is more to a lipid than just being a fat: sphingolipid-guided differentiation of oligodendroglial lineage from embryonic stem cells. Neurochem Res. doi:10.1007/s11064-010-0338-5
6. Wang G, Krishnamurthy K, Chiang YW, Dasgupta S, Bieberich E (2008) Regulation of neural progenitor cell motility by ceramide and potential implications for mouse brain development. J Neurochem 106(2):718–733
7. Bieberich E (2011) Ceramide in stem cell differentiation and embryo development: novel functions of a topological cell-signaling lipid and the concept of ceramide compartments. J Lipids 2011:610306. doi:10.1155/2011/610306
8. Bieberich E (2008) Ceramide signaling in cancer and stem cells. Future Lipidol 3(3):273–300. doi:10.2217/17460875.3.3.273
9. Spassieva S, Bieberich E (2011) The gut-to-breast connection—interdependence of sterols and sphingolipids in multidrug resistance and breast cancer therapy. Anticancer Agents Med Chem 11(9):882–890 doi:BSP/ACAMC/E-Pub/00214
10. He Q, Wang G, Dasgupta S, Dinkins M, Zhu G, Bieberich E (2012) Characterization of an apical ceramide-enriched compartment regulating ciliogenesis. Mol Biol Cell 23(16):3156–3166. doi:10.1091/mbc.E12–02-0079
11. Wang G, Krishnamurthy K, Bieberich E (2009) Regulation of primary cilia formation by ceramide. J Lipid Res 50(10):2103–2110. doi:10.1194/jlr.M900097-JLR200
12. Gulbins E, Kolesnick R (2003) Raft ceramide in molecular medicine. Oncogene 22(45):7070–7077. doi:10.1038/sj.onc.1207146
13. Grassme H, Jekle A, Riehle A, Schwarz H, Berger J, Sandhoff K, Kolesnick R, Gulbins E (2001) CD95 signaling via ceramide-rich membrane rafts. J Biol Chem 276(23):20589–20596. doi:10.1074/jbc.M101207200
14. Harder T, Simons K (1997) Caveolae, DIGs, and the dynamics of sphingolipid-cholesterol microdomains. Curr Opin Cell Biol 9(4):534–542. doi:10.1016/S0955-0674(97)80030-0
15. Sonnino S, Prinetti A (2013) Membrane domains and the "lipid raft" concept. Curr Med Chem 20(1):4–21 doi:CMC-EPUB-20121108-2
16. Yu RK, Tsai YT, Ariga T (2012) Functional roles of gangliosides in neurodevelopment: an overview of recent advances. Neurochem Res 37(6):1230–1244. doi:10.1007/s11064-012-0744-y
17. Simons K, Sampaio JL (2011) Membrane organization and lipid rafts. Cold Spring Harb Perspect Biol 3(10):a004697. doi:10.1101/cshperspect.a004697
18. Lingwood D, Simons K (2010) Lipid rafts as a membrane-organizing principle. Science 327(5961):46–50. doi:10.1126/science.1174621

19. Schulze H, Kolter T, Sandhoff K (2009) Principles of lysosomal membrane degradation: cellular topology and biochemistry of lysosomal lipid degradation. Biochim Biophys Acta 1793(4):674–683. doi:10.1016/j.bbamcr.2008.09.020
20. van Meer G, de Kroon AI (2011) Lipid map of the mammalian cell. J Cell Sci 124(Pt 1):5–8. doi:10.1242/jcs.071233
21. van Meer G, Voelker DR, Feigenson GW (2008) Membrane lipids: where they are and how they behave. Nat Rev Mol Cell Biol 9(2):112–124. doi:10.1038/nrm2330
22. Kobayashi T, Pagano RE (1989) Lipid transport during mitosis. Alternative pathways for delivery of newly synthesized lipids to the cell surface. J Biol Chem 264(10):5966–5973
23. Lipsky NG, Pagano RE (1985) Intracellular translocation of fluorescent sphingolipids in cultured fibroblasts: endogenously synthesized sphingomyelin and glucocerebroside analogues pass through the Golgi apparatus en route to the plasma membrane. J Cell Biol 100(1):27–34
24. Marks DL, Bittman R, Pagano RE (2008) Use of Bodipy-labeled sphingolipid and cholesterol analogs to examine membrane microdomains in cells. Histochem Cell Biol 130(5):819–832. doi:10.1007/s00418-008-0509-5
25. Puri V, Watanabe R, Singh RD, Dominguez M, Brown JC, Wheatley CL, Marks DL, Pagano RE (2001) Clathrin-dependent and -independent internalization of plasma membrane sphingolipids initiates two Golgi targeting pathways. J Cell Biol 154(3):535–547. doi:10.1083/jcb.200102084
26. Chen CS, Martin OC, Pagano RE (1997) Changes in the spectral properties of a plasma membrane lipid analog during the first seconds of endocytosis in living cells. Biophys J 72(1):37–50. doi:10.1016/S0006-3495(97)78645-4
27. van Meer G, Stelzer EH, Wijnaendts-van-Resandt RW, Simons K (1987) Sorting of sphingolipids in epithelial (Madin-Darby canine kidney) cells. J Cell Biol 105(4):1623–1635
28. Surma MA, Klose C, Simons K (2012) Lipid-dependent protein sorting at the trans-Golgi network. Biochim Biophys Acta 1821(8):1059–1067. doi:10.1016/j.bbalip.2011.12.008
29. Kobayashi T, Pimplikar SW, Parton RG, Bhakdi S, Simons K (1992) Sphingolipid transport from the trans-Golgi network to the apical surface in permeabilized MDCK cells. FEBS Lett 300(3):227–231
30. Chinnapen DJ, Hsieh WT, te Welscher YM, Saslowsky DE, Kaoutzani L, Brandsma E, D'Auria L, Park H, Wagner JS, Drake KR, Kang M, Benjamin T, Ullman MD, Costello CE, Kenworthy AK, Baumgart T, Massol RH, Lencer WI (2012) Lipid sorting by ceramide structure from plasma membrane to ER for the cholera toxin receptor ganglioside GM1. Dev Cell 23(3):573–586. doi:10.1016/j.devcel.2012.08.002
31. Kong JN, Hardin K, Dinkins M, Wang G, He Q, Mujadzic T, Zhu G, Bielawski J, Spassieva S, Bieberich E (2015) Regulation of Chlamydomonas flagella and ependymal cell motile cilia by ceramide-mediated translocation of GSK3. Mol Biol Cell 26(24):4451–4465. doi:10.1091/mbc.E15-06-0371
32. Wang G, Dinkins M, He Q, Zhu G, Poirier C, Campbell A, Mayer-Proschel M, Bieberich E (2012) Astrocytes secrete exosomes enriched with proapoptotic ceramide and prostate apoptosis response 4 (PAR-4): potential mechanism of apoptosis induction in Alzheimer disease (AD). J Biol Chem 287(25):21384–21395. doi:10.1074/jbc.M112.340513
33. Pal K, Mukhopadhyay S (2014) Primary cilium and sonic hedgehog signaling during neural tube patterning: role of GPCRs and second messengers. Dev Neurobiol 75(4):337–348. doi:10.1002/dneu.22193
34. Neugebauer JM, Amack JD, Peterson AG, Bisgrove BW, Yost HJ (2009) FGF signalling during embryo development regulates cilia length in diverse epithelia. Nature 458(7238):651–654. doi:10.1038/nature07753
35. Eggenschwiler JT, Anderson KV (2007) Cilia and developmental signaling. Annu Rev Cell Dev Biol 23:345–373
36. Vogel TW, Carter CS, Abode-Iyamah K, Zhang Q, Robinson S (2012) The role of primary cilia in the pathophysiology of neural tube defects. Neurosurg Focus 33(4):E2. doi:10.3171/2012.6.FOCUS12222

37. Cortes CR, Metzis V, Wicking C (2015) Unmasking the ciliopathies: craniofacial defects and the primary cilium. Wiley Interdiscip Rev Dev Biol 4(6):637–653. doi:10.1002/wdev.199
38. May-Simera HL, Kelley MW (2012) Cilia, Wnt signaling, and the cytoskeleton. Cilia 1(1):7. doi:10.1186/2046-2530-1-7
39. Pan J, Seeger-Nukpezah T, Golemis EA (2013) The role of the cilium in normal and abnormal cell cycles: emphasis on renal cystic pathologies. Cell Mol Life Sci 70(11):1849–1874. doi:10.1007/s00018-012-1052-z
40. Cai J, Wu Y, Mirua T, Pierce JL, Lucero MT, Albertine KH, Spangrude GJ, Rao MS (2002) Properties of a fetal multipotent neural stem cell (NEP cell). Dev Biol 251(2):221–240
41. Christensen ST, Pedersen LB, Schneider L, Satir P (2007) Sensory cilia and integration of signal transduction in human health and disease. Traffic 8(2):97–109
42. Noda K, Kitami M, Kitami K, Kaku M, Komatsu Y (2016) Canonical and noncanonical intraflagellar transport regulates craniofacial skeletal development. Proc Natl Acad Sci U S A 113(19):E2589–E2597. doi:10.1073/pnas.1519458113
43. Umberger NL, Caspary T (2015) Ciliary transport regulates PDGF-AA/alphaalpha signaling via elevated mammalian target of rapamycin signaling and diminished PP2A activity. Mol Biol Cell 26(2):350–358. doi:10.1091/mbc.E14-05-0952
44. Falcon-Urrutia P, Carrasco CM, Lois P, Palma V, Roth AD (2015) Shh signaling through the primary cilium modulates rat oligodendrocyte differentiation. PLoS One 10(7):e0133567. doi:10.1371/journal.pone.0133567
45. Zaghloul NA, Brugmann SA (2011) The emerging face of primary cilia. Genesis 49(4):231–246. doi:10.1002/dvg.20728
46. Plotnikova OV, Pugacheva EN, Golemis EA (2009) Primary cilia and the cell cycle. Methods Cell Biol 94:137–160. doi:10.1016/S0091-679X(08)94007-3
47. Wilson SL, Wilson JP, Wang C, Wang B, McConnell SK (2012) Primary cilia and Gli3 activity regulate cerebral cortical size. Dev Neurobiol 72(9):1196–1212. doi:10.1002/dneu.20985
48. Su CY, Bay SN, Mariani LE, Hillman MJ, Caspary T (2012) Temporal deletion of Arl13b reveals that a mispatterned neural tube corrects cell fate over time. Development 139(21):4062–4071. doi:10.1242/dev.082321
49. Ruat M, Roudaut H, Ferent J, Traiffort E (2012) Hedgehog trafficking, cilia and brain functions. Differentiation 83(2):S97–104. doi:10.1016/j.diff.2011.11.011
50. Bay SN, Caspary T (2012) What are those cilia doing in the neural tube? Cilia 1(1):19. doi:10.1186/2046-2530-1-19
51. Han YG, Spassky N, Romaguera-Ros M, Garcia-Verdugo JM, Aguilar A, Schneider-Maunoury S, Alvarez-Buylla A (2008) Hedgehog signaling and primary cilia are required for the formation of adult neural stem cells. Nat Neurosci 11(3):277–284. doi:10.1038/nn2059
52. Breunig JJ, Sarkisian MR, Arellano JI, Morozov YM, Ayoub AE, Sojitra S, Wang B, Flavell RA, Rakic P, Town T (2008) Primary cilia regulate hippocampal neurogenesis by mediating sonic hedgehog signaling. Proc Natl Acad Sci U S A 105(35):13127–13132. doi:10.1073/pnas.0804558105
53. Rohatgi R, Milenkovic L, Scott MP (2007) Patched1 regulates hedgehog signaling at the primary cilium. Science 317(5836):372–376
54. Kim J, Kato M, Beachy PA (2009) Gli2 trafficking links Hedgehog-dependent activation of Smoothened in the primary cilium to transcriptional activation in the nucleus. Proc Natl Acad Sci U S A 106(51):21666–21671. doi:10.1073/pnas.0912180106
55. Valente EM, Rosti RO, Gibbs E, Gleeson JG (2014) Primary cilia in neurodevelopmental disorders. Nat Rev Neurol 10(1):27–36. doi:10.1038/nrneurol.2013.247
56. Koefoed K, Veland IR, Pedersen LB, Larsen LA, Christensen ST (2014) Cilia and coordination of signaling networks during heart development. Organogenesis 10(1):108–125. doi:10.4161/org.27483
57. Guemez-Gamboa A, Coufal NG, Gleeson JG (2014) Primary cilia in the developing and mature brain. Neuron 82(3):511–521. doi:10.1016/j.neuron.2014.04.024
58. Barker AR, Thomas R, Dawe HR (2014) Meckel-Gruber syndrome and the role of primary cilia in kidney, skeleton, and central nervous system development. Organogenesis 10(1):96–107. doi:10.4161/org.27375

59. Komatsu Y, Mishina Y (2013) Establishment of left-right asymmetry in vertebrate development: the node in mouse embryos. Cell Mol Life Sci 70(24):4659–4666. doi:10.1007/s00018-013-1399-9
60. Drummond IA (2012) Cilia functions in development. Curr Opin Cell Biol 24(1):24–30. doi:10.1016/j.ceb.2011.12.007
61. Bodle JC, Loboa EG (2016) Primary cilia: control centers for stem cell lineage specification and potential targets for cell-based therapies. Stem Cells 34(6):1445–1454. doi:10.1002/stem.2341
62. Wang G, Krishnamurthy K, Umapathy NS, Verin AD, Bieberich E (2009) The carboxyl-terminal domain of atypical protein kinase Czeta binds to ceramide and regulates junction formation in epithelial cells. J Biol Chem 284(21):14469–14475. doi:10.1074/jbc.M808909200
63. Wang G, Silva J, Krishnamurthy K, Tran E, Condie BG, Bieberich E (2005) Direct binding to ceramide activates protein kinase Czeta before the formation of a pro-apoptotic complex with PAR-4 in differentiating stem cells. J Biol Chem 280(28):26415–26424
64. Bieberich E, Kawaguchi T, Yu RK (2000) N-acylated serinol is a novel ceramide mimic inducing apoptosis in neuroblastoma cells. J Biol Chem 275(1):177–181
65. Ahn J, Jang J, Choi J, Lee J, Oh SH, Yoon K, Kim S (2014) GSK3beta, but not GSK3alpha, inhibits the neuronal differentiation of neural progenitor cells as a downstream target of mammalian target of rapamycin complex1. Stem Cells Dev 23(10):1121–1133. doi:10.1089/scd.2013.0397
66. Dutta D, Ray S, Home P, Larson M, Wolfe MW, Paul S (2011) Self-renewal versus lineage commitment of embryonic stem cells: protein kinase C signaling shifts the balance. Stem Cells 29(4):618–628. doi:10.1002/stem.605
67. Rajendran G, Dutta D, Hong J, Paul A, Saha B, Mahato B, Ray S, Home P, Ganguly A, Weiss ML, Paul S (2013) Inhibition of protein kinase C signaling maintains rat embryonic stem cell pluripotency. J Biol Chem 288(34):24351–24362. doi:10.1074/jbc.M113.455725
68. Saiz N, Grabarek JB, Sabherwal N, Papalopulu N, Plusa B (2013) Atypical protein kinase C couples cell sorting with primitive endoderm maturation in the mouse blastocyst. Development 140(21):4311–4322. doi:10.1242/dev.093922
69. Sato N, Meijer L, Skaltsounis L, Greengard P, Brivanlou AH (2004) Maintenance of pluripotency in human and mouse embryonic stem cells through activation of Wnt signaling by a pharmacological GSK-3-specific inhibitor. Nat Med 10(1):55–63
70. Dodsworth BT, Flynn R, Cowley SA (2015) The current state of naive human pluripotency. Stem Cells 33(11):3181–3186. doi:10.1002/stem.2085
71. Harwood A, Braga VM (2003) Cdc42 & GSK-3: signals at the crossroads. Nat Cell Biol 5(4):275–277
72. Kim WY, Wang X, Wu Y, Doble BW, Patel S, Woodgett JR, Snider WD (2009) GSK-3 is a master regulator of neural progenitor homeostasis. Nat Neurosci 12(11):1390–1397. doi:10.1038/nn.2408
73. Dobrowsky RT, Hannun YA (1992) Ceramide stimulates a cytosolic protein phosphatase. J Biol Chem 267(8):5048–5051
74. Dobrowsky RT, Kamibayashi C, Mumby MC, Hannun YA (1993) Ceramide activates heterotrimeric protein phosphatase 2A. J Biol Chem 268(21):15523–15530
75. Perry DM, Kitatani K, Roddy P, El-Osta M, Hannun YA (2012) Identification and characterization of protein phosphatase 2C activation by ceramide. J Lipid Res 53(8):1513–1521. doi:10.1194/jlr.M025395
76. Chalfant CE, Szulc Z, Roddy P, Bielawska A, Hannun YA (2004) The structural requirements for ceramide activation of serine-threonine protein phosphatases. J Lipid Res 45(3):496–506
77. Mukhopadhyay A, Saddoughi SA, Song P, Sultan I, Ponnusamy S, Senkal CE, Snook CF, Arnold HK, Sears RC, Hannun YA, Ogretmen B (2008) Direct interaction between the inhibitor 2 and ceramide via sphingolipid-protein binding is involved in the regulation of protein phosphatase 2A activity and signaling. FASEB J 23(3):751–763
78. Yoon BS, Jun EK, Park G, Jun Yoo S, Moon JH, Soon Baik C, Kim A, Kim H, Kim JH, Young Koh G, Taek Lee H, You S (2010) Optimal suppression of protein phosphatase 2A activity is critical for maintenance of human embryonic stem cell self-renewal. Stem Cells 28(5):874–884. doi:10.1002/stem.412

79. Chabu C, Doe CQ (2009) Twins/PP2A regulates aPKC to control neuroblast cell polarity and self-renewal. Dev Biol 330(2):399–405. doi:10.1016/j.ydbio.2009.04.014
80. Lin CF, Chen CL, Chiang CW, Jan MS, Huang WC, Lin YS (2007) GSK-3beta acts downstream of PP2A and the PI 3-kinase-Akt pathway, and upstream of caspase-2 in ceramide-induced mitochondrial apoptosis. J Cell Sci 120(Pt 16):2935–2943. doi:10.1242/jcs.03473
81. Paling NR, Wheadon H, Bone HK, Welham MJ (2004) Regulation of embryonic stem cell self-renewal by phosphoinositide 3-kinase-dependent signaling. J Biol Chem 279(46):48063–48070. doi:10.1074/jbc.M406467200
82. Bieberich E (2008) Smart drugs for smarter stem cells: making SENSe (sphingolipid-enhanced neural stem cells) of ceramide. Neurosignals 16(2–3):124–139
83. Kong JN, He Q, Wang G, Dasgupta S, Dinkins MB, Zhu G, Kim A, Spassieva S, Bieberich E (2015) Guggulsterone and bexarotene induce secretion of exosome-associated breast cancer resistance protein and reduce doxorubicin resistance in MDA-MB-231 cells. Int J Cancer 137(7):1610–1620. doi:10.1002/ijc.29542
84. Trajkovic K, Hsu C, Chiantia S, Rajendran L, Wenzel D, Wieland F, Schwille P, Brugger B, Simons M (2008) Ceramide triggers budding of exosome vesicles into multivesicular endosomes. Science 319(5867):1244–1247. doi:10.1126/science.1153124
85. Shamseddine AA, Airola MV, Hannun YA (2015) Roles and regulation of neutral sphingomyelinase-2 in cellular and pathological processes. Adv Biol Regul 57:24–41. doi:10.1016/j.jbior.2014.10.002
86. Dinkins MB, Dasgupta S, Wang G, Zhu G, Bieberich E (2014) Exosome reduction in vivo is associated with lower amyloid plaque load in the 5XFAD mouse model of Alzheimer's disease. Neurobiol Aging 35(8):1792–1800. doi:10.1016/j.neurobiolaging.2014.02.012
87. Kumar D, Gupta D, Shankar S, Srivastava RK (2015) Biomolecular characterization of exosomes released from cancer stem cells: possible implications for biomarker and treatment of cancer. Oncotarget 6(5):3280–3291. doi:10.18632/oncotarget.2462
88. Fu H, Yang H, Zhang X, Xu W (2016) The emerging roles of exosomes in tumor-stroma interaction. J Cancer Res Clin Oncol. doi:10.1007/s00432-016-2145-0
89. Somasundaram R, Herlyn M (2012) Melanoma exosomes: messengers of metastasis. Nat Med 18(6):853–854. doi:10.1038/nm.2775
90. Yang Q, Diamond MP, Al-Hendy A (2016) The emerging role of extracellular vesicle-derived miRNAs: implication in cancer progression and stem cell related diseases. J Clin Epigenet 2(1)
91. Singla DK (2016) Stem cells and exosomes in cardiac repair. Curr Opin Pharmacol 27:19–23. doi:10.1016/j.coph.2016.01.003
92. Luarte A, Batiz LF, Wyneken U, Lafourcade C (2016) Potential Therapies by stem cell-derived exosomes in CNS diseases: focusing on the neurogenic niche. Stem Cells Int 2016:5736059. doi:10.1155/2016/5736059
93. Jarmalaviciute A, Pivoriunas A (2016) Exosomes as a potential novel therapeutic tools against neurodegenerative diseases. Pharmacol Res. doi:10.1016/j.phrs.2016.02.002
94. Han C, Sun X, Liu L, Jiang H, Shen Y, Xu X, Li J, Zhang G, Huang J, Lin Z, Xiong N, Wang T (2016) Exosomes and their therapeutic potentials of stem cells. Stem Cells Int 2016:7653489. doi:10.1155/2016/7653489
95. Rosca AM, Rayia DM, Tutuianu R (2015) Emerging role of stem cells—derived exosomes as valuable tools for cardiovascular therapy. Curr Stem Cell Res Ther doi:CSCRT-EPUB-71260
96. Alvarez-Erviti L, Seow Y, Yin H, Betts C, Lakhal S, Wood MJ (2011) Delivery of siRNA to the mouse brain by systemic injection of targeted exosomes. Nat Biotechnol 29(4):341–345. doi:10.1038/nbt.1807
97. Pinto SN, Silva LC, de Almeida RF, Prieto M (2008) Membrane domain formation, interdigitation, and morphological alterations induced by the very long chain asymmetric C24:1 ceramide. Biophys J 95(6):2867–2879. doi:10.1529/biophysj.108.129858
98. Pinto SN, Silva LC, Futerman AH, Prieto M (2011) Effect of ceramide structure on membrane biophysical properties: the role of acyl chain length and unsaturation. Biochim Biophys Acta 1808(11):2753–2760. doi:10.1016/j.bbamem.2011.07.023

99. Vicencio JM, Yellon DM, Sivaraman V, Das D, Boi-Doku C, Arjun S, Zheng Y, Riquelme JA, Kearney J, Sharma V, Multhoff G, Hall AR, Davidson SM (2015) Plasma exosomes protect the myocardium from ischemia-reperfusion injury. J Am Coll Cardiol 65(15):1525–1536. doi:10.1016/j.jacc.2015.02.026
100. Zhang H, Xiang M, Meng D, Sun N, Chen S (2016) Inhibition of myocardial ischemia/reperfusion injury by exosomes secreted from mesenchymal stem cells. Stem Cells Int 2016:4328362. doi:10.1155/2016/4328362
101. Lin KC, Yip HK, Shao PL, Wu SC, Chen KH, Chen YT, Yang CC, Sun CK, Kao GS, Chen SY, Chai HT, Chang CL, Chen CH, Lee MS (2016) Combination of adipose-derived mesenchymal stem cells (ADMSC) and ADMSC-derived exosomes for protecting kidney from acute ischemia-reperfusion injury. Int J Cardiol 216:173–185. doi:10.1016/j.ijcard.2016.04.061
102. Akyurekli C, Le Y, Richardson RB, Fergusson D, Tay J, Allan DS (2015) A systematic review of preclinical studies on the therapeutic potential of mesenchymal stromal cell-derived microvesicles. Stem Cell Rev 11(1):150–160. doi:10.1007/s12015-014-9545-9
103. Kunkel GT, Maceyka M, Milstien S, Spiegel S (2013) Targeting the sphingosine-1-phosphate axis in cancer, inflammation and beyond. Nat Rev Drug Discov 12(9):688–702. doi:10.1038/nrd4099
104. Spiegel S, Milstien S (2011) The outs and the ins of sphingosine-1-phosphate in immunity. Nat Rev Immunol 11(6):403–415. doi:10.1038/nri2974
105. Proia RL, Hla T (2015) Emerging biology of sphingosine-1-phosphate: its role in pathogenesis and therapy. J Clin Invest 125(4):1379–1387. doi:10.1172/JCI76369
106. Klyachkin YM, Karapetyan AV, Ratajczak MZ, Abdel-Latif A (2014) The role of bioactive lipids in stem cell mobilization and homing: novel therapeutics for myocardial ischemia. Biomed Res Int 2014:653543. doi:10.1155/2014/653543
107. Mizugishi K, Yamashita T, Olivera A, Miller GF, Spiegel S, Proia RL (2005) Essential role for sphingosine kinases in neural and vascular development. Mol Cell Biol 25(24):11113–11121
108. Bradley E, Dasgupta S, Jiang X, Zhao X, Zhu G, He Q, Dinkins M, Bieberich E, Wang G (2014) Critical role of Spns2, a sphingosine-1-phosphate transporter, in lung cancer cell survival and migration. PLoS One 9(10):e110119. doi:10.1371/journal.pone.0110119
109. Kawahara A, Nishi T, Hisano Y, Fukui H, Yamaguchi A, Mochizuki N (2009) The sphingolipid transporter spns2 functions in migration of zebrafish myocardial precursors. Science 323(5913):524–527. doi:10.1126/science.1167449
110. Nagareddy PR, Asfour A, Klyachkin YM, Abdel-Latif A (2014) A novel role for bioactive lipids in stem cell mobilization during cardiac ischemia: new paradigms in thrombosis: novel mediators and biomarkers. J Thromb Thrombolysis 37(1):24–31. doi:10.1007/s11239-013-1032-7
111. Long J, Darroch P, Wan KF, Kong KC, Ktistakis N, Pyne NJ, Pyne S (2005) Regulation of cell survival by lipid phosphate phosphatases involves the modulation of intracellular phosphatidic acid and sphingosine 1-phosphate pools. Biochem J 391(Pt 1):25–32. doi:10.1042/BJ20050342
112. Sciorra VA, Morris AJ (2002) Roles for lipid phosphate phosphatases in regulation of cellular signaling. Biochim Biophys Acta 1582(1–3):45–51
113. Huang YL, Huang WP, Lee H (2011) Roles of sphingosine 1-phosphate on tumorigenesis. World J Biol Chem 2(2):25–34. doi:10.4331/wjbc.v2.i2.25
114. Kleger A, Busch T, Liebau S, Prelle K, Paschke S, Beil M, Rolletschek A, Wobus A, Wolf E, Adler G, Seufferlein T (2007) The bioactive lipid sphingosylphosphorylcholine induces differentiation of mouse embryonic stem cells and human promyelocytic leukaemia cells. Cell Signal 19(2):367–377. doi:10.1016/j.cellsig.2006.07.015
115. Rodgers A, Mormeneo D, Long JS, Delgado A, Pyne NJ, Pyne S (2009) Sphingosine 1-phosphate regulation of extracellular signal-regulated kinase-1/2 in embryonic stem cells. Stem Cells Dev 18(9):1319–1330. doi:10.1089/scd.2009.0023
116. Pebay A, Wong RC, Pitson SM, Wolvetang EJ, Peh GS, Filipczyk A, Koh KL, Tellis I, Nguyen LT, Pera MF (2005) Essential roles of sphingosine-1-phosphate and platelet-derived

growth factor in the maintenance of human embryonic stem cells. Stem Cells 23(10):1541–1548. doi:10.1634/stemcells.2004-0338
117. Wong RC, Pera MF, Pebay A (2012) Maintenance of human embryonic stem cells by sphingosine-1-phosphate and platelet-derived growth factor. Methods Mol Biol 874:167–175. doi:10.1007/978-1-61779-800-9_13
118. Bradley E, Bieberich E, Mivechi NF, Tangpisuthipongsa D, Wang G (2012) Regulation of embryonic stem cell pluripotency by heat shock protein 90. Stem Cells 30(8):1624–1633. doi:10.1002/stem.1143
119. Burdon T, Smith A, Savatier P (2002) Signalling, cell cycle and pluripotency in embryonic stem cells. Trends Cell Biol 12(9):432–438
120. Smith GS, Kumar A, Saba JD (2013) Sphingosine phosphate lyase regulates murine embryonic stem cell proliferation and pluripotency through an S1P/STAT3 signaling pathway. Biomolecules 3(3):351–368. doi:10.3390/biom3030351
121. Ryu JM, Baek YB, Shin MS, Park JH, Park SH, Lee JH, Han HJ (2014) Sphingosine-1-phosphate-induced Flk-1 transactivation stimulates mouse embryonic stem cell proliferation through S1P1/S1P3-dependent beta-arrestin/c-Src pathways. Stem Cell Res 12(1):69–85. doi:10.1016/j.scr.2013.08.013
122. Arya D, Chang S, DiMuzio P, Carpenter J, Tulenko TN (2014) Sphingosine-1-phosphate promotes the differentiation of adipose-derived stem cells into endothelial nitric oxide synthase (eNOS) expressing endothelial-like cells. J Biomed Sci 21:55. doi:10.1186/1423-0127-21-55
123. Ratajczak MZ, Suszynska M (2016) Emerging strategies to enhance homing and engraftment of hematopoietic stem cells. Stem Cell Rev 12(1):121–128. doi:10.1007/s12015-015-9625-5
124. Adamiak M, Borkowska S, Wysoczynski M, Suszynska M, Kucia M, Rokosh G, Abdel-Latif A, Ratajczak J, Ratajczak MZ (2015) Evidence for the involvement of sphingosine-1-phosphate in the homing and engraftment of hematopoietic stem cells to bone marrow. Oncotarget 6(22):18819–18828. doi:10.18632/oncotarget.4710
125. Xiong Y, Yang P, Proia RL, Hla T (2014) Erythrocyte-derived sphingosine 1-phosphate is essential for vascular development. J Clin Invest 124(11):4823–4828. doi:10.1172/JCI77685
126. Fukui H, Terai K, Nakajima H, Chiba A, Fukuhara S, Mochizuki N (2014) S1P-Yap1 signaling regulates endoderm formation required for cardiac precursor cell migration in zebrafish. Dev Cell 31(1):128–136. doi:10.1016/j.devcel.2014.08.014
127. Marfia G, Campanella R, Navone SE, Di Vito C, Riccitelli E, Hadi LA, Bornati A, de Rezende G, Giussani P, Tringali C, Viani P, Rampini P, Alessandri G, Parati E, Riboni L (2014) Autocrine/paracrine sphingosine-1-phosphate fuels proliferative and stemness qualities of glioblastoma stem cells. Glia 62(12):1968–1981. doi:10.1002/glia.22718
128. Hirata N, Yamada S, Shoda T, Kurihara M, Sekino Y, Kanda Y (2014) Sphingosine-1-phosphate promotes expansion of cancer stem cells via S1PR3 by a ligand-independent Notch activation. Nat Commun 5:4806. doi:10.1038/ncomms5806
129. Kang H, Kim KH, Lim J, Kim YS, Heo J, Choi J, Jeong J, Kim Y, Kim SW, Oh YM, Choo MS, Son J, Kim SJ, Yoo HJ, Oh W, Choi SJ, Lee SW, Shin DM (2015) The therapeutic effects of human mesenchymal stem cells primed with sphingosine-1 phosphate on pulmonary artery hypertension. Stem Cells Dev 24(14):1658–1671. doi:10.1089/scd.2014.0496
130. Maceyka M, Harikumar KB, Milstien S, Spiegel S (2012) Sphingosine-1-phosphate signaling and its role in disease. Trends Cell Biol 22(1):50–60. doi:10.1016/j.tcb.2011.09.003
131. Park K, Ikushiro H, Seo HS, Shin KO, Kim YI, Kim JY, Lee YM, Yano T, Holleran WM, Elias P, Uchida Y (2016) ER stress stimulates production of the key antimicrobial peptide, cathelicidin, by forming a previously unidentified intracellular S1P signaling complex. Proc Natl Acad Sci U S A 113(10):E1334–E1342. doi:10.1073/pnas.1504555113
132. Junttila I, Bourette RP, Rohrschneider LR, Silvennoinen O (2003) M-CSF induced differentiation of myeloid precursor cells involves activation of PKC-delta and expression of Pkare. J Leukoc Biol 73(2):281–288
133. Kim MJ, Moon CH, Kim MY, Kim MH, Lee SH, Baik EJ, Jung YS (2004) Role of PKC-delta during hypoxia in heart-derived H9c2 cells. Jpn J Physiol 54(4):405–414

134. Zhu F, Sweetwyne MT, Hankenson KD (2013) PKCdelta is required for Jagged-1 induction of human mesenchymal stem cell osteogenic differentiation. Stem Cells 31(6):1181–1192. doi:10.1002/stem.1353
135. Lee HJ, Jeong CH, Cha JH, Kim KW (2010) PKC-delta inhibitors sustain self-renewal of mouse embryonic stem cells under hypoxia in vitro. Exp Mol Med 42(4):294–301. doi:10.3858/emm.2010.42.4.028
136. Kinehara M, Kawamura S, Tateyama D, Suga M, Matsumura H, Mimura S, Hirayama N, Hirata M, Uchio-Yamada K, Kohara A, Yanagihara K, Furue MK (2013) Protein kinase C regulates human pluripotent stem cell self-renewal. PLoS One 8(1):e54122. doi:10.1371/journal.pone.0054122
137. Liang G, Zhang Y (2013) Embryonic stem cell and induced pluripotent stem cell: an epigenetic perspective. Cell Res 23(1):49–69. doi:10.1038/cr.2012.175
138. Aloia L, Demajo S, Di Croce L (2015) ZRF1: a novel epigenetic regulator of stem cell identity and cancer. Cell Cycle 14(4):510–515. doi:10.4161/15384101.2014.988022
139. Jamaladdin S, Kelly RD, O'Regan L, Dovey OM, Hodson GE, Millard CJ, Portolano N, Fry AM, Schwabe JW, Cowley SM (2014) Histone deacetylase (HDAC) 1 and 2 are essential for accurate cell division and the pluripotency of embryonic stem cells. Proc Natl Acad Sci U S A 111(27):9840–9845. doi:10.1073/pnas.1321330111
140. Kowno M, Watanabe-Susaki K, Ishimine H, Komazaki S, Enomoto K, Seki Y, Wang YY, Ishigaki Y, Ninomiya N, Noguchi TA, Kokubu Y, Ohnishi K, Nakajima Y, Kato K, Intoh A, Takada H, Yamakawa N, Wang PC, Asashima M, Kurisaki A (2014) Prohibitin 2 regulates the proliferation and lineage-specific differentiation of mouse embryonic stem cells in mitochondria. PLoS One 9(4):e81552. doi:10.1371/journal.pone.0081552
141. Rossi L, Bonuccelli L, Iacopetti P, Evangelista M, Ghezzani C, Tana L, Salvetti A (2014) Prohibitin 2 regulates cell proliferation and mitochondrial cristae morphogenesis in planarian stem cells. Stem Cell Rev 10(6):871–887. doi:10.1007/s12015-014-9540-1
142. Yang R, Tang Q, Miao F, An Y, Li M, Han Y, Wang X, Wang J, Liu P, Chen R (2015) Inhibition of heat-shock protein 90 sensitizes liver cancer stem-like cells to magnetic hyperthermia and enhances anti-tumor effect on hepatocellular carcinoma-burdened nude mice. Int J Nanomed 10:7345–7358. doi:10.2147/IJN.S93758
143. White PT, Subramanian C, Zhu Q, Zhang H, Zhao H, Gallagher R, Timmermann BN, Blagg BS, Cohen MS (2016) Novel HSP90 inhibitors effectively target functions of thyroid cancer stem cell preventing migration and invasion. Surgery 159(1):142–151. doi:10.1016/j.surg.2015.07.050
144. Wanderling S, Simen BB, Ostrovsky O, Ahmed NT, Vogen SM, Gidalevitz T, Argon Y (2007) GRP94 is essential for mesoderm induction and muscle development because it regulates insulin-like growth factor secretion. Mol Biol Cell 18(10):3764–3775. doi:10.1091/mbc.E07-03-0275
145. Chen WT, Tseng CC, Pfaffenbach K, Kanel G, Luo B, Stiles BL, Lee AS (2014) Liver-specific knockout of GRP94 in mice disrupts cell adhesion, activates liver progenitor cells, and accelerates liver tumorigenesis. Hepatology 59(3):947–957. doi:10.1002/hep.26711
146. Lim J, Kim Y, Heo J, Kim KH, Lee S, Lee SW, Kim K, Kim IG, Shin DM (2016) Priming with ceramide-1 phosphate promotes the therapeutic effect of mesenchymal stem/stromal cells on pulmonary artery hypertension. Biochem Biophys Res Commun 473(1):35–41. doi:10.1016/j.bbrc.2016.03.046
147. Marycz K, Smieszek A, Jelen M, Chrzastek K, Grzesiak J, Meissner J (2015) The effect of the bioactive sphingolipids S1P and C1P on multipotent stromal cells—new opportunities in regenerative medicine. Cell Mol Biol Lett 20(3):510–533. doi:10.1515/cmble-2015-0029
148. Kim C, Schneider G, Abdel-Latif A, Mierzejewska K, Sunkara M, Borkowska S, Ratajczak J, Morris AJ, Kucia M, Ratajczak MZ (2013) Ceramide-1-phosphate regulates migration of multipotent stromal cells and endothelial progenitor cells—implications for tissue regeneration. Stem Cells 31(3):500–510. doi:10.1002/stem.1291

149. Lamour NF, Subramanian P, Wijesinghe DS, Stahelin RV, Bonventre JV, Chalfant CE (2009) Ceramide 1-phosphate is required for the translocation of group IVA cytosolic phospholipase A2 and prostaglandin synthesis. J Biol Chem 284(39):26897–26907. doi:10.1074/jbc.M109.001677
150. Lamour NF, Chalfant CE (2008) Ceramide kinase and the ceramide-1-phosphate/cPLA2alpha interaction as a therapeutic target. Curr Drug Targets 9(8):674–682
151. Lamour NF, Chalfant CE (2005) Ceramide-1-phosphate: the "missing" link in eicosanoid biosynthesis and inflammation. Mol Interv 5(6):358–367. doi:10.1124/mi.5.6.8
152. Marcus J, Popko B (2002) Galactolipids are molecular determinants of myelin development and axo-glial organization. Biochim Biophys Acta 1573(3):406–413
153. Ngamukote S, Yanagisawa M, Ariga T, Ando S, Yu RK (2007) Developmental changes of glycosphingolipids and expression of glycogenes in mouse brains. J Neurochem 103(6):2327–2341. doi:10.1111/j.1471-4159.2007.04910.x
154. Zhang SC, Ge B, Duncan ID (2000) Tracing human oligodendroglial development in vitro. J Neurosci Res 59(3):421–429. doi:10.1002/(SICI)1097-4547(20000201)59:3<421::AID-JNR17>3.0.CO;2-C
155. Sommer I, Schachner M (1981) Monoclonal antibodies (O1 to O4) to oligodendrocyte cell surfaces: an immunocytological study in the central nervous system. Dev Biol 83(2):311–327. doi:10.1016/0012-1606(81)90477-2
156. Bansal R, Winkler S, Bheddah S (1999) Negative regulation of oligodendrocyte differentiation by galactosphingolipids. J Neurosci 19(18):7913–7924
157. Bansal R, Pfeiffer SE (1989) Reversible inhibition of oligodendrocyte progenitor differentiation by a monoclonal antibody against surface galactolipids. Proc Natl Acad Sci U S A 86(16):6181–6185
158. Stoffel W, Bosio A (1997) Myelin glycolipids and their functions. Curr Opin Neurobiol 7(5):654–661. doi:10.1016/S0959-4388(97)80085-2
159. Popko B, Dupree JL, Coetzee T, Suzuki K (1999) Genetic analysis of myelin galactolipid function. Adv Exp Med Biol 468:237–244
160. Dupree JL, Coetzee T, Blight A, Suzuki K, Popko B (1998) Myelin galactolipids are essential for proper node of Ranvier formation in the CNS. J Neurosci 18(5):1642–1649
161. Hayashi T, Su TP (2004) Sigma-1 receptors at galactosylceramide-enriched lipid microdomains regulate oligodendrocyte differentiation. Proc Natl Acad Sci U S A 101(41):14949–14954. doi:10.1073/pnas.0402890101
162. Wang TY, Silvius JR (2000) Different sphingolipids show differential partitioning into sphingolipid/cholesterol-rich domains in lipid bilayers. Biophys J 79(3):1478–1489. doi:10.1016/S0006-3495(00)76399-5
163. Moyano AL, Li G, Lopez-Rosas A, Mansson JE, van Breemen RB, Givogri MI (2014) Distribution of C16:0, C18:0, C24:1, and C24:0 sulfatides in central nervous system lipid rafts by quantitative ultra-high-pressure liquid chromatography tandem mass spectrometry. Anal Biochem 467:31–39. doi:10.1016/j.ab.2014.08.033
164. Wang J, Yu RK (2013) Interaction of ganglioside GD3 with an EGF receptor sustains the self-renewal ability of mouse neural stem cells in vitro. Proc Natl Acad Sci U S A 110(47):19137–19142. doi:10.1073/pnas.1307224110
165. Nakatani Y, Yanagisawa M, Suzuki Y, Yu RK (2010) Characterization of GD3 ganglioside as a novel biomarker of mouse neural stem cells. Glycobiology 20(1):78–86. doi:10.1093/glycob/cwp149
166. Yanagisawa M, Liour SS, Yu RK (2004) Involvement of gangliosides in proliferation of immortalized neural progenitor cells. J Neurochem 91(4):804–812. doi:10.1111/j.1471-4159.2004.02750.x
167. Goldman JE, Hirano M, Yu RK, Seyfried TN (1984) GD3 ganglioside is a glycolipid characteristic of immature neuroectodermal cells. J Neuroimmunol 7(2–3):179–192
168. Liour SS, Kapitonov D, Yu RK (2000) Expression of gangliosides in neuronal development of P19 embryonal carcinoma stem cells. J Neurosci Res 62(3):363–373. doi:10.1002/1097-4547(20001101)62:3<363::AID-JNR6>3.0.CO;2-E

169. Zurita AR, Maccioni HJ, Daniotti JL (2001) Modulation of epidermal growth factor receptor phosphorylation by endogenously expressed gangliosides. Biochem J 355(Pt 2):465–472
170. Ledeen R, Wu G (2011) New findings on nuclear gangliosides: overview on metabolism and function. J Neurochem 116(5):714–720. doi:10.1111/j.1471-4159.2010.07115.x
171. Wu G, Xie X, Lu ZH, Ledeen RW (2009) Sodium-calcium exchanger complexed with GM1 ganglioside in nuclear membrane transfers calcium from nucleoplasm to endoplasmic reticulum. Proc Natl Acad Sci U S A 106(26):10829–10834. doi:10.1073/pnas.0903408106
172. Ledeen R, Wu G (2007) GM1 in the nuclear envelope regulates nuclear calcium through association with a nuclear sodium-calcium exchanger. J Neurochem 103(Suppl 1):126–134. doi:10.1111/j.1471-4159.2007.04722.x
173. Xie X, Wu G, Lu ZH, Rohowsky-Kochan C, Ledeen RW (2004) Presence of sodium-calcium exchanger/GM1 complex in the nuclear envelope of non-neural cells: nature of exchanger-GM1 interaction. Neurochem Res 29(11):2135–2146
174. Wu G, Lu ZH, Ledeen RW (1995) GM1 ganglioside in the nuclear membrane modulates nuclear calcium homeostasis during neurite outgrowth. J Neurochem 65(3):1419–1422
175. Yu RK, Bieberich E, Xia T, Zeng G (2004) Regulation of ganglioside biosynthesis in the nervous system. J Lipid Res 45(5):783–793. doi:10.1194/jlr.R300020-JLR200
176. Yu RK, Macala LJ, Taki T, Weinfield HM, Yu FS (1988) Developmental changes in ganglioside composition and synthesis in embryonic rat brain. J Neurochem 50(6):1825–1829
177. Bieberich E, MacKinnon S, Silva J, Yu RK (2001) Regulation of apoptosis during neuronal differentiation by ceramide and b-series complex gangliosides. J Biol Chem 276(48):44396–44404
178. Allende ML, Proia RL (2002) Lubricating cell signaling pathways with gangliosides. Curr Opin Struct Biol 12(5):587–592
179. Kawai H, Allende ML, Wada R, Kono M, Sango K, Deng C, Miyakawa T, Crawley JN, Werth N, Bierfreund U, Sandhoff K, Proia RL (2001) Mice expressing only monosialoganglioside GM3 exhibit lethal audiogenic seizures. J Biol Chem 276(10):6885–6888. doi:10.1074/jbc.C000847200
180. Chiavegatto S, Sun J, Nelson RJ, Schnaar RL (2000) A functional role for complex gangliosides: motor deficits in GM2/GD2 synthase knockout mice. Exp Neurol 166(2):227–234. doi:10.1006/exnr.2000.7504
181. Sheikh KA, Sun J, Liu Y, Kawai H, Crawford TO, Proia RL, Griffin JW, Schnaar RL (1999) Mice lacking complex gangliosides develop Wallerian degeneration and myelination defects. Proc Natl Acad Sci U S A 96(13):7532–7537
182. Wang J, Cheng A, Wakade C, Yu RK (2014) Ganglioside GD3 is required for neurogenesis and long-term maintenance of neural stem cells in the postnatal mouse brain. J Neurosci 34(41):13790–13800. doi:10.1523/JNEUROSCI.2275-14.2014
183. Furukawa K, Ohmi Y, Ohkawa Y, Tajima O (2014) Glycosphingolipids in the regulation of the nervous system. Adv Neurobiol 9:307–320. doi:10.1007/978-1-4939-1154-7_14
184. Wang K, Xu R, Schrandt J, Shah P, Gong YZ, Preston C, Wang L, Yi JK, Lin CL, Sun W, Spyropoulos DD, Rhee S, Li M, Zhou J, Ge S, Zhang G, Snider AJ, Hannun YA, Obeid LM, Mao C (2015) Alkaline ceramidase 3 deficiency results in purkinje cell degeneration and cerebellar ataxia due to dyshomeostasis of sphingolipids in the brain. PLoS Genet 11(10):e1005591. doi:10.1371/journal.pgen.1005591
185. Ginkel C, Hartmann D, vom Dorp K, Zlomuzica A, Farwanah H, Eckhardt M, Sandhoff R, Degen J, Rabionet M, Dere E, Dormann P, Sandhoff K, Willecke K (2012) Ablation of neuronal ceramide synthase 1 in mice decreases ganglioside levels and expression of myelin-associated glycoprotein in oligodendrocytes. J Biol Chem 287(50):41888–41902. doi:10.1074/jbc.M112.413500
186. Imgrund S, Hartmann D, Farwanah H, Eckhardt M, Sandhoff R, Degen J, Gieselmann V, Sandhoff K, Willecke K (2009) Adult ceramide synthase 2 (CERS2)-deficient mice exhibit myelin sheath defects, cerebellar degeneration, and hepatocarcinomas. J Biol Chem 284(48):33549–33560. doi:10.1074/jbc.M109.031971

187. Jennemann R, Sandhoff R, Wang S, Kiss E, Gretz N, Zuliani C, Martin-Villalba A, Jager R, Schorle H, Kenzelmann M, Bonrouhi M, Wiegandt H, Grone HJ (2005) Cell-specific deletion of glucosylceramide synthase in brain leads to severe neural defects after birth. Proc Natl Acad Sci U S A 102(35):12459–12464. doi:10.1073/pnas.0500893102
188. Zhao L, Spassieva SD, Jucius TJ, Shultz LD, Shick HE, Macklin WB, Hannun YA, Obeid LM, Ackerman SL (2011) A deficiency of ceramide biosynthesis causes cerebellar purkinje cell neurodegeneration and lipofuscin accumulation. PLoS Genet 7(5):e1002063. doi:10.1371/journal.pgen.1002063
189. Pewzner-Jung Y, Park H, Laviad EL, Silva LC, Lahiri S, Stiban J, Erez-Roman R, Brugger B, Sachsenheimer T, Wieland F, Prieto M, Merrill AH Jr, Futerman AH (2010) A critical role for ceramide synthase 2 in liver homeostasis: I. Alterations in lipid metabolic pathways. J Biol Chem 285(14):10902–10910. doi:10.1074/jbc.M109.077594
190. Spassieva SD, Ji X, Liu Y, Gable K, Bielawski J, Dunn TM, Bieberich E, Zhao L (2016) Ectopic expression of ceramide synthase 2 in neurons suppresses neurodegeneration induced by ceramide synthase 1 deficiency. Proc Natl Acad Sci U S A 113(21):5928–5933. doi:10.1073/pnas.1522071113
191. Wang E, Norred WP, Bacon CW, Riley RT, Merrill AH Jr (1991) Inhibition of sphingolipid biosynthesis by fumonisins. Implications for diseases associated with Fusarium moniliforme. J Biol Chem 266(22):14486–14490
192. Merrill AH Jr, Schmelz EM, Dillehay DL, Spiegel S, Shayman JA, Schroeder JJ, Riley RT, Voss KA, Wang E (1997) Sphingolipids—the enigmatic lipid class: biochemistry, physiology, and pathophysiology. Toxicol Appl Pharmacol 142(1):208–225
193. Sadler TW, Merrill AH, Stevens VL, Sullards MC, Wang E, Wang P (2002) Prevention of fumonisin B1-induced neural tube defects by folic acid. Teratology 66(4):169–176
194. Marasas WF, Riley RT, Hendricks KA, Stevens VL, Sadler TW, Gelineau-van Waes J, Missmer SA, Cabrera J, Torres O, Gelderblom WC, Allegood J, Martinez C, Maddox J, Miller JD, Starr L, Sullards MC, Roman AV, Voss KA, Wang E, Merrill AH Jr (2004) Fumonisins disrupt sphingolipid metabolism, folate transport, and neural tube development in embryo culture and in vivo: a potential risk factor for human neural tube defects among populations consuming fumonisin-contaminated maize. J Nutr 134(4):711–716
195. Missmer SA, Suarez L, Felkner M, Wang E, Merrill AH Jr, Rothman KJ, Hendricks KA (2006) Exposure to fumonisins and the occurrence of neural tube defects along the Texas-Mexico border. Environ Health Perspect 114(2):237–241
196. De Marco P, Merello E, Mascelli S, Capra V (2006) Current perspectives on the genetic causes of neural tube defects. Neurogenetics 7(4):201–221
197. Yu K, McGlynn S, Matise MP (2013) Floor plate-derived sonic hedgehog regulates glial and ependymal cell fates in the developing spinal cord. Development 140(7):1594–1604. doi:10.1242/dev.090845
198. Murdoch JN, Copp AJ (2010) The relationship between sonic Hedgehog signaling, cilia, and neural tube defects. Birth Defects Res A Clin Mol Teratol 88(8):633–652. doi:10.1002/bdra.20686
199. Herberts CA, Kwa MS, Hermsen HP (2011) Risk factors in the development of stem cell therapy. J Transl Med 9:29. doi:10.1186/1479-5876-9-29
200. Bieberich E, Silva J, Wang G, Krishnamurthy K, Condie BG (2004) Selective apoptosis of pluripotent mouse and human stem cells by novel ceramide analogues prevents teratoma formation and enriches for neural precursors in ES cell-derived neural transplants. J Cell Biol 167(4):723–734
201. Gonsette RE (2004) New immunosuppressants with potential implication in multiple sclerosis. J Neurol Sci 223(1):87–93. doi:10.1016/j.jns.2004.04.025
202. Miron VE, Schubart A, Antel JP (2008) Central nervous system-directed effects of FTY720 (fingolimod). J Neurol Sci 274(1–2):13–17. doi:10.1016/j.jns.2008.06.031
203. Groves A, Kihara Y, Chun J (2013) Fingolimod: direct CNS effects of sphingosine 1-phosphate (S1P) receptor modulation and implications in multiple sclerosis therapy. J Neurol Sci 328(1–2):9–18. doi:10.1016/j.jns.2013.02.011

204. Miron VE, Jung CG, Kim HJ, Kennedy TE, Soliven B, Antel JP (2008) FTY720 modulates human oligodendrocyte progenitor process extension and survival. Ann Neurol 63(1):61–71. doi:10.1002/ana.21227
205. Jung CG, Kim HJ, Miron VE, Cook S, Kennedy TE, Foster CA, Antel JP, Soliven B (2007) Functional consequences of S1P receptor modulation in rat oligodendroglial lineage cells. Glia 55(16):1656–1667. doi:10.1002/glia.20576
206. Berdyshev EV, Gorshkova I, Skobeleva A, Bittman R, Lu X, Dudek SM, Mirzapoiazova T, Garcia JG, Natarajan V (2009) FTY720 inhibits ceramide synthases and up-regulates dihydrosphingosine 1-phosphate formation in human lung endothelial cells. J Biol Chem 284(9):5467–5477
207. Huwiler A, Pfeilschifter J (2008) New players on the center stage: sphingosine 1-phosphate and its receptors as drug targets. Biochem Pharmacol 75(10):1893–1900. doi:10.1016/j.bcp.2007.12.018
208. Brunkhorst R, Vutukuri R, Pfeilschifter W (2014) Fingolimod for the treatment of neurological diseases-state of play and future perspectives. Front Cell Neurosci 8:283. doi:10.3389/fncel.2014.00283
209. Saddoughi SA, Gencer S, Peterson YK, Ward KE, Mukhopadhyay A, Oaks J, Bielawski J, Szulc ZM, Thomas RJ, Selvam SP, Senkal CE, Garrett-Mayer E, De Palma RM, Fedarovich D, Liu A, Habib AA, Stahelin RV, Perrotti D, Ogretmen B (2013) Sphingosine analogue drug FTY720 targets I2PP2A/SET and mediates lung tumour suppression via activation of PP2A-RIPK1-dependent necroptosis. EMBO Mol Med 5(1):105–121. doi:10.1002/emmm.201201283
210. Lahiri S, Park H, Laviad EL, Lu X, Bittman R, Futerman AH (2009) Ceramide synthesis is modulated by the sphingosine analog FTY720 via a mixture of uncompetitive and noncompetitive inhibition in an Acyl-CoA chain length-de pend ent manner. J Biol Chem 284(24):16090–16098
211. Kourembanas S (2015) Exosomes: vehicles of intercellular signaling, biomarkers, and vectors of cell therapy. Annu Rev Physiol 77:13–27. doi:10.1146/annurev-physiol-021014-071641
212. Liour SS, Yu RK (2002) Differential effects of three inhibitors of glycosphingolipid biosynthesis on neuronal differentiation of embryonal carcinoma stem cells. Neurochem Res 27(11):1507–1512
213. Inokuchi J (2009) Neurotrophic and neuroprotective actions of an enhancer of ganglioside biosynthesis. Int Rev Neurobiol 85:319–336. doi:10.1016/S0074-7742(09)85022-8
214. Schneider JS, Bradbury KA, Anada Y, Inokuchi J, Anderson DW (2006) The synthetic ceramide analog L-PDMP partially protects striatal dopamine levels but does not promote dopamine neuron survival in murine models of parkinsonism. Brain Res 1099(1):199–205. doi:10.1016/j.brainres.2006.04.114
215. Yamagishi K, Mishima K, Ohgami Y, Iwasaki K, Jimbo M, Masuda H, Igarashi Y, Inokuchi J, Fujiwara M (2003) A synthetic ceramide analog ameliorates spatial cognition deficit and stimulates biosynthesis of brain gangliosides in rats with cerebral ischemia. Eur J Pharmacol 462(1–3):53–60. doi:10.1016/S0014299903013256
216. Lewis PM, Dunn MP, McMahon JA, Logan M, Martin JF, St-Jacques B, McMahon AP (2001) Cholesterol modification of sonic hedgehog is required for long-range signaling activity and effective modulation of signaling by Ptc1. Cell 105(5):599–612. doi:10.1016/S0092-8674(01)00369-5
217. Li Y, Zhang H, Litingtung Y, Chiang C (2006) Cholesterol modification restricts the spread of Shh gradient in the limb bud. Proc Natl Acad Sci U S A 103(17):6548–6553. doi:10.1073/pnas.0600124103
218. Krahn MP, Wodarz A (2012) Phosphoinositide lipids and cell polarity: linking the plasma membrane to the cytocortex. Essays Biochem 53:15–27. doi:10.1042/bse0530015
219. O'Neill C, Li Y, Jin XL (2015) Survival signalling in the preimplantation embryo. Adv Exp Med Biol 843:129–149. doi:10.1007/978-1-4939-2480-6_5
220. Silva LC, de Almeida RF, Castro BM, Fedorov A, Prieto M (2007) Ceramide-domain formation and collapse in lipid rafts: membrane reorganization by an apoptotic lipid. Biophys J 92(2):502–516

221. Castro BM, Silva LC, Fedorov A, de Almeida RF, Prieto M (2009) Cholesterol-rich fluid membranes solubilize ceramide domains: implications for the structure and dynamics of mammalian intracellular and plasma membranes. J Biol Chem 284(34):22978–22987. doi:10.1074/jbc.M109.026567
222. Sonnino S, Aureli M, Grassi S, Mauri L, Prioni S, Prinetti A (2014) Lipid rafts in neurodegeneration and neuroprotection. Mol Neurobiol 50(1):130–148. doi:10.1007/s12035-013-8614-4
223. Aureli M, Grassi S, Prioni S, Sonnino S, Prinetti A (2015) Lipid membrane domains in the brain. Biochim Biophys Acta 1851(8):1006–1016. doi:10.1016/j.bbalip.2015.02.001
224. Castro BM, Prieto M, Silva LC (2014) Ceramide: a simple sphingolipid with unique biophysical properties. Prog Lipid Res 54:53–67. doi:10.1016/j.plipres.2014.01.004
225. Falconer MM, Vielkind U, Brown DL (1989) Establishment of a stable, acetylated microtubule bundle during neuronal commitment. Cell Motil Cytoskeleton 12(3):169–180. doi:10.1002/cm.970120306
226. Suon S, Jin H, Donaldson AE, Caterson EJ, Tuan RS, Deschennes G, Marshall C, Iacovitti L (2004) Transient differentiation of adult human bone marrow cells into neuron-like cells in culture: development of morphological and biochemical traits is mediated by different molecular mechanisms. Stem Cells Dev 13(6):625–635. doi:10.1089/scd.2004.13.625
227. Falconer MM, Vielkind U, Brown DL (1989) Association of acetylated microtubules, vimentin intermediate filaments, and MAP 2 during early neural differentiation in EC cell culture. Biochem Cell Biol 67(9):537–544
228. Chiu FC, Feng L, Chan SO, Padin C, Federoff JH (1995) Expression of neurofilament proteins during retinoic acid-induced differentiation of P19 embryonal carcinoma cells. Brain Res Mol Brain Res 30(1):77–86
229. Herget T, Esdar C, Oehrlein SA, Heinrich M, Schutze S, Maelicke A, van Echten-Deckert G (2000) Production of ceramides causes apoptosis during early neural differentiation in vitro. J Biol Chem 275(39):30344–30354. doi:10.1074/jbc.M000714200
230. Otaegi G, Yusta-Boyo MJ, Vergano-Vera E, Mendez-Gomez HR, Carrera AC, Abad JL, Gonzalez M, de la Rosa EJ, Vicario-Abejon C, de Pablo F (2006) Modulation of the PI 3-kinase-Akt signalling pathway by IGF-I and PTEN regulates the differentiation of neural stem/precursor cells. J Cell Sci 119(Pt 13):2739–2748. doi:10.1242/jcs.03012
231. Chen Y, Li X, Eswarakumar VP, Seger R, Lonai P (2000) Fibroblast growth factor (FGF) signaling through PI 3-kinase and Akt/PKB is required for embryoid body differentiation. Oncogene 19(33):3750–3756. doi:10.1038/sj.onc.1203726
232. Hu JG, Wang YX, Wang HJ, Bao MS, Wang ZH, Ge X, Wang FC, Zhou JS, Lu HZ (2011) PDGF-AA mediates B104CM-induced oligodendrocyte precursor cell differentiation of embryonic neural stem cells through Erk, PI3K, and p38 signaling. J Mol Neurosci. doi:10.1007/s12031-011-9652-x
233. Cartwright P, McLean C, Sheppard A, Rivett D, Jones K, Dalton S (2005) LIF/STAT3 controls ES cell self-renewal and pluripotency by a Myc-dependent mechanism. Development 132(5):885–896. doi:10.1242/dev.01670

Chapter 3
Autotaxin in Stem Cell Biology and Neurodevelopment

Babette Fuss

Abbreviations

ATX	Autotaxin
CNS	Central nervous system
ENPP2	Ecto-nucleotide pyrophosphatase/phosphodiesterase 2
LPA	Lysophosphatidic acid
LPC	Lysophosphatidylcholine
LysoLPD	Lysophospholipase D
MORFO	Modulator of oligodendrocyte differentiation and focal adhesion organization
PD-Iα	Phosphodiesterase Iα

3.1 Introduction

Autotaxin, also designated ecto-nucleotide pyrophosphatase/phosphodiesterase 2 (ENPP2), phosphodiesterase Iα/autotaxin (PD-Iα/ATX), or lysophospholipase D (lysoPLD), was originally discovered as an autocrine motility-stimulating factor released by human melanoma cells [1]. This functional property was the foundation for its name autotaxin, which remains its most commonly used designation despite the realization that most of the functions assigned to autotaxin are mediated by its enzymatic activity, now known to generate the lipid signaling molecule lysophosphatidic acid (LPA) [2–5]. The gene encoding autotaxin has been described to give rise to five alternatively spliced protein products, referred to as autotaxin α, β, γ, δ, and ε

B. Fuss, Ph.D. (✉)
Department of Anatomy and Neurobiology, Virginia Commonwealth University School of Medicine, 1101 East Marshall Street, Richmond, VA 23298, USA
e-mail: Babette.Fuss@vcuhealth.org

A. Pébay, R.C.B. Wong (eds.), *Lipidomics of Stem Cells*, Stem Cell Biology and Regenerative Medicine, DOI 10.1007/978-3-319-49343-5_3

[6–8]. These isoforms of autotaxin display characteristic expression patterns, whereby autotaxin γ, originally identified as phosphodiesterase Iα (PD-Iα), is considered a brain-specific isoform [6, 9, 10]. The functional consequences of autotaxin's alternative splicing events are largely unknown; even though, an insertion uniquely found present in autotaxin α and ε has been described to mediate recruitment to the cell membrane through interaction with heparin sulfate proteoglycans [11]. Unless stated otherwise, the term autotaxin is used in this chapter collectively for all isoforms.

Autotaxin has been established as a secreted protein that is expressed by a large variety of tumor cells and within a number of different tissues during normal development and in the adult [6, 9, 10, 12–16]. Furthermore, high protein levels of autotaxin have been observed in biological fluids such as plasma and cerebrospinal fluid [3, 4, 17]. Given the historical discovery of autotaxin as a tumor cell motility-stimulating factor, it is of no surprise that a major focus in the research related to autotaxin has long remained in tumor cell biology. More recently, however, a plethora of additional functions have emerged including roles in stem cell biology and neurodevelopment. The following paragraphs will review the major characteristics of autotaxin from a historical perspective, autotaxin's major structure–function relationships, and autotaxin's evolving roles in stem cell biology and neurodevelopment.

3.2 A Historical Perspective

Autotaxin was originally characterized as a secreted "autocrine motility factor" of 100–130 kDa due to its ascribed function in stimulating both random and directed migration of human melanoma cells at picomolar concentrations [1]. Subsequent cDNA cloning revealed homology to PC-1, a pyrophosphatase/type I phosphodiesterase, and led to the classification of autotaxin as an ecto-/exo-enzyme possessing 5′-nucleotide phosphodiesterase/ATP pyrophosphatase activity [18–20]. Recognition of a conserved structural relationship between the catalytic domains of B10/gp130RB13-6/PD-Ibeta [21, 22], PC-1, and autotaxin prompted a change in nomenclature and led to the creation of the family of nucleotide pyrophosphatases/phosphodiesterases (NPPs) or ecto-NPPs (ENPPs) [23, 24]. To date, this protein family encompasses seven human genes, which are numbered according to the order of their discovery, whereby autotaxin is referred to as ENPP2 [25].

As eluded to above, ENPPs were originally characterized to hydrolyze pyrophosphate or phosphodiester bonds in (di)nucleotides and their derivatives. However, it is becoming increasingly apparent that some of the family members prefer other substrates. In this context, autotaxin has been uncovered to act primarily as an extracellular lysophospholipase-D (lysoPLD) generating the lipid signaling molecule lysophosphatidic acid (LPA) [2–5], and it is now broadly accepted that autotaxin's role as nucleotide phosphodiesterase is, if at all, of minor physiological importance [3–5, 26, 27]. Consistent with this point of view, autotaxin's initially discovered tumor cell motility-stimulating function has been found to require not only its catalytic activity but to also be largely mediated via the generation of LPA [4, 13, 28] and activation of one or more of LPA's cognate G protein-coupled receptors [29, 30]. Interestingly, a critical role of G protein-coupled receptors in mediating motility responses stimulated by autotaxin had already been indicated in the initial studies in which the protein was discovered [1]. A prominent role of autotaxin as LPA-producing enzyme could be further corroborated through findings made in genetically modified mice, in which plasma LPA levels were seen critically affected by deletion of the gene encoding autotaxin or by transgenic overexpression of the enzymatically active protein [27, 31, 32]. Next to LPA, sphingosine-1-phosphate [33] and cyclic phosphatidic acid [34] have been reported as enzymatic products of autotaxin's catalytic activity; their physiological importance, however, remains unclear.

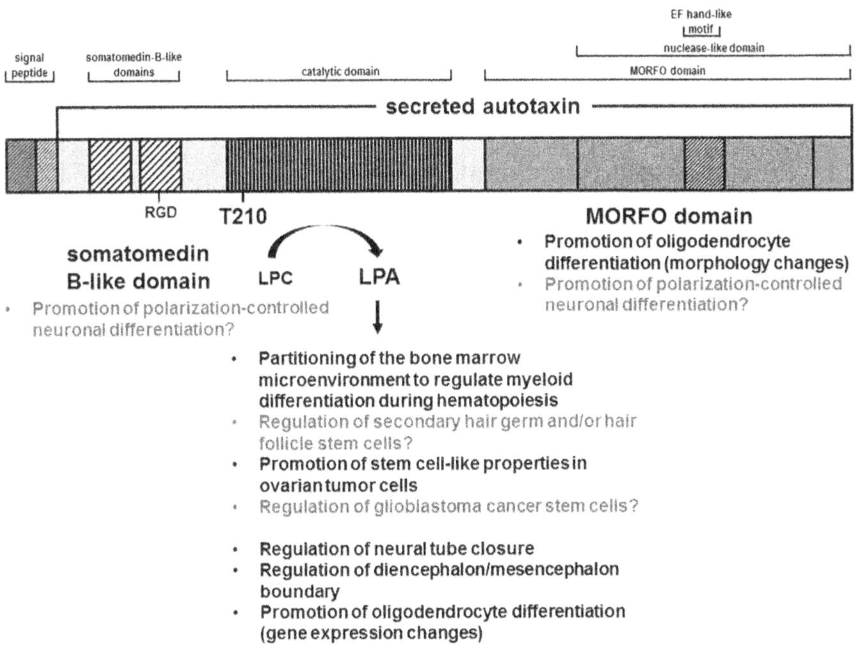

Fig. 3.1 Scheme of the structure–function domains of autotaxin and assigned functions in stem cell biology and neurodevelopment. The N-terminal hydrophobic sequence of autotaxin is a signal peptide, thus resulting in the secretion of the protein. Two somatomedin B-like domains are located at the N-terminal end of the protein. The catalytic domain of autotaxin functions as lysoPLD generating LPA, which, in turn, exerts its effects through binding to its cognate G protein-coupled receptors. Catalytic activity is dependent on the catalytic site residue T210. At the C-terminal end, the Modulator of Oligodendrocyte Remodeling and Focal adhesion Organization (MORFO) domain entails the nuclease-like domain, which is enzymatically inactive, and an EF hand-like motif. Functions assigned to autotaxin's domains are listed at the bottom. *?* indicates functions for which solid experimental evidence is lacking or for which the respective domain has not yet been identified

From a functional point of view, autotaxin has been implicated in a number of physiological and pathophysiological processes (Fig. 3.1). In light of its long established role in cancer development and metastasis [8], it is of no surprise that the autotaxin-LPA axis has gained much attention as a potential therapeutic target for cancer treatment [35–43]. With regard to more physiological functions, first insights came from the characterization of autotaxin knockout mice, i.e., mice in which critical parts of the gene encoding autotaxin are deleted leading to an ubiquitous lack of autotaxin expression [27, 31, 44]. These mice display an embryonically lethal phenotype with most noticeable deficits in blood vessel maturation and/or stabilization, which are caused primarily by a lack of autotaxin's catalytic and LPA-producing activity [27, 44–46]. Additional studies in the zebrafish substantiate such a role of the autotaxin-LPA axis [47], and they provide support for evolutionary conservation of autotaxin's physiological functions. An interesting logistic follow-up of these findings lies in the projection that autotaxin's angiogenic role may also contribute to its tumor progression enhancing effects [48, 49]. Physiological roles in addition to the above have emerged for autotaxin [50, 51], and they point toward critical functions during stem cell biology and nervous system development.

3.3 Structure and Functional Domains

Autotaxin has been characterized as a multifunctional and multi-modular protein consisting, next to its catalytic domain, of two repetitive N-terminal somatomedin B-like domains and a domain involved in the regulation of adhesion, the so-called modulator of oligodendrocyte differentiation and focal adhesion organization (MORFO) domain, which entails an inert (catalytically inactive) nuclease-like domain and a single EF hand-like motif [8, 52–54] (Fig. 3.1). Autotaxin represents a phylogenetically conserved protein with 93% sequence identity between human and mouse and a high conservation of functionally important residues [55]. As introduced above, autotaxin is best known for its catalytic activity, whereby a threonine residue located in the middle of the catalytic domain (T210 in human; T209 in rat and mouse) serves as a point of transient binding of reaction intermediates and is, thereby, essential for catalytic activity. Structurally, this site is located within a hydrophilic binding groove shown to accommodate the glycerol backbone of the lipid products [55, 56]. The catalytically active domain also includes two essential zinc ions for which the coordination shell is constructed by conserved aspartate and histidine residues [26, 33, 57]. The multitude of residues involved in the catalytic domain fold point toward a very compact 3D architecture of autotaxin. Indeed, crystal structure analysis revealed that the central catalytic domain interacts with both the N-terminal somatomedin B-like domains and the C-terminal nuclease-like domain [55, 56, 58–60]. This arrangement is stabilized by an N-linked glycan chain and a disulfide bridge between the catalytic and nuclease-like domains both of which were found to be essential for enzymatic activity [61–63]. Such insights gained from the structural analysis of autotaxin explain the involvement of large parts of the protein in its function as lysoPLD as well as the observed preference of autotaxin for acyl chain length and saturation of its lysophosphatidylcholine (LPC) substrates (18:0 << 16:0 < 14:0 and 18:0 << 18:1) [3, 55, 64]. In addition, they provide new inspiration for the design and selection of modulators with high selectivity and potency for autotaxin's enzymatic activity [65–72].

It has been suggested that LPA is generated by autotaxin in the proximity of the LPA-activated cell surface receptor [73–75]. Such a localized synthesis would be compatible with the observed low concentrations of LPA in plasma and could be involved in targeting LPA to distinct LPA receptors. Integrin-dependent association of plasma autotaxin with activated platelets initially pointed toward the integrin-binding RGD motif located within the second somatomedin B-like domain [12] as an underlying mechanism. Interestingly, however, mutation analysis revealed that binding occurs in an RGD-independent manner and involves the solvent-exposed surface of autotaxin's second somatomedin B-like domain [56, 76]. Functionally, autotaxin's ability to bind to integrins has been proposed to critically contribute to the protein's effects on rapid directional cell migration [77].

Autotaxin has initially been thought to represent a type II transmembrane protein; however, it is now well recognized to be synthesized as a pre-pro-protein that is secreted upon removal of the N-terminal signal peptide and further trimming by a furin-type protease [78, 79]. Interestingly, there is evidence that autotaxin secretion requires N-glycosylation and is dependent on amino acid residues located within autotaxin's C-terminal nuclease-like domain [62, 80]. Despite the likely physiological significance of regulatory mechanisms controlling autotaxin secretion and expression, a comprehensive picture still needs to be developed [81–88].

As becomes evident from the above, a major focus has been on the enzymatic activity of autotaxin. Notwithstanding its critical role in a plethora of biological activities assigned to autotaxin, there is increasing evidence for nonenzymatic functions of the protein. For example, cell positioning in the ventricular zone of the CNS has been found regulated by non-catalytic activities of autotaxin [89]. In addition, the MORFO domain was found to be critical for promoting the morphological maturation of oligodendrocytes [90–93]. As such nonenzymatic functions have been primarily reported to affect neurodevelopmental processes in the CNS, they will be discussed in more detail below.

3.4 Functional Roles in Stem Cell Biology

In a multicellular organism, different cell types have to be established during development and maintained in the adult in well-balanced numbers and proportions. A critical component of this intricate process lies in the ability of stem cells to undergo symmetric divisions to expand the stem cell pool and asymmetric divisions to produce, in addition to self-renewed stem cells, daughter cells that will differentiate [94–96]. The decision of whether to undergo symmetric or asymmetric division and the cell fate identity of the differentiating daughter cell is controlled by a number of intrinsic as well as extrinsic cues [97]. One of the latter that has emerged as a critical factor in stem cell regulation is LPA [98, 99]. Given autotaxin's well-established identity as a prominent LPA generating enzyme, functional roles of autotaxin in stem cell biology are seemingly apparent; yet, we are only at the beginning of understanding the potential importance of autotaxin in regulating stem cells and their progenies.

3.4.1 Embryonic Stem Cells

Embryonic stem cells have the ability to proliferate and replicate themselves indefinitely (self-renewal) while still maintaining the developmental potential to form any cell type of the body; they are thus considered pluripotent [100, 101]. Their projected promise in cellular therapy has prompted intense research into their regulation and biology [102, 103]. In this context, it is of note that totipotency, defined as the ability to develop into a complete organism by not only generating all the cells of the body but also organizing them in a specific temporal and spatial sequence, is not a typical feature of an embryonic stem cell [104–107].

As mentioned above, recent evidence suggests that LPA and its downstream signaling pathways play significant roles in the regulation of various aspects of stem cell biology. With regard to embryonic stem cells, the expression of LPA receptors by these cells is highly suggestive of LPA as a physiologically relevant signal [108–110]. This point of view may be further supported by the observation that LPA can induce calcium signals and early gene *c-fos* and *c-myc* expression in embryonic stem cells [110–112]. Consistent with a thus anticipated role of LPA signaling during early development, LPA was shown to promote preimplantation development of mouse embryos from the four-cell to blastocyst stage possibly via the stimulation of cell division at the pronuclear and/or 2-cell stage [113]. In accord with the above, LPA treatment of mouse embryonic stem cells was described to increase their proliferation and DNA synthesis rate [110]. Somewhat controversially, however, LPA was found to not affect the size or morphology of human embryonic stem cell colonies [108]. The mechanistic underpinnings for these apparent differences still need to be determined.

In contrast to the receptors of the LPA signaling axis, the LPA generating enzyme autotaxin does not appear to be expressed by embryonic stem cells [114], and its reported developmental expression pattern does not include the early preimplantation stages [14]. Thus, any LPA effects at these early developmental stages likely involve autotaxin expression in the uterus during early pregnancy [115–117] and/or may rely on an autotaxin-independent LPA synthesis pathway [116, 118].

3.4.2 Tissue-Specific Stem Cells

Tissue-specific stem cells, i.e., stem cells that produce only a limited set of specialized cells characteristic of a particular tissue, are found in the developing and adult organism within specialized microenvironments, so-called niches [119–121].

Such niches are considered anatomically defined sites of communication between stem cells and their respective tissues, and they are thought to enable a coordinated control over various stem cell activities, ranging from dormancy and activation to migration and differentiation. Related to tissue-specific stem cells, roles for autotaxin have been investigated in association with hematopoietic and hair follicle stem cell niches. In addition, autotaxin functions have been implicated to regulate neural stem and progenitor cells; these will be discussed below under the topic of neurodevelopment.

Within the hematopoietic stem cell niche located in the bone marrow, populations of "long-term" hematopoietic stem cells and "short-term" hematopoietic stem/progenitor cells interact with a complex multicellular microenvironment that includes hematopoietic stem cell progenies and non-hematopoietic cell types [122]. Importantly, the hematopoietic niche is considered to be dynamic and able to respond to bone marrow stress, such as cell loss induced by for example toxic substances, including chemotherapeutic agents [123, 124]. In a study published by Ortlepp et al. [125], autotaxin was described to be expressed by human hematopoietic stem/progenitor cells, as defined by the expression of the cell surface marker CD34. In addition, autotaxin was found to stimulate proliferation and motility of these CD34-positive cells via the generation of LPA and a likely autocrine mechanism [125]. While CD34 is predominantly regarded as a marker for hematopoietic stem/progenitor cells, it is also present on myeloid/erythroid progenitor cells [126–128]. Importantly, recent studies indicate a more widespread expression of CD34, including an expression by mesenchymal stem/stromal cells [129]. Mesenchymal stem/stromal cells are considered part of the hematopoietic stem cell niche. They can in vitro be stimulated to differentiate into osteoblasts, chondrocytes, and adipocytes; the in vivo differentiation hierarchy, however, appears to still be only hypothetical [130, 131]. Thus, the primary in vivo function of mesenchymal stem/stromal cells may lie in their ability to provide a microenvironment for other stem cells. In the context of the abovementioned effects of autotaxin on CD34-positive cells, it is of interest that primary bone marrow-derived mesenchymal stem/stromal cells have been described to express autotaxin [132]. In addition, autotaxin was found to enhance the migration of human umbilical cord blood-derived mesenchymal stem/stromal cells in a wound repair model system [133], and LPA was shown to function as a chemoattractant for bone marrow-derived mouse mesenchymal stem/stromal cells [134]. These findings would be consistent with the previously described motility-stimulating effects toward CD34-positive cells [125]. On the other hand, LPA treatment of culture-expanded mesenchymal stem/stromal cells from human bone marrow has been described to inhibit cell migration through the activation of intracellular Rho and increased actin stress fiber formation [135]. The role of the autotaxin-LPA signaling axis in regulating mesenchymal stem/stromal cells appears thus rather confusing, a situation that may be explained by cellular heterogeneity within the utilized mesenchymal stem/stromal cell populations and/or a lack of well-established and reliable surface markers [136, 137]. With regard to hematopoietic stem cells, a critical role of autotaxin is supported by the finding that LPA, potentially generated by autotaxin [132], triggers enhanced cell motility and invasion of primitive hematopoietic cells into stromal cell layers [138, 139]. More recent data revealed that LPA-mediated in vitro stimulation of CD34-positive human hematopoietic progenitors induces myeloid but not lymphoid differentiation [132]. In these studies, LPA was additionally characterized as an enhancer of myeloid progenitor cell migration and proliferation. Given the relative high expression of autotaxin by perivascular stromal cells, it was proposed that autotaxin-derived LPA may mediate anatomical partitioning of the bone marrow microenvironment and, thereby, regulate myeloid differentiation during hematopoiesis. Thus, the current data point toward a role of autotaxin and its enzymatic activity in regulating hematopoietic stem cells and hematopoiesis via mostly paracrine effects through the expression of autotaxin by stromal cells.

The hair follicle stem cell niche has emerged as an important paradigm to study stem cells in quiescence and in action due to the unique synchronized cycles of extended periods of rest and brief bouts of activation [121, 140]. Hair (re)generation is fueled by hair follicle stem cells, which are located in the outer layer of the bulge at the very bottom of each hair follicle, and a small cluster of cells beneath it, known as the secondary hair germ. During the phase of activation, specialized mesenchymal cells, referred to as the dermal papilla, stimulate cells of the secondary hair germ. This event leads to the generation of the hair follicle transit amplifying cell matrix, which consists of cells that divide a finite number of times until they become differentiated [141, 142]. Interestingly, autotaxin has been described to represent one of the highest expressed signature genes in the dermal papilla of growing hair follicles [143], thus suggesting a potential critical role of autotaxin during hair follicle morphogenesis. Surprisingly, however, conditional autotaxin knockout studies revealed no effect on follicle numbers, lengths, and sizes, but rather identified lipase H, also known as phosphatidic acid (PA)-selective phospholipase $A_1\alpha$, as an LPA-producing enzyme that could functionally compensate for the genetic deletion of autotaxin and/or represent the physiologically more prominent LPA generating enzyme in the hair follicle [144]. The latter may be supported by studies demonstrating a critical role of lipase H in regulating the formation of the inner root sheath [145, 146]. However, there may also still be room for functionally redundant roles of lipase H and autotaxin in the regulation of for example secondary hair germ and/or hair follicle stem cells.

3.4.3 Cancer Stem Cells

Even though still to some extent controversial and possibly not applicable to all tumors, the cancer stem cell model has provided a conceptual framework for explaining functional and phenotypic heterogeneity among cells within a tumor [147–151]. In this model, cancers are organized into a hierarchy of subpopulations of tumorigenic cancer stem cells and their non-tumorigenic progeny, whereby it is the cancer stem cell that is thought to drive tumor growth and disease progression, possibly through therapy resistance and metastasis [148, 149, 152].

Given autotaxin's well-documented role in tumorigenesis, it is of no surprise that autotaxin has been implicated in regulating cancer stem cell biology. In this context, cancer stem cells have been described present within primary human ovarian tumors [153], and the expression of autotaxin has been associated with chemoresistance of ovarian tumor cells [154, 155]. More recently, it was shown that autotaxin can confer stem cell-like properties to ovarian carcinoma cells and that silencing of autotaxin expression can lead to increased susceptibility to chemotherapy drugs [156]. These findings suggest that the autotaxin-LPA axis could present a promising target for the development of therapeutic strategies directed at ovarian cancer stem cells within epithelial ovarian tumors.

The other type of cancer stem cell for which the autotaxin-LPA axis has been proposed as a therapeutic target are those present in glioblastoma multiforme, the most highly malignant type of brain tumor [43]. The evidence here, however, is indirect and complicated by the complexity of LPA signaling in glioblastoma [157]. Autotaxin has been found highly expressed in glioblastoma multiforme, and its expression and enzymatic activity have been implicated in facilitating tumor cell invasion and possibly neovascularization [28, 158, 159]. In addition, inhibition of autotaxin's enzymatic activity has been proposed to enhance radiosensitivity of glioblastoma multiforme cells [159]. The idea of a potential role of autotaxin-LPA signaling in glioblastoma cancer stem cells comes from the observation that the LPA receptor LPA1 is highly expressed in CD133 (prominin1)-positive glioblastoma cancer stem cells presumably leading to enhanced migratory responses to LPA [43, 160]. Even though cancer stem cells have been described present in glioblastoma multiforme, tumorigenic cells can be found in both CD133-positive as well as negative cell population [161, 162]; CD133's role as sole marker for cancer stem cells has thus been questioned [148, 161, 162]. Consequently, much more research will be necessary to establish the extent to which there may be a role for autotaxin-LPA signaling in regulating what is thought to represent a cancer stem cell.

3.5 Functional Roles in Neurodevelopment

During development of the vertebrate embryo, the process of neural induction, which occurs at gastrulation, leads to the generation of the neural plate, which consists of cells that are derived from the ectoderm and restricted to giving rise to neural tissue [163]. Folding of the neural plate and subsequent closure at the dorsal end leads to the formation of the neural tube in a process that is referred to as neurulation. At this early stage, four primary brain regions can be distinguished; these will give rise to the spinal cord, hindbrain, midbrain, and forebrain. First evidence for a critical role of autotaxin in these early neurodevelopmental events came from the characterization of autotaxin knockout mice. Next to the previously mentioned vascular defects, large cavities or effusions in the future forebrain region and a lack of proper neural tube closure were observed in the majority of autotaxin knockout embryos at embryonic day 8.5 and 9.5, respectively [27, 31, 44]. These neurodevelopmental phenotypes were found associated with a decrease in proliferation and an increase in apoptosis [44, 164]. Importantly, they are thought to be caused primarily due to a lack of autotaxin's enzymatic function. First, mice expressing enzymatically inactive but not fully functional autotaxin were reported to display early embryonic lethality similar to autotaxin knockout mice [46]. Second, addition of LPA to autotaxin knockout embryonic explant cultures was found to restore explant size to wild-type levels [44]. Third, inhibition of LPA receptors in ex vivo whole embryo cultures has been described to lead to head cavity formations similar to the ones seen in autotaxin knockout mice [165]. The latter study also provides some mechanistic insight by demonstrating that inhibition of Rho/ROCK as well as actin polymerization also results in the formation of head cavities. A crucial role of autotaxin in brain development could be further underscored by studies done in the developing chick embryo in which silencing of endogenous autotaxin expression was found to affect the integrity of the diencephalon (posterior part of the forebrain)-mesencephalon (midbrain) boundary and the proliferation of caudal diencephalon-mesencephalon neuroepithelial cells. Autotaxin expression has been observed during early development in both mouse and chick embryos [14, 164, 165]. This expression includes cells of the floor plate, a region located at the ventral midline of the embryonic neural tube and known to release factors important for the formation of a fully functional nervous system [14, 15, 165]. Thus, it has been proposed that the neurodevelopmental defects seen upon genetic deletion of autotaxin are due to local autotaxin deficiency and not secondary to circulatory failure [27]. Such a vascular deficiency independent effect of autotaxin knockout on neurodevelopment may be supported by a lack of neural tube defects in Gα13 knockout mice, which have been described to phenotypically display vascular deficiencies similar to the ones seen in autotaxin knockout embryos [44, 45]. Nevertheless, to clearly define the roles of autotaxin during neurodevelopment and to unequivocally dissect neural and vascular phenotypes, conditional deletion of autotaxin in well-defined cell types and/or tissues will be necessary.

3.5.1 Neural Stem Cells and Neurogenesis

Neurogenesis, the generation of neurons from neuroepithelial stem cells, occurs during early vertebrate development in the embryonic neural tube. Receptors for LPA have been described to be expressed by neural stem/progenitor cells, and LPA has been implicated in regulating their morphological rearrangements, proliferation, and differentiation. However, the effects of LPA on neural stem/progenitor cells seem to be dependent on a complex set of factors, including the region of origin, species, and developmental stage [98, 99]. Thus, correlating any of the early developmental phenotypes seen upon autotaxin knockout in the mouse with existing data about LPA signaling in neural stem/progenitor cells is currently challenging.

At the onset of neurogenesis, the neural tube wall of the developing cerebral cortex of the forebrain, termed the ventricular zone, is occupied by neural stem/progenitor cells which are referred to as radial glia [166–168]. These cells initially span the entire thickness of the wall by extending processes to both the ventricular (inner) and pial (outer) surfaces of the developing brain. Once neurons are generated from these stem cells, they migrate along the processes of radial glia toward the pial surface to form a series of distinct layers, ultimately forming the cerebral cortex [163]. During this developmental time period, autotaxin has been described to be expressed by cells located within the ventricular zone as well as the cortical plate occupied by newly generated neurons [15, 16, 89]. Interestingly, downregulation (knockdown) of autotaxin expression or conditional autotaxin gene deletion specifically at this developmental stage was found to lead to a distorted cellular morphology with most cells appearing round instead of elongated and in some cases lacking an association of their processes or endfeet with the ventricular surface [89]. It is of note that this developmental stage is past the survival time for conventional autotaxin knockout embryos. In addition, at this developmental stage, cells located within the ventricular zone are characterized by apical (ventricular) and basal (pial) subcellular regions leading to what is being referred to as polarization [169]. This polarization has been proposed to be critical for regulating proliferation and neurogenesis. In this context, downregulation of autotaxin expression was found to disrupt cellular polarity and to result in an increase in the number of cells located within the ventricular zone associated with a decrease in the percentage of cells expressing a postmitotic neuronal marker (Tuj1). Most remarkably, these phenotypes could be rescued by the expression of catalytically inactive autotaxin [89]. These findings highlight that during neurodevelopment at least some functions of autotaxin are mediated by domains other than its catalytically active site. In future studies it will be important to establish which of autotaxin's domains are involved in the regulation of neurogenesis and the establishment of neuronal stem/progenitor cell polarity. Potential candidates may be the integrin binding somatomedin B-like domain or the MORFO domain. In addition, it will be crucial to dissect functions mediated by non-catalytic and catalytic activities. In this context, it is worth mentioning that the role of LPA in regulating early developmental neurogenesis is still puzzling. On the one hand, LPA has been found to inhibit neuronal differentiation of neural stem/progenitor cells derived from human embryonic stem cells or induced pluripotent stem cells [109, 170], while on the other it has been described to induce neuronal differentiation from mouse cortical neuroblasts and rat embryonic neural stem cells [171, 172].

3.5.2 *Oligodendrogenesis*

Oligodendrocytes, the myelinating cells of the CNS, originate from distinct regions of the ventricular zone as well as from the so-called subventricular zone located postnatally within the developing cerebral cortex of the forebrain [173, 174].

Temporally, several waves of oligodendrocyte differentiation have been described, whereby the early wave typically involves more ventral sources. Interestingly, soluble factors derived from the floorplate have been implicated in driving oligodendrogenesis in the developing spinal cord and hindbrain during the first wave [175]. Thus, the high expression of autotaxin observed in the floorplate prompted studies toward the elucidation of autotaxin in oligodendrogenesis. These studies were done in the developing zebrafish, and they revealed that indeed downregulation of autotaxin expression or mutation of the autotaxin gene inhibits the appearance of early stages of the oligodendrocyte lineage in the hindbrain [176, 177]. Importantly, the oligodendrogenesis promoting function of autotaxin was found to be mediated by its enzymatic activity [177]. In addition, and from a mechanistic point of view, autotaxin, via its LPA generating activity, was found to induce epigenetic changes that had previously been shown to be crucial for oligodendrogenesis [178]. Based on these findings, the following model has been proposed: LPA, generated via the lyso-PLD activity of autotaxin, activates one (or more) of its cognate LPA receptors, which have been found expressed by cells of the oligodendrocyte lineage [179–183]. Activation of the above autotaxin-LPA axis initiates a downstream signaling cascade leading to the activation of histone deacetylation, which in turn mediates repression of transcriptional inhibitors of oligodendrocyte differentiation, thereby promoting gene expression changes that are associated with the transition from an oligodendrocyte progenitor to an early-stage differentiating oligodendrocyte [177]. Notably, while this model was initially characterized in the zebrafish, its validity for rodent oligodendrogenesis could be established, suggesting an evolutionarily conserved mechanism [177].

Once oligodendrocyte progenitor cells have been generated in the respective regions of the ventricular zone or the postnatal subventricular zone, they migrate into prospective white matter regions where they undergo discrete steps of differentiation, which are characterized by typical changes in morphology and gene expression [184, 185]. Autotaxin has been found expressed by oligodendrocytes during this process of differentiation/maturation, thus suggesting the existence of potential autocrine regulatory functions. In this regard, autotaxin has been described to facilitate the morphological maturation of post-migratory, premyelinating oligodendrocytes [92], a process that is characterized by the transition of cells that extend a few processes to cells that generate a highly complex process network. Remarkably, this functional property of autotaxin was found to be independent of its enzymatic activity and to be mediated by its MORFO domain [90–92, 186]. In early studies, the MORFO domain was described to antagonize adhesion of oligodendrocytes to naturally occurring extracellular matrix molecules such as fibronectin in an active fashion involving pertussis toxin-sensitive G-proteins and a reorganized assembly of focal adhesions, i.e., macromolecular complexes linking the extracellular matrix with the cell surface and the underlying actin cytoskeleton [90–92]. This finding classified autotaxin as a matricellular protein, i.e., a protein that mediates an intermediate adhesive state and, thereby, supports cellular remodeling [187]. Subsequently, expression of the purinergic receptor P2Y12 was identified as a critical component of the mechanism mediating the effects of autotaxin's MORFO domain on the morphological maturation of differentiating oligodendrocytes. Taken together, the picture is emerging that autotaxin regulates oligodendrocyte differentiation via the concerted action of its enzymatic activity and its functions mediated by the MORFO domain. This dual domain mechanism may be crucial for the complex coordination of gene expression and morphological changes as seen during a well-controlled maturation from an oligodendrocyte progenitor cell to a fully functional oligodendrocyte.

3.6 Conclusion

Much research has been conducted since the initial discovery of autotaxin. Nevertheless, there are a number of critical outstanding questions. With regard to autotaxin's enzymatic activity for example, little is known about how its activity is regulated and targeted to specific cell surface receptors of the LPA receptor family. In addition, few details are known about the regulation of autotaxin expression and secretion. Such questions become particularly relevant in the context of pathological conditions in which autotaxin expression and secretion are upregulated. Related to stem cells and the nervous system, glioblastoma multiforme comes into mind [43]. However, other pathologies have been identified. For example, it has been shown that autotaxin is strongly upregulated in reactive astrocytes following neurotrauma [16]. In addition, an upregulation of autotaxin expression has been implicated in nerve injury-induced neuropathic pain [188–190]. Interestingly, in contrast to autotaxin's critical roles during development, its expression appears largely dispensable for homeostatic maintenance in the adult [191], an observation that is encouraging in the context of therapeutic interventions designed to inhibit autotaxin and its enzymatic activity.

Most functions of autotaxin, whether physiologic or pathological, have been assigned to its enzymatic activity. However, there is increasing evidence for non-catalytic functions of autotaxin. Interestingly, these have so far been identified primarily in mechanisms related to nervous system development and targeting central nervous system cells [89, 186]. These findings raise new questions related to the extent of such non-enzymatic functions and their interrelationship with autotaxin's ability to generate LPA. In the long term, it is the hope that better understanding the biology of autotaxin will not only advance our understanding of developmental and homeostatic processes, including those involving stem cells and the nervous system, but also aid in the development of innovative strategies to counteract pathological effects of autotaxin as seen under conditions where its expression is upregulated.

Acknowledgements This work was supported by grants from the National Institute of Health (B.F) and the National Multiple Sclerosis Society (B.F.).

References

1. Stracke ML, Krutzsch HC, Unsworth EJ, Arestad A, Cioce V, Schiffmann E, Liotta LA (1992) Identification, purification, and partial sequence analysis of autotaxin, a novel motility-stimulating protein. J Biol Chem 267(4):2524–2529
2. Moolenaar WH (2002) Lysophospholipids in the limelight: autotaxin takes center stage. J Cell Biol 158(2):197–199
3. Tokumura A, Majima E, Kariya Y, Tominaga K, Kogure K, Yasuda K, Fukuzawa K (2002) Identification of human plasma lysophospholipase D, a lysophosphatidic acid-producing enzyme, as autotaxin, a multifunctional phosphodiesterase. J Biol Chem 277(42):39436–39442
4. Umezu-Goto M, Kishi Y, Taira A, Hama K, Dohmae N, Takio K, Yamori T, Mills GB, Inoue K, Aoki J, Arai H (2002) Autotaxin has lysophospholipase D activity leading to tumor cell growth and motility by lysophosphatidic acid production. J Cell Biol 158(2):227–233
5. Ferry G, Tellier E, Try A, Gres S, Naime I, Simon MF, Rodriguez M, Boucher J, Tack I, Gesta S, Chomarat P, Dieu M, Raes M, Galizzi JP, Valet P, Boutin JA, Saulnier-Blache JS (2003) Autotaxin is released from adipocytes, catalyzes lysophosphatidic acid synthesis, and activates preadipocyte proliferation. Up-regulated expression with adipocyte differentiation and obesity. J Biol Chem 278(20):18162–18169
6. Giganti A, Rodriguez M, Fould B, Moulharat N, Coge F, Chomarat P, Galizzi JP, Valet P, Saulnier-Blache JS, Boutin JA, Ferry G (2008) Murine and human autotaxin alpha, beta, and gamma isoforms: gene organization, tissue distribution, and biochemical characterization. J Biol Chem 283(12):7776–7789
7. Hashimoto T, Okudaira S, Igarashi K, Hama K, Yatomi Y, Aoki J (2012) Identification and biochemical characterization of a novel autotaxin isoform, ATXdelta, with a four-amino acid deletion. J Biochem 151(1):89–97
8. Federico L, Jeong KJ, Vellano CP, Mills GB (2016) Autotaxin, a lysophospholipase D with pleomorphic effects in oncogenesis and cancer progression. J Lipid Res 57(1):25–35
9. Narita M, Goji J, Nakamura H, Sano K (1994) Molecular cloning, expression, and localization of a brain-specific phosphodiesterase I/nucleotide pyrophosphatase (PD-I alpha) from rat brain. J Biol Chem 269(45):28235–28242
10. Fuss B, Baba H, Phan T, Tuohy VK, Macklin WB (1997) Phosphodiesterase I, a novel adhesion molecule and/or cytokine involved in oligodendrocyte function. J Neurosci 17(23):9095–9103
11. Houben AJ, van Wijk XM, van Meeteren LA, van Zeijl L, van de Westerlo EM, Hausmann J, Fish A, Perrakis A, van Kuppevelt TH, Moolenaar WH (2013) The polybasic insertion in autotaxin alpha confers specific binding to heparin and cell surface heparan sulfate proteoglycans. J Biol Chem 288(1):510–519
12. Kawagoe H, Soma O, Goji J, Nishimura N, Narita M, Inazawa J, Nakamura H, Sano K (1995) Molecular cloning and chromosomal assignment of the human brain-type phosphodiesterase I/nucleotide pyrophosphatase gene (PDNP2). Genomics 30(2):380–384
13. Lee HY, Clair T, Mulvaney PT, Woodhouse EC, Aznavoorian S, Liotta LA, Stracke ML (1996) Stimulation of tumor cell motility linked to phosphodiesterase catalytic site of autotaxin. J Biol Chem 271(40):24408–24412
14. Bachner D, Ahrens M, Betat N, Schroder D, Gross G (1999) Developmental expression analysis of murine autotaxin (ATX). Mech Dev 84(1–2):121–125
15. Ohuchi H, Hayashibara Y, Matsuda H, Onoi M, Mitsumori M, Tanaka M, Aoki J, Arai H, Noji S (2007) Diversified expression patterns of autotaxin, a gene for phospholipid-generating enzyme during mouse and chicken development. Dev Dyn 236(4):114–1143
16. Savaskan NE, Rocha L, Kotter MR, Baer A, Lubec G, van Meeteren LA, Kishi Y, Aoki J, Moolenaar WH, Nitsch R, Brauer AU (2006) Autotaxin (NPP-2) in the brain: cell type-specific expression and regulation during development and after neurotrauma. Cell Mol Life Sci 64(2):230–243
17. Sato K, Malchinkhuu E, Muraki T, Ishikawa K, Hayashi K, Tosaka M, Mochiduki A, Inoue K, Tomura H, Mogi C, Nochi H, Tamoto K, Okajima F (2005) Identification of autotaxin as

a neurite retraction-inducing factor of PC12 cells in cerebrospinal fluid and its possible sources. J Neurochem 92(4):904–914
18. Murata J, Lee HY, Clair T, Krutzsch HC, Arestad AA, Sobel ME, Liotta LA, Stracke ML (1994) cDNA cloning of the human tumor motility-stimulating protein, autotaxin, reveals a homology with phosphodiesterases. J Biol Chem 269(48):30479–30484
19. Stracke ML, Clair T, Liotta LA (1997) Autotaxin, tumor motility-stimulating exophosphodiesterase. Adv Enzyme Regul 37:135–144
20. Clair T, Lee HY, Liotta LA, Stracke ML (1997) Autotaxin is an exoenzyme possessing 5'-nucleotide phosphodiesterase/ATP pyrophosphatase and ATPase activities. J Biol Chem 272(2):996–1001
21. Deissler H, Lottspeich F, Rajewsky MF (1995) Affinity purification and cDNA cloning of rat neural differentiation and tumor cell surface antigen gp130RB13-6 reveals relationship to human and murine PC-1. J Biol Chem 270(17):9849–9855
22. Jin-Hua P, Goding JW, Nakamura H, Sano K (1997) Molecular cloning and chromosomal localization of PD-Ibeta (PDNP3), a new member of the human phosphodiesterase I genes. Genomics 45(2):412–415
23. Bollen M, Gijsbers R, Ceulemans H, Stalmans W, Stefan C (2000) Nucleotide pyrophosphatases/phosphodiesterases on the move. Crit Rev Biochem Mol Biol 35(6):393–432
24. Zimmermann H, Beaudoin AR, Bollen M, Goding JW, Guidotti G, Kirley TL, Robson SC, Sano K (2000) Proposed nomenclature for two novel nucleotide hydrolyzing enzyme families expressed on the cell surface. In: Vanduffel L, Lemmens R (eds) Ecto-ATPases and related ectonucleotidases. Shaker Publishing B.V, Maastrich, pp 1–8
25. Stefan C, Jansen S, Bollen M (2005) NPP-type ectophosphodiesterases: unity in diversity. Trends Biochem Sci 30(10):542–550
26. Gijsbers R, Aoki J, Arai H, Bollen M (2003) The hydrolysis of lysophospholipids and nucleotides by autotaxin (NPP2) involves a single catalytic site. FEBS Lett 538(1–3):60–64
27. van Meeteren LA, Ruurs P, Stortelers C, Bouwman P, van Rooijen MA, Pradere JP, Pettit TR, Wakelam MJ, Saulnier-Blache JS, Mummery CL, Moolenaar WH, Jonkers J (2006) Autotaxin, a secreted lysophospholipase D, is essential for blood vessel formation during development. Mol Cell Biol 26(13):5015–5022
28. Kishi Y, Okudaira S, Tanaka M, Hama K, Shida D, Kitayama J, Yamori T, Aoki J, Fujimaki T, Arai H (2006) Autotaxin is overexpressed in glioblastoma multiforme and contributes to cell motility of glioblastoma by converting lysophosphatidylcholine to lysophosphatidic acid. J Biol Chem 281(25):17492–17500
29. Hama K, Aoki J, Fukaya M, Kishi Y, Sakai T, Suzuki R, Ohta H, Yamori T, Watanabe M, Chun J, Arai H (2004) Lysophosphatidic acid and autotaxin stimulate cell motility of neoplastic and non-neoplastic cells through LPA1. J Biol Chem 279(17):17634–17639
30. Gaetano CG, Samadi N, Tomsig JL, Macdonald TL, Lynch KR, Brindley DN (2009) Inhibition of autotaxin production or activity blocks lysophosphatidylcholine-induced migration of human breast cancer and melanoma cells. Mol Carcinog 48(9):801–809
31. Tanaka M, Okudaira S, Kishi Y, Ohkawa R, Iseki S, Ota M, Noji S, Yatomi Y, Aoki J, Arai H (2006) Autotaxin stabilizes blood vessels and is required for embryonic vasculature by producing lysophosphatidic acid. J Biol Chem 281(35):25822–25830
32. Yukiura H, Kano K, Kise R, Inoue A, Aoki J (2015) Autotaxin overexpression causes embryonic lethality and vascular defects. PLoS One 10(5):e0126734
33. Clair T, Aoki J, Koh E, Bandle RW, Nam SW, Ptaszynska MM, Mills GB, Schiffmann E, Liotta LA, Stracke ML (2003) Autotaxin hydrolyzes sphingosylphosphorylcholine to produce the regulator of migration, sphingosine-1-phosphate. Cancer Res 63(17):5446–5453
34. Tsuda S, Okudaira S, Moriya-Ito K, Shimamoto C, Tanaka M, Aoki J, Arai H, Murakami-Murofushi K, Kobayashi T (2006) Cyclic phosphatidic acid is produced by autotaxin in blood. J Biol Chem 281(36):26081–26088
35. Parrill AL, Baker DL (2008) Autotaxin inhibition: challenges and progress toward novel anticancer agents. Anticancer Agents Med Chem 8(8):917–923

36. Xu X, Yang G, Zhang H, Prestwich GD (2009) Evaluating dual activity LPA receptor pan-antagonist/autotaxin inhibitors as anti-cancer agents in vivo using engineered human tumors. Prostaglandins Other Lipid Mediat 89(3–4):140–146
37. Gotoh M, Fujiwara Y, Yue J, Liu J, Lee S, Fells J, Uchiyama A, Murakami-Murofushi K, Kennel S, Wall J, Patil R, Gupte R, Balazs L, Miller DD, Tigyi GJ (2012) Controlling cancer through the autotaxin-lysophosphatidic acid receptor axis. Biochem Soc Trans 40(1):31–36
38. Houben AJ, Moolenaar WH (2011) Autotaxin and LPA receptor signaling in cancer. Cancer Metastasis Rev 30(3–4):557–565
39. Teo K, Brunton VG (2014) The role and therapeutic potential of the autotaxin-lysophosphatidate signalling axis in breast cancer. Biochem J 463(1):157–165
40. Barbayianni E, Kaffe E, Aidinis V, Kokotos G (2015) Autotaxin, a secreted lysophospholipase D, as a promising therapeutic target in chronic inflammation and cancer. Prog Lipid Res 58:76–96
41. Benesch MG, Tang X, Venkatraman G, Bekele RT, Brindley DN (2015) Recent advances in targeting the autotaxin-lysophosphatidate-lipid phosphate phosphatase axis in vivo. J Biomed Res 30. [Epub ahead of print]
42. Leblanc R, Peyruchaud O (2015) New insights into the autotaxin/LPA axis in cancer development and metastasis. Exp Cell Res 333(2):183–189
43. Tabuchi S (2015) The autotaxin-lysophosphatidic acid-lysophosphatidic acid receptor cascade: proposal of a novel potential therapeutic target for treating glioblastoma multiforme. Lipids Health Dis 14:56
44. Fotopoulou S, Oikonomou N, Grigorieva E, Nikitopoulou I, Paparountas T, Thanassopoulou A, Zhao Z, Xu Y, Kontoyiannis DL, Remboutsika E, Aidinis V (2010) ATX expression and LPA signalling are vital for the development of the nervous system. Dev Biol 339(2):451–464
45. Offermanns S, Mancino V, Revel JP, Simon MI (1997) Vascular system defects and impaired cell chemokinesis as a result of Galpha13 deficiency. Science 275(5299):533–536
46. Ferry G, Giganti A, Coge F, Bertaux F, Thiam K, Boutin JA (2007) Functional invalidation of the autotaxin gene by a single amino acid mutation in mouse is lethal. FEBS Lett 581(18):3572–3578
47. Yukiura H, Hama K, Nakanaga K, Tanaka M, Asaoka Y, Okudaira S, Arima N, Inoue A, Hashimoto T, Arai H, Kawahara A, Nishina H, Aoki J (2011) Autotaxin regulates vascular development via multiple lysophosphatidic acid (LPA) receptors in zebrafish. J Biol Chem 286(51):43972–43983
48. Nam SW, Clair T, Kim YS, McMarlin A, Schiffmann E, Liotta LA, Stracke ML (2001) Autotaxin (NPP-2), a metastasis-enhancing motogen, is an angiogenic factor. Cancer Res 61(18):6938–6944
49. Xu X, Prestwich GD (2010) Inhibition of tumor growth and angiogenesis by a lysophosphatidic acid antagonist in an engineered three-dimensional lung cancer xenograft model. Cancer 116(7):1739–1750
50. Okudaira S, Yukiura H, Aoki J (2010) Biological roles of lysophosphatidic acid signaling through its production by autotaxin. Biochimie 92(6):698–706
51. Moolenaar WH, Houben AJ, Lee SJ, van Meeteren LA (2013) Autotaxin in embryonic development. Biochim Biophys Acta 1831(1):13–19
52. Yuelling LM, Fuss B (2008) Autotaxin (ATX): a multi-functional and multi-modular protein possessing enzymatic lysoPLD activity and matricellular properties. Biochim Biophys Acta 1781(9):525–530
53. Nakanaga K, Hama K, Aoki J (2010) Autotaxin--an LPA producing enzyme with diverse functions. J Biochem 148(1):13–24
54. Perrakis A, Moolenaar WH (2014) Autotaxin: structure-function and signaling. J Lipid Res 55(6):1010–1018
55. Nishimasu H, Okudaira S, Hama K, Mihara E, Dohmae N, Inoue A, Ishitani R, Takagi J, Aoki J, Nureki O (2011) Crystal structure of autotaxin and insight into GPCR activation by lipid mediators. Nat Struct Mol Biol 18(2):205–212

56. Hausmann J, Kamtekar S, Christodoulou E, Day JE, Wu T, Fulkerson Z, Albers HM, van Meeteren LA, Houben AJ, van Zeijl L, Jansen S, Andries M, Hall T, Pegg LE, Benson TE, Kasiem M, Harlos K, Kooi CW, Smyth SS, Ovaa H, Bollen M, Morris AJ, Moolenaar WH, Perrakis A (2011) Structural basis of substrate discrimination and integrin binding by autotaxin. Nat Struct Mol Biol 18(2):198–204
57. Koh E, Clair T, Woodhouse EC, Schiffmann E, Liotta L, Stracke M (2003) Site-directed mutations in the tumor-associated cytokine, autotaxin, eliminate nucleotide phosphodiesterase, lysophospholipase D, and motogenic activities. Cancer Res 63(9):2042–2045
58. Tabchy A, Tigyi G, Mills GB (2011) Location, location, location: a crystal-clear view of autotaxin saturating LPA receptors. Nat Struct Mol Biol 18(2):117–118
59. Nishimasu H, Ishitani R, Aoki J, Nureki O (2012) A 3D view of autotaxin. Trends Pharmacol Sci 33(3):138–145
60. Hausmann J, Perrakis A, Moolenaar WH (2013) Structure-function relationships of autotaxin, a secreted lysophospholipase D. Adv Biol Regul 53(1):112–117
61. Jansen S, Callewaert N, Dewerte I, Andries M, Ceulemans H, Bollen M (2007) An essential oligomannosidic glycan chain in the catalytic domain of autotaxin, a secreted lysophospholipase-D. J Biol Chem 282(15):11084–11091
62. Jansen S, Andries M, Derua R, Waelkens E, Bollen M (2009) Domain interplay mediated by an essential disulfide linkage is critical for the activity and secretion of the metastasis-promoting enzyme autotaxin. J Biol Chem 284(21):14296–14302
63. Koyama M, Nishimasu H, Ishitani R, Nureki O (2012) Molecular dynamics simulation of autotaxin: roles of the nuclease-like domain and the glycan modification. J Phys Chem B 116(39):11798–11808
64. Cimpean A, Stefan C, Gijsbers R, Stalmans W, Bollen M (2004) Substrate-specifying determinants of the nucleotide pyrophosphatases/phosphodiesterases NPP1 and NPP2. Biochem J 381(Pt 1):71–77
65. Albers HM, Hendrickx LJ, van Tol RJ, Hausmann J, Perrakis A, Ovaa H (2011) Structure-based design of novel boronic acid-based inhibitors of autotaxin. J Med Chem 54(13):4619–4626
66. Mize CD, Abbott AM, Gacasan SB, Parrill AL, Baker DL (2011) Ligand-based autotaxin pharmacophore models reflect structure-based docking results. J Mol Graph Model 31:76–86
67. Kawaguchi M, Okabe T, Okudaira S, Nishimasu H, Ishitani R, Kojima H, Nureki O, Aoki J, Nagano T (2013) Screening and X-ray crystal structure-based optimization of autotaxin (ENPP2) inhibitors, using a newly developed fluorescence probe. ASC Chem Biol 8(8):1713–1721
68. Norman DD, Ibezim A, Scott WE, White S, Parrill AL, Baker DL (2013) Autotaxin inhibition: development and application of computational tools to identify site-selective lead compounds. Bioorg Med Chem 21(17):5548–5560
69. Fells JI, Lee SC, Fujiwara Y, Norman DD, Lim KG, Tsukahara R, Liu J, Patil R, Miller DD, Kirby RJ, Nelson S, Seibel W, Papoian R, Parrill AL, Baker DL, Bittman R, Tigyi G (2013) Hits of a high-throughput screen identify the hydrophobic pocket of autotaxin/lysophospholipase D as an inhibitory surface. Mol Pharmacol 84(3):415–424
70. Fells JI, Lee SC, Norman DD, Tsukahara R, Kirby JR, Nelson S, Seibel W, Papoian R, Patil R, Miller DD, Parrill AL, Pham TC, Baker DL, Bittman R, Tigyi G (2014) Targeting the hydrophobic pocket of autotaxin with virtual screening of inhibitors identifies a common aromatic sulfonamide structural motif. FEBS J 281(4):1017–1028
71. Stein AJ, Bain G, Prodanovich P, Santini AM, Darlington J, Stelzer NM, Sidhu RS, Schaub J, Goulet L, Lonergan D, Calderon I, Evans JF, Hutchinson JH (2015) Structural basis for inhibition of human autotaxin by four potent compounds with distinct modes of binding. Mol Pharmacol 88(6):982–992
72. Kato K, Ikeda H, Miyakawa S, Futakawa S, Nonaka Y, Fujiwara M, Okudaira S, Kano K, Aoki J, Morita J, Ishitani R, Nishimasu H, Nakamura Y, Nureki O (2016) Structural basis for specific inhibition of Autotaxin by a DNA aptamer. Nat Struct Mol Biol 23(5):395–401

73. Kanda H, Newton R, Klein R, Morita Y, Gunn MD, Rosen SD (2008) Autotaxin, an ectoenzyme that produces lysophosphatidic acid, promotes the entry of lymphocytes into secondary lymphoid organs. Nat Immunol 9(4):415–423
74. Pamuklar Z, Federico L, Liu S, Umezu-Goto M, Dong A, Panchatcharam M, Fulkerson Z, Berdyshev E, Natarajan V, Fang X, van Meeteren LA, Moolenaar WH, Mills GB, Morris AJ, Smyth SS (2009) Autotaxin/lysopholipase D and lysophosphatidic acid regulate murine hemostasis and thrombosis. J Biol Chem 284(11):7385–7394
75. Zhao J, He D, Berdyshev E, Zhong M, Salgia R, Morris AJ, Smyth SS, Natarajan V, Zhao Y (2011) Autotaxin induces lung epithelial cell migration through lysoPLD activity-dependent and -independent pathways. Biochem J 439(1):45–55
76. Fulkerson Z, Wu T, Sunkara M, Kooi CV, Morris AJ, Smyth SS (2011) Binding of autotaxin to integrins localizes lysophosphatidic acid production to platelets and mammalian cells. J Biol Chem 286(40):34654–34663
77. Wu T, Kooi CV, Shah P, Charnigo R, Huang C, Smyth SS, Morris AJ (2014) Integrin-mediated cell surface recruitment of autotaxin promotes persistent directional cell migration. FASEB J 28(2):861–870
78. Jansen S, Stefan C, Creemers JW, Waelkens E, Van Eynde A, Stalmans W, Bollen M (2005) Proteolytic maturation and activation of autotaxin (NPP2), a secreted metastasis-enhancing lysophospholipase D. J Cell Sci 118(Pt 14):3081–3089
79. Koike S, Keino-Masu K, Ohto T, Masu M (2006) The N-terminal hydrophobic sequence of autotaxin (ENPP2) functions as a signal peptide. Genes Cells 11(2):133–142
80. Pradere JP, Tarnus E, Gres S, Valet P, Saulnier-Blache JS (2007) Secretion and lysophospholipase D activity of autotaxin by adipocytes are controlled by N-glycosylation and signal peptidase. Biochim Biophys Acta 1771(1):93–102
81. Santos AN, Riemann D, Santos AN, Kehlen A, Thiele K, Langner J (1996) Treatment of fibroblast-like synoviocytes with IFN-gamma results in the down-regulation of autotaxin mRNA. Biochem Biophys Res Commun 229(2):419–424
82. Kawagoe H, Stracke ML, Nakamura H, Sano K (1997) Expression and transcriptional regulation of the PD-Ialpha/autotaxin gene in neuroblastoma. Cancer Res 57(12):2516–2521
83. Kehlen A, Lauterbach R, Santos AN, Thiele K, Kabisch U, Weber E, Riemann D, Langner J (2001) IL-1 beta- and IL-4-induced down-regulation of autotaxin mRNA and PC-1 in fibroblast-like synoviocytes of patients with rheumatoid arthritis (RA). Clin Exp Immunol 123(1):147–154
84. Chen M, O'Connor KL (2005) Integrin alpha6beta4 promotes expression of autotaxin/ENPP2 autocrine motility factor in breast carcinoma cells. Oncogene 24(32):5125–5130
85. Farina AR, Cappabianca L, Ruggeri P, Di Ianni N, Ragone M, Merolle S, Sano K, Stracke ML, Horowitz JM, Gulino A, Mackay AR (2012) Constitutive autotaxin transcription by Nmyc-amplified and non-amplified neuroblastoma cells is regulated by a novel AP-1 and SP-mediated mechanism and abrogated by curcumin. FEBS Lett 586(20):3681–3691
86. Sioletic S, Czaplinski J, Hu L, Fletcher JA, Fletcher CD, Wagner AJ, Loda M, Demetri GD, Sicinska ET, Snyder EL (2014) c-Jun promotes cell migration and drives expression of the motility factor ENPP2 in soft tissue sarcomas. J Pathol 234(2):190–202
87. Benesch MG, Zhao YY, Curtis JM, McMullen TP, Brindley DN (2015) Regulation of autotaxin expression and secretion by lysophosphatidate and sphingosine 1-phosphate. J Lipid Res 56(6):1134–1144
88. Song J, Guan M, Zhao Z, Zhang J (2015) Type I Interferons Function as Autocrine and Paracrine Factors to Induce Autotaxin in Response to TLR Activation. PLoS One 10(8):e0136629
89. Greenman R, Gorelik A, Sapir T, Baumgart J, Zamor V, Segal-Salto M, Levin-Zaidman S, Aidinis V, Aoki J, Nitsch R, Vogt J, Reiner O (2015) Non-cell autonomous and non-catalytic activities of ATX in the developing brain. Front Neurosci 9:53
90. Fox MA, Colello RJ, Macklin WB, Fuss B (2003) Phosphodiesterase-Ialpha/autotaxin: a counteradhesive protein expressed by oligodendrocytes during onset of myelination. Mol Cell Neurosci 23(3):507–519

91. Fox MA, Alexander JK, Afshari FS, Colello RJ, Fuss B (2004) Phosphodiesterase-I alpha/autotaxin controls cytoskeletal organization and FAK phosphorylation during myelination. Mol Cell Neurosci 27(2):140–150
92. Dennis J, White MA, Forrest AD, Yuelling LM, Nogaroli L, Afshari FS, Fox MA, Fuss B (2008) Phosphodiesterase-Ialpha/autotaxin's MORFO domain regulates oligodendroglial process network formation and focal adhesion organization. Mol Cell Neurosci 37(2):412–424
93. Dennis J, Morgan MK, Graf MR, Fuss B (2012) P2Y(12) receptor expression is a critical determinant of functional responsiveness to ATX's MORFO domain. Purinergic Signal 8:181–190
94. Morrison SJ, Kimble J (2006) Asymmetric and symmetric stem-cell divisions in development and cancer. Nature 441(7097):1068–1074
95. Knoblich JA (2008) Mechanisms of asymmetric stem cell division. Cell 132(4):583–597
96. Januschke J, Nathke I (2014) Stem cell decisions: a twist of fate or a niche market? Semin Cell Dev Biol 34:116–123
97. Watt FM, Hogan BL (2000) Out of Eden: stem cells and their niches. Science 287(5457):1427–1430
98. Pebay A, Bonder CS, Pitson SM (2007) Stem cell regulation by lysophospholipids. Prostaglandins Other Lipid Mediat 84(3–4):83–97
99. Pitson SM, Pebay A (2009) Regulation of stem cell pluripotency and neural differentiation by lysophospholipids. Neurosignals 17(4):242–254
100. Bradley A, Evans M, Kaufman MH, Robertson E (1984) Formation of germ-line chimaeras from embryo-derived teratocarcinoma cell lines. Nature 309(5965):255–256
101. Nagy A, Gocza E, Diaz EM, Prideaux VR, Ivanyi E, Markkula M, Rossant J (1990) Embryonic stem cells alone are able to support fetal development in the mouse. Development 110(3):815–821
102. Ilic D, Ogilvie C (2016) Human embryonic stem cells-what have we done? What are we doing? Where are we going? Stem Cells. [Epub ahead of print]
103. Avior Y, Sagi I, Benvenisty N (2016) Pluripotent stem cells in disease modelling and drug discovery. Nat Rev Mol Cell Biol 17(3):170–182
104. Verfaillie CM, Pera MF, Lansdorp PM (2002) Stem cells: hype and reality. Hematol Am Soc Hematol Educ Program 2002:369–391
105. Cauffman G, De Rycke M, Sermon K, Liebaers I, Van de Velde H (2009) Markers that define stemness in ESC are unable to identify the totipotent cells in human preimplantation embryos. Hum Reprod 24(1):63–70
106. Galan A, Diaz-Gimeno P, Poo ME, Valbuena D, Sanchez E, Ruiz V, Dopazo J, Montaner D, Conesa A, Simon C (2013) Defining the genomic signature of totipotency and pluripotency during early human development. PLoS One 8(4):e62135
107. Condic ML (2014) Totipotency: what it is and what it is not. Stem Cells Dev 23(8):796–812
108. Pebay A, Wong RC, Pitson SM, Wolvetang EJ, Peh GS, Filipczyk A, Koh KL, Tellis I, Nguyen LT, Pera MF (2005) Essential roles of sphingosine-1-phosphate and platelet-derived growth factor in the maintenance of human embryonic stem cells. Stem Cells 23(10):1541–1548
109. Dottori M, Leung J, Turnley AM, Pebay A (2008) Lysophosphatidic acid inhibits neuronal differentiation of neural stem/progenitor cells derived from human embryonic stem cells. Stem Cells 26(5):1146–1154
110. Todorova MG, Fuentes E, Soria B, Nadal A, Quesada I (2009) Lysophosphatidic acid induces Ca2+ mobilization and c-Myc expression in mouse embryonic stem cells via the phospholipase C pathway. Cell Signal 21(4):523–528
111. Schuck S, Soloaga A, Schratt G, Arthur JS, Nordheim A (2003) The kinase MSK1 is required for induction of c-fos by lysophosphatidic acid in mouse embryonic stem cells. BMC Mol Biol 4:6
112. Apati A, Paszty K, Hegedus L, Kolacsek O, Orban TI, Erdei Z, Szebenyi K, Pentek A, Enyedi A, Sarkadi B (2013) Characterization of calcium signals in human embryonic stem cells and in their differentiated offspring by a stably integrated calcium indicator protein. Cell Signal 25(4):752–759

113. Kobayashi T, Yamano S, Murayama S, Ishikawa H, Tokumura A, Aono T (1994) Effect of lysophosphatidic acid on the preimplantation development of mouse embryos. FEBS Lett 351(1):38–40
114. Ahn JI, Lee KH, Shin DM, Shim JW, Kim CM, Kim H, Lee SH, Lee YS (2004) Temporal expression changes during differentiation of neural stem cells derived from mouse embryonic stem cell. J Cell Biochem 93(3):563–578
115. Liszewska E, Reinaud P, Billon-Denis E, Dubois O, Robin P, Charpigny G (2009) Lysophosphatidic acid signaling during embryo development in sheep: involvement in prostaglandin synthesis. Endocrinology 150(1):422–434
116. Boruszewska D, Kowalczyk-Zieba I, Piotrowska-Tomala K, Saulnier-Blache JS, Acosta T, Skarzynski DJ, Woclawek-Potocka I (2013) Which bovine endometrial cells are the source of and target for lysophosphatidic acid? Reprod Biol 13(1):100–103
117. Brunnert D, Sztachelska M, Bornkessel F, Treder N, Wolczynski S, Goyal P, Zygmunt M (2014) Lysophosphatidic acid and sphingosine 1-phosphate metabolic pathways and their receptors are differentially regulated during decidualization of human endometrial stromal cells. Mol Hum Reprod 20(10):1016–1025
118. Aoki J, Inoue A, Okudaira S (2008) Two pathways for lysophosphatidic acid production. Biochim Biophys Acta 1781(9):513–518
119. Schofield R (1978) The relationship between the spleen colony-forming cell and the haemopoietic stem cell. Blood Cells 4(1–2):7–25
120. Morrison SJ, Spradling AC (2008) Stem cells and niches: mechanisms that promote stem cell maintenance throughout life. Cell 132(4):598–611
121. Lander AD, Kimble J, Clevers H, Fuchs E, Montarras D, Buckingham M, Calof AL, Trumpp A, Oskarsson T (2012) What does the concept of the stem cell niche really mean today? BMC Biol 10:19
122. Sanchez-Aguilera A, Mendez-Ferrer S (2016) The hematopoietic stem-cell niche in health and leukemia. Cell Mol Life Sci. [Epub ahead of print]
123. Wilson A, Laurenti E, Oser G, van der Wath RC, Blanco-Bose W, Jaworski M, Offner S, Dunant CF, Eshkind L, Bockamp E, Lio P, Macdonald HR, Trumpp A (2008) Hematopoietic stem cells reversibly switch from dormancy to self-renewal during homeostasis and repair. Cell 135(6):1118–1129
124. Brenet F, Scandura JM (2015) Cutting the brakes on hematopoietic regeneration by blocking TGFbeta to limit chemotherapy-induced myelosuppression. Mol Cell Oncol 2(3):e978703
125. Ortlepp C, Steudel C, Heiderich C, Koch S, Jacobi A, Ryser M, Brenner S, Bornhauser M, Brors B, Hofmann WK, Ehninger G, Thiede C (2013) Autotaxin is expressed in FLT3-ITD positive acute myeloid leukemia and hematopoietic stem cells and promotes cell migration and proliferation. Exp Hematol 41(5):444–461
126. Lansdorp PM, Sutherland HJ, Eaves CJ (1990) Selective expression of CD45 isoforms on functional subpopulations of CD34+ hemopoietic cells from human bone marrow. J Exp Med 172(1):363–366
127. Lansdorp PM, Dragowska W (1992) Long-term erythropoiesis from constant numbers of CD34+ cells in serum-free cultures initiated with highly purified progenitor cells from human bone marrow. J Exp Med 175(6):1501–1509
128. Mayani H, Dragowska W, Lansdorp PM (1993) Cytokine-induced selective expansion and maturation of erythroid versus myeloid progenitors from purified cord blood precursor cells. Blood 81(12):3252–3258
129. Sidney LE, Branch MJ, Dunphy SE, Dua HS, Hopkinson A (2014) Concise review: evidence for CD34 as a common marker for diverse progenitors. Stem Cells 32(6):1380–1389
130. Lindner U, Kramer J, Rohwedel J, Schlenke P (2010) Mesenchymal stem or stromal cells: toward a better understanding of their biology? Transfus Med Hemother 37(2):75–83
131. Mendez-Ferrer S, Scadden DT, Sanchez-Aguilera A (2015) Bone marrow stem cells: current and emerging concepts. Ann N Y Acad Sci 1335:32–44
132. Evseenko D, Latour B, Richardson W, Corselli M, Sahaghian A, Cardinal S, Zhu Y, Chan R, Dunn B, Crooks GM (2013) Lysophosphatidic acid mediates myeloid differentiation within the human bone marrow microenvironment. PLoS One 8(5):e63718

133. Ryu JM, Han HJ (2015) Autotaxin-LPA axis regulates hMSC migration by adherent junction disruption and cytoskeletal rearrangement via LPAR1/3-dependent PKC/GSK3beta/beta-catenin and PKC/Rho GTPase pathways. Stem Cells 33(3):819–832
134. Annabi B, Thibeault S, Lee YT, Bousquet-Gagnon N, Eliopoulos N, Barrette S, Galipeau J, Beliveau R (2003) Matrix metalloproteinase regulation of sphingosine-1-phosphate-induced angiogenic properties of bone marrow stromal cells. Exp Hematol 31(7):640–649
135. Jaganathan BG, Ruester B, Dressel L, Stein S, Grez M, Seifried E, Henschler R (2007) Rho inhibition induces migration of mesenchymal stromal cells. Stem Cells 25(8):1966–1974
136. Krause DS, Theise ND, Collector MI, Henegariu O, Hwang S, Gardner R, Neutzel S, Sharkis SJ (2001) Multi-organ, multi-lineage engraftment by a single bone marrow-derived stem cell. Cell 105(3):369–377
137. Rostovskaya M, Anastassiadis K (2012) Differential expression of surface markers in mouse bone marrow mesenchymal stromal cell subpopulations with distinct lineage commitment. PLoS One 7(12):e51221
138. Yanai N, Matsui N, Furusawa T, Okubo T, Obinata M (2000) Sphingosine-1-phosphate and lysophosphatidic acid trigger invasion of primitive hematopoietic cells into stromal cell layers. Blood 96(1):139–144
139. Whetton AD, Lu Y, Pierce A, Carney L, Spooncer E (2003) Lysophospholipids synergistically promote primitive hematopoietic cell chemotaxis via a mechanism involving Vav 1. Blood 102(8):2798–2802
140. Fuchs E (2009) The tortoise and the hair: slow-cycling cells in the stem cell race. Cell 137(5):811–819
141. Hsu YC, Li L, Fuchs E (2014) Emerging interactions between skin stem cells and their niches. Nat Med 20(8):847–856
142. Hsu YC, Li L, Fuchs E (2014) Transit-amplifying cells orchestrate stem cell activity and tissue regeneration. Cell 157(4):935–949
143. Rendl M, Lewis L, Fuchs E (2005) Molecular dissection of mesenchymal-epithelial interactions in the hair follicle. PLoS Biol 3(11):e331
144. Grisanti L, Rezza A, Clavel C, Sennett R, Rendl M (2013) Enpp2/Autotaxin in dermal papilla precursors is dispensable for hair follicle morphogenesis. J Invest Dermatol 133(10):2332–2339
145. Yanagida K, Masago K, Nakanishi H, Kihara Y, Hamano F, Tajima Y, Taguchi R, Shimizu T, Ishii S (2009) Identification and characterization of a novel lysophosphatidic acid receptor, p2y5/LPA6. J Biol Chem 284(26):17731–17741
146. Inoue A, Arima N, Ishiguro J, Prestwich GD, Arai H, Aoki J (2011) LPA-producing enzyme PA-PLA(1)alpha regulates hair follicle development by modulating EGFR signalling. EMBO J 30(20):4248–4260
147. Rosen JM, Jordan CT (2009) The increasing complexity of the cancer stem cell paradigm. Science (New York, NY) 324(5935):1670–1673. doi:10.1126/science.1171837
148. Meacham CE, Morrison SJ (2013) Tumour heterogeneity and cancer cell plasticity. Nature 501(7467):328–337
149. Carnero A, Garcia-Mayea Y, Mir C, Lorente J, Rubio IT, Me LL (2016) The cancer stem-cell signaling network and resistance to therapy. Cancer Treat Rev 49:25–36
150. Chen W, Dong J, Haiech J, Kilhoffer MC, Zeniou M (2016) Cancer Stem Cell Quiescence and Plasticity as Major Challenges in Cancer Therapy. Stem Cells Int 2016:1740936
151. Woodward WA, Hill RP (2016) Cancer Stem Cells. Recent Results Cancer Res 198:25–44
152. Gkountela S, Aceto N (2016) Stem-like features of cancer cells on their way to metastasis. Biol Direct 11:33
153. Zhang S, Balch C, Chan MW, Lai HC, Matei D, Schilder JM, Yan PS, Huang TH, Nephew KP (2008) Identification and characterization of ovarian cancer-initiating cells from primary human tumors. Cancer Res 68(11):4311–4320
154. Jazaeri AA, Awtrey CS, Chandramouli GV, Chuang YE, Khan J, Sotiriou C, Aprelikova O, Yee CJ, Zorn KK, Birrer MJ, Barrett JC, Boyd J (2005) Gene expression profiles associated with response to chemotherapy in epithelial ovarian cancers. Clin Cancer Res 11(17):6300–6310

155. Vidot S, Witham J, Agarwal R, Greenhough S, Bamrah HS, Tigyi GJ, Kaye SB, Richardson A (2010) Autotaxin delays apoptosis induced by carboplatin in ovarian cancer cells. Cell Signal 22(6):926–935
156. Seo EJ, Kwon YW, Jang IH, Kim DK, Lee SI, Choi EJ, Kim KH, Suh DS, Lee JH, Choi KU, Lee JW, Mok HJ, Kim KP, Matsumoto H, Aoki J, Kim JH (2016) Autotaxin regulates maintenance of ovarian cancer stem cells through lysophosphatidic acid-mediated autocrine mechanism. Stem Cells 34(3):551–564
157. Ng W, Pebay A, Drummond K, Burgess A, Kaye AH, Morokoff A (2014) Complexities of lysophospholipid signalling in glioblastoma. J Clin Neurosci 21(6):893–898
158. Hoelzinger DB, Mariani L, Weis J, Woyke T, Berens TJ, McDonough WS, Sloan A, Coons SW, Berens ME (2005) Gene expression profile of glioblastoma multiforme invasive phenotype points to new therapeutic targets. Neoplasia 7(1):7–16
159. Bhave SR, Dadey DY, Karvas RM, Ferraro DJ, Kotipatruni RP, Jaboin JJ, Hallahan AN, Dewees TA, Linkous AG, Hallahan DE, Thotala D (2013) Autotaxin inhibition with PF-8380 enhances the radiosensitivity of human and murine glioblastoma cell lines. Front Oncol 3:236
160. Annabi B, Lachambre MP, Plouffe K, Sartelet H, Beliveau R (2009) Modulation of invasive properties of CD133+ glioblastoma stem cells: a role for MT1-MMP in bioactive lysophospholipid signaling. Mol Carcinog 48(10):910–919
161. Lathia JD, Mack SC, Mulkearns-Hubert EE, Valentim CL, Rich JN (2015) Cancer stem cells in glioblastoma. Genes Dev 29(12):1203–1217
162. Bradshaw A, Wickremsekera A, Tan ST, Peng L, Davis PF, Itinteang T (2016) Cancer Stem Cell Hierarchy in Glioblastoma Multiforme. Front Surg 3:21
163. Sanes DH, Reh TA, Harris WA (2011) Development of the nervous system, 3rd edn. Academic Press, Cambridge, MA
164. Koike S, Yutoh Y, Keino-Masu K, Noji S, Masu M, Ohuchi H (2011) Autotaxin is required for the cranial neural tube closure and establishment of the midbrain-hindbrain boundary during mouse development. Dev Dyn 240(2):413–421
165. Koike S, Keino-Masu K, Masu M (2010) Deficiency of autotaxin/lysophospholipase D results in head cavity formation in mouse embryos through the LPA receptor-Rho-ROCK pathway. Biochem Biophys Res Commun 400(1):66–71
166. Campbell K, Gotz M (2002) Radial glia: multi-purpose cells for vertebrate brain development. Trends Neurosci 25(5):235–238
167. Fishell G, Kriegstein AR (2003) Neurons from radial glia: the consequences of asymmetric inheritance. Curr Opin Neurobiol 13(1):34–41
168. Anthony TE, Klein C, Fishell G, Heintz N (2004) Radial glia serve as neuronal progenitors in all regions of the central nervous system. Neuron 41(6):881–890
169. Wilsch-Brauninger M, Florio M, Huttner WB (2016) Neocortex expansion in development and evolution—from cell biology to single genes. Curr Opin Neurobiol 39:122–132
170. Frisca F, Crombie DE, Dottori M, Goldshmit Y, Pebay A (2013) Rho/ROCK pathway is essential to the expansion, differentiation, and morphological rearrangements of human neural stem/progenitor cells induced by lysophosphatidic acid. J Lipid Res 54(5):1192–1206
171. Cui HL, Qiao JT (2006) Promotive action of lysophosphatidic acid on proliferation of rat embryonic neural stem cells and their differentiation to cholinergic neurons in vitro. Sheng Li Xue Bao 58(6):547–555
172. Fukushima N, Shano S, Moriyama R, Chun J (2007) Lysophosphatidic acid stimulates neuronal differentiation of cortical neuroblasts through the LPA1-G(i/o) pathway. Neurochem Int 50(2):302–307
173. Miller RH (2005) Dorsally derived oligodendrocytes come of age. Neuron 45(1):1–3
174. Richardson WD, Kessaris N, Pringle N (2006) Oligodendrocyte wars. Nat Rev Neurosci 7(1):11–18
175. Miller RH (2002) Regulation of oligodendrocyte development in the vertebrate CNS. Prog Neurobiol 67(6):451–467
176. Yuelling LW, Waggener CT, Afshari FS, Lister JA, Fuss B (2012) Autotaxin/ENPP2 regulates oligodendrocyte differentiation in vivo in the developing zebrafish hindbrain. Glia 60(10):1605–1618

177. Wheeler NA, Lister JA, Fuss B (2015) The Autotaxin-Lysophosphatidic Acid Axis Modulates Histone Acetylation and Gene Expression during Oligodendrocyte Differentiation. J Neurosci 35(32):11399–11414
178. Liu J, Moyon S, Hernandez M, Casaccia P (2016) Epigenetic control of oligodendrocyte development: adding new players to old keepers. Curr Opin Neurobiol 39:133–138
179. Weiner JA, Hecht JH, Chun J (1998) Lysophosphatidic acid receptor gene vzg-1/lpA1/edg-2 is expressed by mature oligodendrocytes during myelination in the postnatal murine brain. J Comp Neurol 398(4):587–598
180. Stankoff B, Barron S, Allard J, Barbin G, Noel F, Aigrot MS, Premont J, Sokoloff P, Zalc B, Lubetzki C (2002) Oligodendroglial expression of Edg-2 receptor: developmental analysis and pharmacological responses to lysophosphatidic acid. Mol Cell Neurosci 20(3):415–428
181. Dawson J, Hotchin N, Lax S, Rumsby M (2003) Lysophosphatidic acid induces process retraction in CG-4 line oligodendrocytes and oligodendrocyte precursor cells but not in differentiated oligodendrocytes. J Neurochem 87(4):947–957
182. Nogaroli L, Yuelling LM, Dennis J, Gorse K, Payne SG, Fuss B (2009) Lysophosphatidic acid can support the formation of membranous structures and an increase in MBP mRNA levels in differentiating oligodendrocytes. Neurochem Res 34(1):182–193
183. Zhang Y, Chen K, Sloan SA, Bennett ML, Scholze AR, O'Keeffe S, Phatnani HP, Guarnieri P, Caneda C, Ruderisch N, Deng S, Liddelow SA, Zhang C, Daneman R, Maniatis T, Barres BA, Wu JQ (2014) An RNA-sequencing transcriptome and splicing database of glia, neurons, and vascular cells of the cerebral cortex. J Neurosci 34(36):11929–11947
184. Pfeiffer SE, Warrington AE, Bansal R (1993) The oligodendrocyte and its many cellular processes. Trends Cell Biol 3:191–197
185. Dugas JC, Tai YC, Speed TP, Ngai J, Barres BA (2006) Functional genomic analysis of oligodendrocyte differentiation. J Neurosci 26(43):10967–10983
186. Dennis J, Nogaroli L, Fuss B (2005) Phosphodiesterase-Ialpha/autotaxin (PD-Ialpha/ATX): a multifunctional protein involved in central nervous system development and disease. J Neurosci Res 82(6):737–742
187. Murphy-Ullrich JE (2001) The de-adhesive activity of matricellular proteins: is intermediate cell adhesion an adaptive state? J Clin Invest 107(7):785–790
188. Nagai J, Uchida H, Matsushita Y, Yano R, Ueda M, Niwa M, Aoki J, Chun J, Ueda H (2010) Autotaxin and lysophosphatidic acid1 receptor-mediated demyelination of dorsal root fibers by sciatic nerve injury and intrathecal lysophosphatidylcholine. Mol Pain 6:78
189. Inoue M, Xie W, Matsushita Y, Chun J, Aoki J, Ueda H (2008) Lysophosphatidylcholine induces neuropathic pain through an action of autotaxin to generate lysophosphatidic acid. Neuroscience 152(2):296–298
190. Ma L, Uchida H, Nagai J, Inoue M, Aoki J, Ueda H (2010) Evidence for de novo synthesis of lysophosphatidic acid in the spinal cord through phospholipase A2 and autotaxin in nerve injury-induced neuropathic pain. J Pharmacol Exp Ther 333(2):540–546
191. Katsifa A, Kaffe E, Nikolaidou-Katsaridou N, Economides AN, Newbigging S, McKerlie C, Aidinis V (2015) The Bulk of Autotaxin Activity Is Dispensable for Adult Mouse Life. PLoS One 10(11):e0143083

Chapter 4
Lysophosphatidic Acid (LPA) Signaling in Neurogenesis

Whitney S. McDonald and Jerold Chun

Abbreviations

5-HT	Serotonin
ATX	Autotaxin
Ca^{2+}	Calcium
cAMP	Cyclic adenosine monophosphate
CNS	Central nervous system
CSF	Cerebrospinal fluid
DAG	Diacylglycerol
E	Embryonic day
EDG	Endothelial differentiation gene
EGF	Epidermal growth factor
Enpp2	Epidermal growth factor receptor phosphodiesterase family member 2
FABP	Fatty acid binding protein
GPAT	Glycerophosphate acyltransferase
GPCR	G protein-coupled receptor
GRK2	G protein-coupled receptor kinase 2
HIF-1α	Hypoxia inducible factor-1 alpha
IZ	Intermediate zone
LCAT	Lecithin cholesterol acyltransferase
LP	Lysophospholipids

W.S. McDonald, Ph.D. • J. Chun, M.D., Ph.D. (✉)
Sanford Burnham Prebys Medical Discovery Institute, 10901 North Torrey Pines Road, La Jolla, CA 92037, USA
e-mail: jchun@sbpdiscovery.org

A. Pébay, R.C.B. Wong (eds.), *Lipidomics of Stem Cells*, Stem Cell Biology and Regenerative Medicine, DOI 10.1007/978-3-319-49343-5_4

LPA	Lysophosphatidic acid
LPA_{1-6}	Lysophosphatidic acid GPCR 1–6
LPAAT	Lysophosphatidic acid acyltransferase
LPAR1–6	Human Lysophosphatidic acid GPCR genes 1–6
Lpar1–6	Mouse Lysophosphatidic acid GPCR genes 1–6
LPC	Lysophosphatidylcholine
LPP	Lipid phosphate phosphatases
MAG	Monoacylglycerol
MAP	Microtubule-associated protein
NGF	Nerve growth factor
NPC	Neuroprogenitor cell
P2Y	Purinergic family genes
PA	Phosphatidic acid
PC	Phosphatidylcholine
PCD	Programmed cell death
PHH	Post-hemorrhagic hydrocephalus
PL	Phospholipids
PLA1	Phospholipase A1
PLC	Phospholipase-C
PS	Phosphatidylserine
PSA-NCAM	Polysialylated neural cell adhesion protein
PS-PLA1	Phosphatidylserine-specific phospholipase A1
S1P	Sphingosine-1-phosphate
SOX2	Sex determining region Y-box 2
sPLA2	Secretory phospholipase A2
SVZ	Subventricular zone
TrkA	Tyrosine kinase receptor type 1
VEGF	Vascular endothelial growth factor
VZ	Ventricular zone

4.1 Introduction

Lysophosphatidic acid (LPA) is a simple glycerophospholipid (molecular weight: 430–480 Da) present at low levels in all major cell types as well as in blood. LPA has intracellular and extracellular metabolic pathways and signals through class A, rhodopsin-like G protein-coupled receptors (GPCRs). Identification of the first high-affinity GPCR (LPA_1) in 1996 [1] quickly leads to deorphanization of other LPA as well as sphingosine-1-phosphate (S1P) receptors, particularly those of the endothelial differentiation gene (EDG) cluster and later, those of the P2Y purinergic family genes [2]. Each LPA receptor couples with one or more of four heterotrimeric G_α ($G_{12/13}$, $G_{q/11}$, $G_{I/O}$, and G_s) proteins and has distinct tissue expression and patterning. In view of the heterogeneity of receptor expression, G proteins activated, and downstream signaling cascades, LPA receptor activation can lead to diverse, pervasive, redundant, and sometimes antagonizing outcomes in biological processes.

4.2 LPA Structure, Distribution, and Metabolism

4.2.1 *Structure*

All LPA molecules have a phosphate head group attached to a glycerol backbone ester linked to a single aliphatic chain; each species of LPA has a distinct length and saturation of the acyl chain. Species with a phosphate monoester and an aliphatic chain of more than 12 carbons are biologically active LPA [3, 4]. Common LPA species detected in the brain and biological fluids include palmitoyl (16:0), stearoyl (18:0), oleoyl (18:1), and arachidonoyl (20:4). Acyl groups LPA 18:1 is most commonly used in laboratory settings [5–10]. Although the structure of LPA is fairly simplistic, studies found that structural differences in LPA species underlie the relative potency of LPA receptor activation and downstream signaling effects; unsaturated LPA (18:1) species are more bioactive than saturated LPA (18:0) species [11–13].

4.2.2 *Distribution*

LPA is ubiquitously present in most fluids including plasma, serum, and cerebrospinal fluid (CSF) at bioactively potent concentrations [9, 14]. In physiological conditions, blood LPA levels are detected at ~10 μM in serum, ~0.1 μM in plasma [15, 16], and low nanomolar to micromolar levels in the CSF [9]. The brain is reported to have the highest concentration of LPA [17] expressed at low micromolar levels in the CSF, choroid plexus, neural tube, meninges, and blood vessels of the developing brain [9]. Although the spatiotemporal distribution of LPA during neurogenesis remains elusive, advanced technologies in chromatography, mass spectrometry, and laser capture can be useful diagnostics to map LPA expression throughout CNS development [16, 18].

4.2.3 *Metabolism*

LPA is produced by many different cell types including neurons, activated platelets, tumor cells, and adipocytes [19–23].The species of LPA that are generated reflect the structure of the precursor phospholipid (e.g., lysophosphatidylcholine 18:1 creates LPA 18:1) [14]. LPA metabolism has intracellular and extracellular enzymatic pathways. The extracellular metabolic pathway for LPA is mediated by Autotaxin (ATX, also known as lysophospholipase D, gene name Enpp2) activity in the blood and is perhaps the most well-defined mechanism for LPA production. In plasma, phospholipids such as phosphatidylcholine (PC) and phosphatidylserine (PS) are converted to their lysophospholipid (LP) forms through phospholipase A1 (PLA1)/lecithin cholesterol acyltransferase (LCAT) activity and through secretory phospholipase A2 (sPLA2), calcium-independent phospholipase A2 (cPLA2), or phosphatidylserine-specific phospholipase A1 (PS-PLA1) activity, respectively [24, 25]. ATX converts those LPs to LPA [26–28] and it can act as a signaling molecule through LPA receptors on the plasma membrane [29] (Fig. 4.1). ATX activity is crucial for maintaining vascular and neuronal embryonic development by inducing vascular endothelial growth factor (VEGF), endothelial migration and proliferation [30, 31], and matrix remodeling in angiogenesis [20, 32]. Knockout of the ATX gene (Enpp2) produces major neural and vascular deficits and subsequent death at embryonic day (E) 9.5 [33, 34]. Conditional deletion of Enpp2 in Sox2-positive epiblasts results in neural tube deficits [35], and Enpp2 heterozygous null mice survive into adulthood with ~50% lower LPA levels in the plasma as compared to the wild type [33, 35]. These studies suggest that extracellular LPA metabolism through ATX is crucial for CNS development.

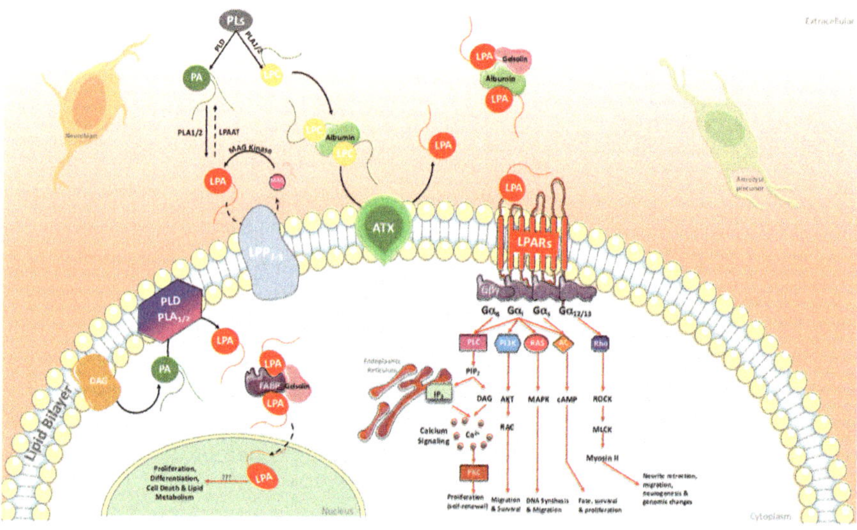

Fig. 4.1 Intracellular and extracellular LPA metabolism and signaling. *LPA* lysophosphatidic acid, *PLs* phospholipids, *PLD* phospholipase D, *PLA* phospholipase A, *PA* phosphatidic acid, *LPAAT* LPA acyltransferase, *MAG* monoacylglycerol, *LPP* lipid phosphate phosphatase, *FABP* fatty acid binding protein, *LPC* lysophosphatidylcholine, *ATX* autotaxin, *DAG* diacylglycerol, *LPARs* lysophosphatidic acid receptors 1–6, *PLC* phospholipase C, *PIP2* phosphatidylinositol 4,5-bisphosphate, *IP3* inositol 1,4,5-trisphosphate, *DAG* diacyl glycerol, *AC* adenylyl cyclase, *PKC* protein kinase C, *PI3K* phosphatidylinositol-4,5-bisphosphate 3-kinase, *AKT* protein kinase B, *RAC* Ras-related C3 botulinum toxin substrate, *MAPK* mitogen-activated protein kinases, *AC* adenylyl cyclase, *cAMP* cyclic adenosine monophosphate, *ROCK* Rho-associated protein kinase, *MLCK* myosin light chain kinase. *Solid black arrow* indicates metabolism; *dashed black arrow* indicates catabolism; *solid red arrows* indicate signaling cascades

Intracellular LPA is an intermediate in the synthesis of glycerolipids [29]. In cancer cells, intracellular LPA appears to be transported extracellularly by fatty acid binding proteins (FABP) as well as gelsolin to activate cell surface receptors and produce morphological changes [26] (Fig. 4.1). Intracellular LPA is produced by making phospholipids (PLs) like phosphatidic acid (PA), from diacylglycerol (DAG) and the actions of diacylglycerol kinase or phospholipase D. LPA is then generated from hydrolysis of PA by phospholipase A1 and A2 (PLA1 and PLA2) [24]. LPA is also synthesized by monoacylglycerol (MAG) kinase phosphorylating MAG. The intracellular metabolic pathway for LPA occurs in neurons [36] within the endoplasmic reticulum and the mitochondria by the acetylation of glycerol-3-phosphate through glycerophosphate acyltransferase (GPAT) [14, 24, 37, 38] and at the leading edge of migrating monocytes through calcium-independent PLA2 activity [39, 40]. LPA metabolism thus influences DNA synthesis, progenitor population expansion, and migration.

4.2.4 Catabolism

Dephosphorylation of intracellular LPA by LPA acyltransferase (LPAAT) is a major pathway that terminates LPA's signaling processes and synthesizes complex glycerophospholipids [14, 29]. Extracellular LPA is hydrolyzed by lipid phosphate phosphatases 1–3 (LPP1–3) and phospholipid phosphatase and is converted into MAG (Fig. 4.1). This process can be reversed when MAG is rephosphorylated by MAG kinase to produce LPA and subsequent LPA metabolites and signaling. The myriad mechanisms for LPA synthesis and degradation suggest that the presence of LPA is tightly regulated and disruption of this system may initiate and exacerbate cellular pathologies.

4.3 Lysophosphatidic Acid Receptor Signaling and Downstream Pathways

The first LPA GPCR (LPA_1) was discovered in 1996, which was cloned as a gene (Lpar1) in the proliferative zone of the developing brain [1, 41, 42]. Since this discovery, five other LPA receptor genes (human: LPAR1–6; mouse: Lpar1–6) have been cloned and these six receptors (LPA_{1-6}) play a plethora of roles in embryonic cortical development. These include cellular apoptosis, proliferation, migration, adhesion, differentiation, morphology, electrophysiological changes, and signaling. Each LPAR couples to one or more of the four heterotrimeric G_α ($G_{12/13}$, $G_{q/11}$, $G_{i/o}$, and G_s) proteins that initiate an array of cascades with diverse effects in physiology and pathophysiology [9, 23, 43–45]. Normal brain development is dependent on the intricate spatiotemporal patterning of LPARs [8] and disrupted LPAR expression results in major phenotypes including embryonic and postnatal fatalities.

4.3.1 LPA_1

LPA_1 (previously known as VZG-1 and EDG-2) couples with three G_α proteins ($G_{12/13}$, $G_{q/11}$, and $G_{I/o}$); $G_{12/13}$ activates the Rho/ROCK pathway, $G_{q/11}$ activates the phospholipase-C (PLC) pathway, and $G_{I/o}$ activates the RAS/mitogen-activated protein kinase (MAPK) pathway. During CNS development, LPA_1 spatiotemporal patterning is highly regulated. In particular, LPA_1 is highly expressed in the neurogenic ventricular zone and was thus discovered to mediate NPC proliferation and differentiation [46–48]. LPA_1 activation is considered one of the primary signaling systems in cortical development [180]; Lpar1 knockout caused ~50% perinatal lethality and the surviving mice demonstrated significant neurodevelopmental deficits such as craniofacial defects [23], smaller bodies [46, 49–51], altered pain sensation [52, 53], and increased death of Schwann cells [54] and neurons of the cortex [55, 56] and hippocampus [57]. Lpar1 null neuronal cultures have impaired synaptic transmission and altered neurotransmitter release indicating that Lpar1 expression is also important for neuronal function [58]. Lpar1 signaling can be antagonized by receptor tyrosine kinases and GPCR complexes such as a neurotrophic tyrosine kinase receptor type 1 (TrkA) and LPAR1 integrated signaling complex [59, 60]. TrkA typically binds nerve growth factor (NGF) which activates the beta-arrestin-dependent ERK1/2 pathway [59, 61] and causes neurite extension. Active LPA_1 in the TrkA–LPAR1 complex enhances the ERK1/2 response by creating more Gαβ dimers for TrkA to use. LPA binding to LPA_1 can dissociate this complex [60], thus antagonizing the effects of NGF on cells and causing neurite retraction, counteracting NGF-induced neurite extension [62]. These studies suggest that tightly regulated Lpar1 expression and LPA signaling is crucial for normal CNS development and indicates a possible link between aberrations in the LPA system with neurodevelopmental disorders.

4.3.2 LPA_2

LPA_2 (previously known as EDG-4) is highly expressed in the embryonic brain and couples with $G_{12/13}$, $G_{q/11}$, and $G_{I/O}$ to produce cellular responses such as neuronal differentiation [48, 63, 64], cell migration [65, 66], survival, and altered immune function [8, 67–69]. Although Lpar2 null mice are phenotypically normal at the pre- and postnatal stages, Lpar1 and Lpar2 double knockout mice have an exacerbated phenotype of Lpar1 null mice [70, 71]. In addition, embryonic exposure to LPA ex vivo causes increased neurogenesis in an LPA_1- and LPA_2-dependent manner [68], suggesting functional redundancy within the Lpar1 and Lpar2 signaling system.

4.3.3 LPA_3

LPA_3 (previously known as EDG-7) is less sensitive to LPA species with saturated acyl chains and prefers unsaturated fatty acid chains in the SN-2 position. LPA_3 couples to $G_{i/o}$ and G_q [42, 72, 73]. LPA_3 activation is associated with neurite elongation,

PLC activation, Ca^{2+} mobilization, and MAPK activation [23, 74, 75]. During embryonic cortical development, Lpar3 is primarily expressed in the lateral nasal and maxillary process as well as the optic vesicle [76]. Lpar3 null mice develop normally with viable litters. However, the reproductive system of female Lpar3 null mice is affected, resulting in substantially delayed embryo implantation and spacing defects [77, 78].

4.3.4 *LPA₄*

LPA_4 (previously known as purinergic G protein-coupled receptor 9; p2y9/GPCR orphan receptor 23; GPR23) is a non-EDG receptor that interacts with $G\alpha_s$ proteins in addition to $G_{12/13}$, $G_{q/11}$, and $G_{i/o}$ [79, 80]. There is no clear phenotypical distinction between LPA_4 knockout and wild-type mice although Lpar4 knockout mice have decreased prenatal survival [81]. Signaling through LPA_4 mediates Ca^{2+} mobilization and cAMP accumulation, and modulates cell morphology, migration, aggregation, and angiogenesis [2, 8, 9]. LPA_4 signaling can also antagonize traditional LPA signaling, such as LPA_1- and LPA_2-induced cell motility by enhancing chemorepulsive cues [82].

4.3.5 *LPA₅*

LPA_5 (previously known as GPR92/GPR93) is expressed throughout the developing CNS and signals via $G_{12/13}$ and G_q proteins [76, 79]. LPA_5 activation mediates neurite retraction, stress fiber formation, and increased intracellular Ca^{2+} levels [83]. Lpar5 is also involved in neuropathic pain [84], as well as immune function [10, 85].

4.3.6 *LPA₆*

LPA_6 (previously known as P2Y5) is the latest deorphanized LPAR that utilizes $G_{12/13}$ and signals through the Rho pathway. *LPAR6* is the first gene found to mediate human hair growth and was found to preferentially respond to 2-acyl-LPA rather than 1-acyl-LPA [86]. *LPAR6* activation also produces increased intracellular Ca^{2+} through G_s stimulation of cAMP and ERK1/2 pathways [87].

4.4 Lysophosphatidic Acid Signaling in Embryonic Corticogenesis and Neurodevelopmental Disease

Studies of LPAR expression and LPA bioactivity in the developing CNS have shown myriad effects on LPA signaling during fetal corticogenesis and neural progenitor cell (NPC) survival and function. Altered LPA signaling has been identified as a potent mediator of NPC function and one of the primary mechanisms in the resulting neurodegenerative and neuropsychiatric disorders, such as hydrocephalus [55, 88], gliomas [7, 89], Alzheimer's disease [90, 91], neuropsychiatric disorders [8, 92–94], neuropathic pain [95–97], and hypoxia [49, 98, 99].

4.4.1 *Fetal Corticogenesis*

During early embryonic development, neuroepithelial (NE) cells of the ectoderm proliferate and invaginate to form the neural tube. The highly regulated process of symmetric and asymmetric divisions of NE cells forms several distinct embryonic layers: the ventricular zone (VZ), subventricular zone (SVZ), intermediate zone (IZ), cortical plate (CP), and marginal zone (MZ). Radial glial cells in the VZ migrate and proliferate to form distinct pools of NPCs, to provide migratory support for nascent neurons through the cortical layers and to form functional connections in the cortex [100]. The basic cellular components and functional connections of the cortex are generated from NPC proliferation, migration, differentiation, and programmed cell death. Astrocytes, oligodendrocytes, and other supportive cell types such as ependymal cells, microglia, and meninges are formed in late embryogenesis to early postnatal stages [101–103].

4.4.2 *LPA Is a Potent Neuromodulator*

LPA has neurotransmitter-like effects on NPCs and preferentially modulates calcium and chloride conductance in cortical neuroblasts [104, 105]. Whole-cell patch clamp of E11 cortical neuroblasts demonstrates the striking effect of altered membrane potential (depolarization) immediately after LPA exposure and no response to L-glutamate or GABA, suggesting that embryonic NPCs are preferentially responsive to LPA [104]. Even after embryonic cortical NPC growth in culture for up to 12 h, the majority of cells produced were LPA responsive; some GABA and L-glutamate responsive cells were also produced but those cells preferentially depolarized to extracellular LPA even after a short refractory period of GABA or L-glutamate activation [104]. Calcium conductance changes [104, 105] and inward chloride current from RhoA activation [106] underlies the LPA-induced changes in the membrane potential of cortical NPCs. Fluctuations in membrane potential are known to influence NPC survival, proliferation, differentiation, morphology, and migration, all of which are also mediated by LPA signaling [104–111]. These studies provide compelling evidence that anomalies in LPA signaling during cortical development may have significant physiological consequences.

4.4.3 LPA in Mitogenesis and Neurogenesis

Mitogenesis is typically triggered by proteins; however, LPA signaling through its GPCRs promotes cell cycle progression and survival of mitogenic NPCs to increase cellular output. LPA_1 was originally identified in the major proliferative and neurogenic regions of the embryonic brain: the cortex, ventricular zones, and olfactory bulbs [1, 41]. Wild-type embryonic mouse cortices exogenously exposed to LPA had enhanced cortical folding and thickness through terminal mitosis of NPCs; this cortical response was lost in LPA_1/LPA_2 double knockout mice [68]. Enhanced proliferation and neural cluster formation was also observed in embryonic cortical cell cultures exposed to LPA. The effects of LPA signaling on NPC population expansion and neuronal fate commitment during corticogenesis suggest that multiple factors such as developmental stage, species, LPA levels, and LPA receptor expression may contribute to neurogenesis and cortical organization. In the rat, high concentrations of LPA (10 μM) inhibit proliferation [112] and lower concentrations (up to 1.0 μM) promote proliferation of cortical NPCs [113]; similar effects are also observed in human embryonic stem cells mediated by the Rho/ROCK pathway [114]. Alternatively, high concentrations (10 μM) of LPA promote proliferation of mouse NPCs [48]. Similar controversies are also evident with regard to LPA-induced neuronal differentiation. NPC differentiation is caused by LPA_1-G_i-dependent activation [115]. LPA_1 and LPA_3 activation of PI3K/AKT and the Rho/ROCK pathway also inhibits neuronal differentiation [114], suggesting LPAR functional antagonism as was observed in the cortex [116] and distinct expression profiles may modulate progenitor cell fate determination. Compelling studies using a spontaneously occurring Lpar1 null mutant mouse, termed maLPA [46], displayed loss of cortical layers, altered neuronal marker expression, increased cell death, and a reduced VZ population, suggesting that reducing LPA signaling through Lpar1 may also attenuate neurogenesis [56, 57].

4.4.4 LPA in NPC Survival and Programmed Cell Death

NPCs express multiple LPAR subtypes [6]; the effects of LPA signaling on NPC survival are dependent on receptor specificity and G protein activation. Some studies suggest that LPA's pro-survival activity is dependent on LPA_1- or LPA_2-mediated G_i activation of GSK-3, Akt, and beta-catenin [54, 117]. Conversely, LPA's pro-apoptotic effects are due to LPA_1-, LPA_2-, and LPA_4-mediated $G_{12/13}$ activation of GSK-3 [118]. Ex vivo culture of embryonic cortices with LPA shows an LPA_1- and LPA_2-dependent abrogation of NPC death and increased terminal mitosis of NPCs, resulting in an expanded neuronal population and gyri-like formations of the murine cortex [68]. During corticogenesis, the spatiotemporal expression of LPAR1 coincides with programmed cell death [119, 120], suggesting that LPA signaling may modulate programmed cell death pathways in the developing brain.

4.4.5 LPA and NPC Migration

One integral facet of embryonic cortical development is the movement of NPC nuclei between the apical ventricular surface and more basal positions, referred to as interkinetic nuclear migration (INM) [121, 122]. Within the highly compact brain, INM is thought to provide adequate physical space for mitosis, which permits the expansion of progenitor pools that later populate each cortical layer. During this process, NPCs elongate and extend lamellipodia to sense the surrounding environment for chemoattractive, chemorepulsive, and growth factor cues. Through an LPAR1-dependent manner, exogenous addition of LPA can disrupt INM [181], inhibit fiber extension, and cause mitotic displacement [99]. Consistently, studies have also shown that LPA is a chemorepulsive signaling molecule and causes neurite retraction, cell rounding, and modulates neuronal migration in a Rho-dependent manner [22, 113, 123–125]. Under certain conditions, LPA indirectly affects NPC function and causes axonal extension [126–128]. In co-cultures of cortical neuroblasts and astrocytes primed with LPA, the astrocytes induced neuronal differentiation and axonal outgrowth in the neuroblasts, a process mediated by epidermal growth factor (EGF) and MAPK pathways in an Lpar1- and Lpar2-dependent manner [126, 127, 129]. Although further research is needed, the evidence suggests that LPA signaling may have significant control of NPC migration during cortical development, effectively causing cortical disorganization when the spatiotemporal patterning of LPA is altered.

4.5 LPA Signaling and Neurodevelopmental Diseases

Hypoxic and hemorrhagic injury in the developing brain are two major risk factors for neurodevelopmental pathology that can result in neurological and neuropsychiatric disorders [99, 116, 130–132]. Malfunctioning and reduced progenitor cell populations in the developing brain are among the primary pathologies in neurodevelopmental diseases. Abnormal LPA signaling during cortical development is correlated with abnormal NPC function, hypoxic or hemorrhagic events, and downstream effects such as hydrocephalus and schizophrenia. Further study is needed to gain mechanistic insight into the role of LPA signaling in developmental injury regarding changes in neurogenesis and NPC function in these disease states.

4.5.1 Hypoxic Injury

Hypoxic injury is commonly associated with developmental nervous system disorders. In the developing brain, hypoxia causes mitotic displacement, inflammation, and decreased cholinergic and serotonergic fiber formation [133, 134]. Cortical exposure to hypoxia causes overactivation of Lpar1 and downregulation of G protein-coupled receptor kinase 2 (GRK2) [99], both of which are also caused by LPA signaling in NPCs [68, 99]. Hypoxic injury enhances expression of LPA-induced hypoxia inducible factor-1 alpha (HIF-1α) in cancer cells and VEGF in the vasculature [98, 135], indicating that LPA signaling is directly linked to injury mechanisms produced by the hypoxic brain.

4.5.2 Schizophrenia

Neuropsychiatric disorders like schizophrenia are often initiated by fetal exposure to hypoxia, hemorrhage, and/or infection [99, 116, 136]. The pathology of schizophrenia includes deviations in cellular, molecular, and neurotransmitter pathways that result in psychological and cognitive deficits, most of which are also linked to LPA signaling [58, 99, 116]. Perturbed glutamatergic and serotonergic (5-HT) signaling are major hallmarks of a schizophrenic brain [137, 138]; LPA signaling through LPA_1 is also known to attenuate glutamatergic signaling pathways [139]. In addition, LPAR1 null mice have altered 5-HT levels and reduced glutamate synapses [58, 140]. Fetal intraventricular exposure to LPA was found to recapitulate neurochemical, behavioral, and genetic hallmarks of schizophrenia [116] and altered Lpar expression is linked to the cellular pathologies in schizophrenia including loss of parvalbumin-positive cells in the frontal cortex, reduced neurogenesis in the hippocampus, and behavior related to anxiety, depression, and cognitive decline [92, 93, 141]. Interestingly, an LPA or serum-dependent model of neuropsychiatric disease produced by fetal brain exposure recapitulated nearly 50% of genes previously identified in schizophrenia study [116]. Overall, this provides compelling evidence that abnormal LPA signaling during fetal brain development can have persistent detrimental outcomes evident in adult brain function.

4.5.3 Hemorrhagic Injury and Hydrocephalus

Fetal intracranial hemorrhage (ICH) is a major risk for hydrocephalus (post-hemorrhagic hydrocephalus, PHH) [131, 142, 143]. PHH is a common neurological disorder characterized by increased head size, cortical thinning, ventricular cerebrospinal fluid accumulation, and ventriculomegaly. The cellular pathologies of PHH include ependymal cell denudation, neurorosette formation, and NPC displacement [144–146]. The etiology of PHH remains unclear and the prognosis reflects a range of debilitating neurodevelopmental and psychiatric sequelae. The pathogenesis of PHH is possibly mediated by enhanced LPA signaling from blood exposure in the developing brain. Blood is known to contain LPA many fold over the Kd of various LPARs; these high levels can be exacerbated by ATX-mediated conversion of lysophosphatidylcholine (LPC) into LPA and/or degranulation of platelets during a hemorrhagic event [49, 50, 147]. The cellular pathologies of hydrocephalus can be initiated by intraventricular exposure to LPA or blood components in the fetal mouse brain [55]. The effects of blood or LPA exposure were prevented using an LPA_1/LPA_3 selective antagonist as well as in Lpar1 and Lpar2 double null mutant mouse.

The basic mechanisms behind PHH remain unclear. However, clinical evidence indicated that chromosome segregation deficits—aneuploidies—predict a higher risk for poor clinical outcomes of hydrocephalus and development of other associated neurological disorders (i.e., schizophrenia, Down syndrome, and brain tumors) [148–157]. In the healthy developing brain, aneuploidy is associated with programmed cell death (PCD), synaptic transmission, differentiation, and gene expression [155, 158–162]. While large deviations in chromosome number undergo PCD, smaller aneuploidies evade this process, integrate into the brain circuitry, and become functional neurons [162]. The effects of these abnormal cell karyotypes on brain function remain unknown although evidence suggests that genomic mosaicism in the brain is associated with neurodegenerative [163–167] and neuropsychiatric [166, 168–170] diseases. LPA signaling, such as through RhoA, alters normal NPC mitosis and consequently enhances aneuploidies [171–176]. LPA signaling often results in mitotic deregulation, altered survival, migration, and fate of developing neurons [22, 88, 177, 178]. We speculate that intraventricular exposure to LPA may alter forms of genomic mosaicism including aneuploidies and CNVs [171–176, 179]. In view of LPA's genomic effects on NPCs, the pathogenesis of LPA-induced PHH may be a consequence of LPA-induced aneuploidies or smaller genomic changes in the progenitor population of the developing cortex, which remains to be examined in future work.

4.6 Conclusion

LPA signaling is highly regulated during corticogenesis which influences a vast array of NPC functions including proliferation, survival, migration, morphology, fate, and karyotype. Perturbations to the normal spatiotemporal expression of LPARs and LPA during development have significant functional consequences in the brain, as demonstrated by the many effects relevant to neurodevelopmental disorders that have been linked to LPA signaling. There are currently no medical treatments for the neurodevelopmental and neuropsychiatric disorders that arise from fetal hypoxic or hemorrhagic insults. Interrogating potential genomic and molecular mechanisms underlying LPA's effect on corticogenesis after fetal brain injury may help develop effective therapeutics for neurodevelopmental disorders that target LPA metabolism and signaling.

Grant Support

R01 NS084398
R01 AA021402

References

1. Hecht JH, Weiner JA, Post SR, Chun J (1996) Ventricular zone gene-1 (vzg-1) encodes a lysophosphatidic acid receptor expressed in neurogenic regions of the developing cerebral cortex. J Cell Biol 135(4):1071–1083
2. Kihara Y, Mizuno H, Chun J (2015) Lysophospholipid receptors in drug discovery. Exp Cell Res 333(2):171–177
3. Lyncha KR, Macdonald TL (2001) Structure activity relationships of lysophospholipid mediators. Prostaglandins Other Lipid Mediat 64(1–4):33–45
4. Lyncha KR, Macdonald TL (2002) Structure-activity relationships of lysophosphatidic acid analogs. Biochim Biophys Acta 1582(1–3):289–294
5. Sugiura T, Nakane S, Kishimoto S, Waku K, Yoshioka Y, Tokumura A et al (1999) Occurrence of lysophosphatidic acid and its alkyl ether-linked analog in rat brain and comparison of their biological activities toward cultured neural cells. Biochim Biophys Acta 1440(2–3):194–204
6. Yung YC, Stoddard NC, Mirendil H, Chun J (2015) Lysophosphatidic acid signaling in the nervous system. Neuron 85(4):669–682
7. Steiner MR, Urso JR, Klein J, Steiner SM (2002) Multiple astrocyte responses to lysophosphatidic acids. Biochim Biophys Acta 1582(1–3):154–160
8. Choi JW, Herr DR, Noguchi K, Yung YC, Lee C-W, Mutoh T et al (2010) LPA receptors: subtypes and biological actions. Annu Rev Pharmacol Toxicol 50:157–186
9. Yung YC, Stoddard NC, Chun J (2014) LPA receptor signaling: pharmacology, physiology, and pathophysiology. J Lipid Res 55(7):1192–1214
10. Hu J, Oda SK, Shotts K, Donovan EE, Strauch P, Pujanauski LM et al (2014) Lysophosphatidic acid receptor 5 inhibits B cell antigen receptor signaling and antibody response. J Immunol 193(1):85–95
11. Ma L, Nagai J, Chun J, Ueda H (2013) An LPA species (18:1 LPA) plays key roles in the self-amplification of spinal LPA production in the peripheral neuropathic pain model. Mol Pain 9(1):29
12. Chan LC, Peters W, Xu Y, Chun J, Farese RV, Cases S (2007) LPA3 receptor mediates chemotaxis of immature murine dendritic cells to unsaturated lysophosphatidic acid (LPA). J Leukoc Biol 82(5):1193–1200
13. Hayashi K, Takahashi M, Nishida W, Yoshida K, Ohkawa Y, Kitabatake A et al (2001) Phenotypic modulation of vascular smooth muscle cells induced by unsaturated lysophosphatidic acids. Circ Res 89(3):251–258
14. Aoki J (2004) Mechanisms of lysophosphatidic acid production. Semin Cell Dev Biol 15(5):477–489
15. Baker DL, Desiderio DM, Miller DD, Tolley B, Tigyi GJ (2001) Direct quantitative analysis of lysophosphatidic acid molecular species by stable isotope dilution electrospray ionization liquid chromatography-mass spectrometry. Anal Biochem 292(2):287–295
16. Hosogaya S, Yatomi Y, Nakamura K, Ohkawa R, Okubo S, Yokota H et al (2008) Measurement of plasma lysophosphatidic acid concentration in healthy subjects: strong correlation with lysophospholipase D activity. Ann Clin Biochem 45(Pt 4):364–368
17. Das AK, Hajra AK (1989) Quantification, characterization and fatty acid composition of lysophosphatidic acid in different rat tissues. Lipids 24(4):329–333
18. Triebl A, Trötzmüller M, Eberl A, Hanel P, Hartler J, Köfeler HC (2014) Quantitation of phosphatidic acid and lysophosphatidic acid molecular species using hydrophilic interaction liquid chromatography coupled to electrospray ionization high resolution mass spectrometry. J Chromatogr A 1347:104–110
19. Eichholtz T, Jalink K, Fahrenfort I, Moolenaar WH (1993) The bioactive phospholipid lysophosphatidic acid is released from activated platelets. Biochem J 291(Pt 3):677–680
20. Siess W, Zangl KJ, Essler M, Bauer M, Brandl R, Corrinth C et al (1999) Lysophosphatidic acid mediates the rapid activation of platelets and endothelial cells by mildly oxidized low

density lipoprotein and accumulates in human atherosclerotic lesions. Proc Natl Acad Sci U S A 96(12):6931–6936

21. Kishi Y, Okudaira S, Tanaka M, Hama K, Shida D, Kitayama J et al (2006) Autotaxin is overexpressed in glioblastoma multiforme and contributes to cell motility of glioblastoma by converting lysophosphatidylcholine to lysophosphatidic acid. J Biol Chem 281(25):17492–17500
22. Fukushima N, Weiner JA, Kaushal D, Contos JJAA, Rehen SK, Kingsbury MA et al (2002) Lysophosphatidic acid influences the morphology and motility of young, postmitotic cortical neurons. Mol Cell Neurosci 20(2):271–282
23. Ishii I, Contos JJ, Fukushima N, Chun J (2000) Functional comparisons of the lysophosphatidic acid receptors, LP(A1)/VZG-1/EDG-2, LP(A2)/EDG-4, and LP(A3)/EDG-7 in neuronal cell lines using a retrovirus expression system. Mol Pharmacol 58(5):895–902
24. Aoki J, Inoue A, Okudaira S (2008) Two pathways for lysophosphatidic acid production. Biochim Biophys Acta 1781(9):513–518
25. Underwood KW, Song C, Kriz RW, Chang XJ, Knopf JL, Lin L-L (1998) A novel calcium-independent phospholipase A2, cPLA2-, that is prenylated and contains homology to cPLA2. J Biol Chem 273(34):21926–21932
26. Tokumura A, Majima E, Kariya Y, Tominaga K, Kogure K, Yasuda K et al (2002) Identification of human plasma lysophospholipase D, a lysophosphatidic acid-producing enzyme, as autotaxin, a multifunctional phosphodiesterase. J Biol Chem 277(42):39436–39442
27. Tokumura A, Kanaya Y, Kitahara M, Miyake M, Yoshioka Y, Fukuzawa K (2002) Increased formation of lysophosphatidic acids by lysophospholipase D in serum of hypercholesterolemic rabbits. J Lipid Res 43(2):307–315
28. Umezu-Goto M, Kishi Y, Taira A, Hama K, Dohmae N, Takio K et al (2002) Autotaxin has lysophospholipase D activity leading to tumor cell growth and motility by lysophosphatidic acid production. J Cell Biol 158(2):227–233
29. Pagès C, Simon M-FF, Valet P, Saulnier-Blache JS (2001) Lysophosphatidic acid synthesis and release. Prostaglandins Other Lipid Mediat 64(1–4):1–10
30. Lee H, Goetzl EJ, An S (2000) Lysophosphatidic acid and sphingosine 1-phosphate stimulate endothelial cell wound healing. Am J Physiol Cell Physiol 278(3):C612–C618
31. Panetti TS, Chen H, Misenheimer TM, Getzler SB, Mosher DF (1997) Endothelial cell mitogenesis induced by LPA: inhibition by thrombospondin-1 and thrombospondin-2. J Lab Clin Med 129(2):208–216
32. Wu WT, Chen C-N, Lin CI, Chen JH, Lee H (2005) Lysophospholipids enhance matrix metalloproteinase-2 expression in human endothelial cells. Endocrinology 146(8):3387–3400
33. van Meeteren LA, Ruurs P, Stortelers C, Bouwman P, van Rooijen MA, Pradère JP et al (2006) Autotaxin, a secreted lysophospholipase D, is essential for blood vessel formation during development. Mol Cell Biol 26(13):5015–5022
34. Tanaka M, Okudaira S, Kishi Y, Ohkawa R, Iseki S, Ota M et al (2006) Autotaxin stabilizes blood vessels and is required for embryonic vasculature by producing lysophosphatidic acid. J Biol Chem 281(35):25822–25830
35. Fotopoulou S, Oikonomou N, Grigorieva E, Nikitopoulou I, Paparountas T, Thanassopoulou A et al (2010) ATX expression and LPA signalling are vital for the development of the nervous system. Dev Biol 339(2):451–464
36. Fukushima N, Weiner JA, Chun J (2000) Lysophosphatidic acid (LPA) is a novel extracellular regulator of cortical neuroblast morphology. Dev Biol 228(1):6–18
37. Bektas M, Payne SG, Liu H, Goparaju S, Milstien S, Spiegel S (2005) A novel acylglycerol kinase that produces lysophosphatidic acid modulates cross talk with EGFR in prostate cancer cells. J Cell Biol 169(5):801–811
38. Gendaszewska-Darmach E (2008) Lysophosphatidic acids, cyclic phosphatidic acids and autotaxin as promising targets in therapies of cancer and other diseases. Acta Biochim Pol 55(2):227–240
39. Mishra RS, Carnevale KA, Cathcart MK (2008) iPLA2beta: front and center in human monocyte chemotaxis to MCP-1. J Exp Med 205(2):347–359

40. Carnevale KA, Cathcart MK (2001) Calcium-independent phospholipase A(2) is required for human monocyte chemotaxis to monocyte chemoattractant protein 1. J Immunol 167(6):3414–3421
41. Weiner JA, Hecht JH, Chun J (1998) Lysophosphatidic acid receptor gene vzg-1/lp(A)1/edg-2 is expressed by mature oligodendrocytes during myelination in the postnatal murine brain. J Comp Neurol 398(4):587–598
42. Contos JJ, Chun J (2001) The mouse lpA3/Edg7 lysophosphatidic acid receptor gene: genomic structure, chromosomal localization, and expression pattern. Gene 267(2):243–253
43. Moolenaar WH (1999) Bioactive lysophospholipids and their G protein-coupled receptors. Exp Cell Res 253(1):230–238
44. Guo Z, Liliom K, Fischer DJ, Bathurst IC, Tomei LD, Kiefer MC et al (1996) Molecular cloning of a high-affinity receptor for the growth factor-like lipid mediator lysophosphatidic acid from Xenopus oocytes. Proc Natl Acad Sci U S A 93(25):14367–14372
45. Choi JW, Chun J (2013) Lysophospholipids and their receptors in the central nervous system. Biochim Biophys Acta 1831(1):20–32
46. Contos JJ, Fukushima N, Weiner JA, Kaushal D, Chun J (2000) Requirement for the lpA1 lysophosphatidic acid receptor gene in normal suckling behavior. Proc Natl Acad Sci U S A 97(24):13384–13389
47. Fukushima N, Kimura Y, Chun J (1998) A single receptor encoded by vzg-1/lpA1/edg-2 couples to G proteins and mediates multiple cellular responses to lysophosphatidic acid. Proc Natl Acad Sci U S A 95(11):6151–6156
48. Svetlov SI, Ignatova TN, Wang KKW, Hayes RL, English D, Kukekov VG (2004) Lysophosphatidic acid induces clonal generation of mouse neurospheres via proliferation of Sca-1- and AC133-positive neural progenitors. Stem Cells Dev 13(6):685–693
49. Aoki J, Taira A, Takanezawa Y, Kishi Y, Hama K, Kishimoto T et al (2011) Autotaxin, a secreted lysophospholipase D, is essential for blood vessel formation during development. J Neurosci 21(1):595–604
50. Yanagida K, Masago K, Nakanishi H, Kihara Y, Hamano F, Tajima Y et al (2011) Proc Natl Acad Sci U S A 21(1):595–604
51. Dusaulcy R, Daviaud D, Pradère JP, Grès S, Valet P, Saulnier-Blache JS (2009) Altered food consumption in mice lacking lysophosphatidic acid receptor-1. J Physiol Biochem 65(4):345–350
52. Halder SK, Yano R, Chun J, Ueda H (2013) Involvement of LPA(1) receptor signaling in cerebral ischemia-induced neuropathic pain. Neuroscience 235:10–15
53. Xie W, Uchida H, Nagai J, Ueda M, Chun J, Ueda H (2010) Calpain-mediated down-regulation of myelin-associated glycoprotein in lysophosphatidic acid-induced neuropathic pain. J Neurochem 113(4):1002–1011
54. Weiner JA, Fukushima N, Contos JJA, Scherer SS, Chun J (2001) Regulation of Schwann cell morphology and adhesion by receptor-mediated lysophosphatidic acid signaling. J Neurosci 21(18):7069–7078
55. Yung YC, Mutoh T, Lin M-EE, Noguchi K, Rivera RR, Choi JW et al (2011) Lysophosphatidic acid signaling may initiate fetal hydrocephalus. Sci Transl Med 3(99):99ra87
56. Estivill-Torrus G (2008) Absence of LPA1 signaling results in defective cortical development. Cereb Cortex 18(4):938–950
57. Matas-Rico E, García-Diaz B, Llebrez-Zayas P, López-Barroso D, Santín L, Pedraza C et al (2008) Deletion of lysophosphatidic acid receptor LPA1 reduces neurogenesis in the mouse dentate gyrus. Mol Cell Neurosci 39(3):342–355
58. Harrison SM, Reavill C, Brown G, Brown JT, Cluderay JE, Crook B et al (2003) LPA1 receptor-deficient mice have phenotypic changes observed in psychiatric disease. Mol Cell Neurosci 24(4):1170–1179
59. Miller WE, Lefkowitz RJ (2001) Expanding roles for beta-arrestins as scaffolds and adapters in GPCR signaling and trafficking. Curr Opin Cell Biol 13(2):139–145
60. Moughal NA, Waters C, Sambi B, Pyne S, Pyne NJ (2004) Nerve growth factor signaling involves interaction between the Trk A receptor and lysophosphatidate receptor 1 systems:

nuclear translocation of the lysophosphatidate receptor 1 and Trk A receptors in pheochromocytoma 12 cells. Cell Signal 16(1):127–136
61. Rakhit S, Pyne S, Pyne NJ (2001) Nerve growth factor stimulation of p42/p44 mitogen-activated protein kinase in PC12 cells: role of G(i/o), G protein-coupled receptor kinase 2, beta-arrestin I, and endocytic processing. Mol Pharmacol 60(1):63–70
62. Moughal NA, Waters CM, Valentine WJ, Connell M, Richardson JC, Tigyi G et al (2006) Protean agonism of the lysophosphatidic acid receptor-1 with Ki16425 reduces nerve growth factor-induced neurite outgrowth in pheochromocytoma 12 cells. J Neurochem 98(6):1920–1929
63. Beer MS, Stanton JA, Salim K, Rigby M, Heavens RP, Smith D et al (2000) EDG receptors as a therapeutic target in the nervous system. Ann N Y Acad Sci 905:118–131
64. Savitz SI, Dhallu MS, Malhotra S, Mammis A, Ocava LC, Rosenbaum PS et al (2006) EDG receptors as a potential therapeutic target in retinal ischemia-reperfusion injury. Brain Res 1118(1):168–175
65. Yan H, Lu D, Rivkees SA (2003) Lysophosphatidic acid regulates the proliferation and migration of olfactory ensheathing cells in vitro. Glia 44(1):26–36
66. Zhang Y, Chen Y-CM, Krummel MF, Rosen SD (2012) Autotaxin through lysophosphatidic acid stimulates polarization, motility, and transendothelial migration of naive T cells. J Immunol 189(8):3914–3924
67. Kihara Y, Maceyka M, Spiegel S, Chun J (2014) Lysophospholipid receptor nomenclature review: IUPHAR Review 8. Br J Pharmacol 171(15):3575–3594
68. Kingsbury MA, Rehen SK, Contos JJA, Higgins CM, Chun J (2003) Non-proliferative effects of lysophosphatidic acid enhance cortical growth and folding. Nat Neurosci 6(12):1292–1299
69. Kuriyama S, Theveneau E, Benedetto A, Parsons M, Tanaka M, Charras G et al (2014) In vivo collective cell migration requires an LPAR2-dependent increase in tissue fluidity. J Cell Biol 206(1):113–127
70. Contos JJ, Ishii I, Fukushima N, Kingsbury MA, Ye X, Kawamura S et al (2002) Characterization of lpa(2) (Edg4) and lpa(1)/lpa(2) (Edg2/Edg4) lysophosphatidic acid receptor knockout mice: signaling deficits without obvious phenotypic abnormality attributable to lpa(2). Mol Cell Biol 22(19):6921–6929
71. Contos JJ, Ishii I, Chun J (2000) Lysophosphatidic acid receptors. Mol Pharmacol 58:1188–1196
72. Hama K, Bandoh K, Kakehi Y, Aoki J, Arai H (2002) Lysophosphatidic acid (LPA) receptors are activated differentially by biological fluids: possible role of LPA-binding proteins in activation of LPA receptors. FEBS Lett 523(1–3):187–192
73. Hama K, Aoki J, Inoue A, Endo T, Amano T, Motoki R et al (2007) Embryo spacing and implantation timing are differentially regulated by LPA3-mediated lysophosphatidic acid signaling in mice. Biol Reprod 77(6):954–959
74. Bandoh K, Aoki J, Hosono H, Kobayashi S, Kobayashi T, Murakami-Murofushi K et al (1999) Molecular cloning and characterization of a novel human G-protein-coupled receptor, EDG7, for lysophosphatidic acid. J Biol Chem 274(39):27776–27785
75. Im DS, Heise CE, Harding MA, George SR, O'Dowd BF, Theodorescu D et al (2000) Molecular cloning and characterization of a lysophosphatidic acid receptor, Edg-7, expressed in prostate. Mol Pharmacol 57(4):753–759
76. Ohuchi H, Hamada A, Matsuda H, Takagi A, Tanaka M, Aoki J et al (2008) Expression patterns of the lysophospholipid receptor genes during mouse early development. Dev Dyn 237(11):3280–3294
77. Ye X, Hama K, Contos JJ, Anliker B, Inoue A, Skinner MK et al (2005) LPA3-mediated lysophosphatidic acid signalling in embryo implantation and spacing. Nature 435(7038):104–108
78. Lai S-L, Yao W-L, Tsao K-C, Houben AJS, Albers HMHG, Ovaa H et al (2012) Autotaxin/Lpar3 signaling regulates Kupffer's vesicle formation and left-right asymmetry in zebrafish. Development 139(23):4439–4448

79. Lee C-WW, Rivera R, Gardell S, Dubin AE, Chun J (2006) GPR92 as a new G12/13- and Gq-coupled lysophosphatidic acid receptor that increases cAMP, LPA5. J Biol Chem 281(33):23589–23597
80. Lee C-W, Rivera R, Dubin AE, Chun J (2007) LPA(4)/GPR23 is a lysophosphatidic acid (LPA) receptor utilizing G(s)-, G(q)/G(i)-mediated calcium signaling and G(12/13)-mediated Rho activation. J Biol Chem 282(7):4310–4317
81. Sumida H, Noguchi K, Kihara Y, Abe M, Yanagida K, Hamano F et al (2010) LPA4 regulates blood and lymphatic vessel formation during mouse embryogenesis. Blood 116(23):5060–5070
82. Lee Z, Cheng C-T, Zhang H, Subler MA, Wu J, Mukherjee A et al (2008) Role of LPA4/p2y9/GPR23 in negative regulation of cell motility. Mol Biol Cell 19(12):5435–5445
83. Lee KJ, Kim SJ, Kim SW, Choi SH, Shin YC, Park SH et al (2006) Chronic mild stress decreases survival, but not proliferation, of new-born cells in adult rat hippocampus. Exp Mol Med 38(1):44–54
84. Lin M-E, Rivera RR, Chun J (2012) Targeted deletion of LPA5 identifies novel roles for lysophosphatidic acid signaling in development of neuropathic pain. J Biol Chem 287(21):17608–17617
85. Oda SK, Strauch P, Fujiwara Y, Al-Shami A, Oravecz T, Tigyi G et al (2013) Lysophosphatidic acid inhibits CD8 T cell activation and control of tumor progression. Cancer Immunol Res 1(4):245–255
86. Yanagida K, Masago K, Nakanishi H, Kihara Y, Hamano F, Tajima Y et al (2009) Identification and characterization of a novel lysophosphatidic acid receptor, p2y5/LPA6. J Biol Chem 284(26):17731–17741
87. Lee M, Choi S, Halldén G, Yo SJ, Schichnes D, Aponte GW (2009) P2Y5 is a G(alpha)i, G(alpha)12/13 G protein-coupled receptor activated by lysophosphatidic acid that reduces intestinal cell adhesion. Am J Physiol Gastrointest Liver Physiol 297(4):G641–G654
88. Park R, Moon UY, Park JY, Hughes LJ, Johnson RL, Cho S-H et al (2016) Yap is required for ependymal integrity and is suppressed in LPA-induced hydrocephalus. Nat Commun 7:10329
89. Cechin SR, Dunkley PR, Rodnight R (2005) Signal transduction mechanisms involved in the proliferation of C6 glioma cells induced by lysophosphatidic acid. Neurochem Res 30(5):603–611
90. Sayas CL (1999) The neurite retraction induced by lysophosphatidic acid increases Alzheimer's disease-like Tau phosphorylation. J Biol Chem 274(52):37046–37052
91. Sun Y, Kim N-H, Yang H, Kim S-H, Huh S-O (2011) Lysophosphatidic acid induces neurite retraction in differentiated neuroblastoma cells via GSK-3β activation. Mol Cells 31(5):483–489
92. Castilla-Ortega E, Sánchez-López J, Hoyo-Becerra C, Matas-Rico E, Zambrana-Infantes E, Chun J et al (2010) Exploratory, anxiety and spatial memory impairments are dissociated in mice lacking the LPA1 receptor. Neurobiol Learn Mem 94(1):73–82
93. Santin LJ, Bilbao A, Pedraza C, Matas-Rico E, López-Barroso D, Castilla-Ortega E et al (2009) Behavioral phenotype of maLPA1-null mice: increased anxiety-like behavior and spatial memory deficits. Genes Brain Behav 8(8):772–784
94. Lin M-E, Herr DR, Chun J (2010) Lysophosphatidic acid (LPA) receptors: signaling properties and disease relevance. Prostaglandins Other Lipid Mediat 91(3–4):130–138
95. Fujita R, Kiguchi N, Ueda H (2007) LPA-mediated demyelination in ex vivo culture of dorsal root. Neurochem Int 50(2):351–355
96. Inoue M, Rashid MH, Fujita R, Contos JJ, Chun J, Ueda H (2004) Initiation of neuropathic pain requires lysophosphatidic acid receptor signaling. Nat Med 10(7):712–718
97. Yano R, Ma L, Nagai J, Ueda H (2013) Interleukin-1β plays key roles in LPA-induced amplification of LPA production in neuropathic pain model. Cell Mol Neurobiol 33(8):1033–1041
98. Lee S-J, No YR, Dang DT, Dang LH, Yang VW, Shim H et al (2013) Regulation of hypoxia-inducible factor 1α (HIF-1α) by lysophosphatidic acid is dependent on interplay between p53 and Krüppel-like factor 5. J Biol Chem 288(35):25244–25253

99. Herr KJ, Herr DR, Lee C-W, Noguchi K, Chun J (2011) Stereotyped fetal brain disorganization is induced by hypoxia and requires lysophosphatidic acid receptor 1 (LPA1) signaling. Proc Natl Acad Sci U S A 108(37):15444–15449
100. Noctor SC, Flint AC, Weissman TA, Dammerman RS, Kriegstein AR (2001) Neurons derived from radial glial cells establish radial units in neocortex. Nature 409(6821):714–720
101. Prinz M, Priller J (2014) Microglia and brain macrophages in the molecular age: from origin to neuropsychiatric disease. Nat Rev Neurosci 15(5):300–312
102. Jakovcevski I, Filipovic R, Mo Z, Rakic S, Zecevic N (2009) Oligodendrocyte development and the onset of myelination in the human fetal brain. Front Neuroanat 3:5
103. Yang Y, Higashimori H, Morel L (2013) Developmental maturation of astrocytes and pathogenesis of neurodevelopmental disorders. J Neurodev Disord 5(1):22
104. Dubin AE, Bahnson T, Weiner JA, Fukushima N, Chun J (1999) Lysophosphatidic acid stimulates neurotransmitter-like conductance changes that precede GABA and L-glutamate in early, presumptive cortical neuroblasts. J Neurosci 19(4):1371–1381
105. Dubin AE, Herr DR, Chun J (2010) Diversity of lysophosphatidic acid receptor-mediated intracellular calcium signaling in early cortical neurogenesis. J Neurosci 30(21):7300–7309
106. Postma FR, Jalink K, Hengeveld T, Offermanns S, Moolenaar WH (2001) Gα13 mediates activation of a depolarizing chloride current that accompanies RhoA activation in both neuronal and nonneuronal cells. Curr Biol 11(2):121–124
107. Apáti Á, Pászty K, Erdei Z, Szebényi K, Homolya L, Sarkadi B (2012) Calcium signaling in pluripotent stem cells. Mol Cell Endocrinol 353(1–2):57–67
108. Schilling T, Stock C, Schwab A, Eder C (2004) Functional importance of Ca^{2+}-activated K^+ channels for lysophosphatidic acid-induced microglial migration. Eur J Neurosci 19(6):1469–1474
109. García-García E, Pino-Barrio MJ, López-Medina L, Martínez-Serrano A (2012) Intermediate progenitors are increased by lengthening of the cell cycle through calcium signaling and p53 expression in human neural progenitors. Mol Biol Cell 23(7):1167–1180
110. Weissman TA, Riquelme PA, Ivic L, Flint AC, Kriegstein AR (2004) Calcium waves propagate through radial glial cells and modulate proliferation in the developing neocortex. Neuron 43(5):647–661
111. Young SZ, Taylor MM, Wu S, Ikeda-Matsuo Y, Kubera C, Bordey A (2012) NKCC1 knockdown decreases neuron production through GABA(A)-regulated neural progenitor proliferation and delays dendrite development. J Neurosci 32(39):13630–13638
112. Cui H-LL, Qiao J-TT (2006) Promotive action of lysophosphatidic acid on proliferation of rat embryonic neural stem cells and their differentiation to cholinergic neurons in vitro. Sheng Li Xue Bao 58(6):547–555
113. Hurst JH, Mumaw J, Machacek DW, Sturkie C, Callihan P, Stice SL et al (2008) Human neural progenitors express functional lysophospholipid receptors that regulate cell growth and morphology. BMC Neurosci 9(1):118
114. Dottori M, Leung J, Turnley AM, Pébay A (2008) Lysophosphatidic acid inhibits neuronal differentiation of neural stem/progenitor cells derived from human embryonic stem cells. Stem Cells 26(5):1146–1154
115. Pitson SM, Pébay A (2009) Regulation of stem cell pluripotency and neural differentiation by lysophospholipids. Neurosignals 17(4):242–254
116. Mirendil H, Thomas EA, De Loera C, Okada K, Inomata Y, Chun J (2015) LPA signaling initiates schizophrenia-like brain and behavioral changes in a mouse model of prenatal brain hemorrhage. Transl Psychiatry 5(4):e541
117. Weiner JA, Chun J (1999) Schwann cell survival mediated by the signaling phospholipid lysophosphatidic acid. Proc Natl Acad Sci U S A 96(9):5233–5238
118. Sun Y, Nam J-S, Han D-H, Kim N-H, Choi H-K, Lee JK et al (2010) Lysophosphatidic acid induces upregulation of Mcl-1 and protects apoptosis in a PTX-dependent manner in H19-7 cells. Cell Signal 22(3):484–494
119. Blaschke AJ, Weiner JA, Chun J (1998) Programmed cell death is a universal feature of embryonic and postnatal neuroproliferative regions throughout the central nervous system. J Comp Neurol 396(1):39–50

120. Blaschke AJ, Staley K, Chun J (1996) Widespread programmed cell death in proliferative and postmitotic regions of the fetal cerebral cortex. Development 122(4):1165–1174
121. Buffo A, Rite I, Tripathi P, Lepier A, Colak D, Horn A-P et al (2008) Origin and progeny of reactive gliosis: a source of multipotent cells in the injured brain. Proc Natl Acad Sci U S A 105(9):3581–3586
122. Hayes NL, Nowakowski RS (2000) Exploiting the dynamics of S-phase tracers in developing brain: interkinetic nuclear migration for cells entering versus leaving the S-phase. Dev Neurosci 22(1–2):44–55
123. Pébay A, Bonder CS, Pitson SM (2007) Stem cell regulation by lysophospholipids. Prostaglandins Other Lipid Mediat 84(3–4):83–97
124. Halstead JR, Savaskan NE, van den Bout I, Van Horck F, Hajdo-Milasinovic A, Snell M, et al (2010) Rac controls PIP5K localisation and PtdIns(4,5)P_2 synthesis, which modulates vinculin localisation and neurite dynamics. J Cell Sci 123(Pt 20):3535–3546
125. Campbell DS, Holt CE (2003) Apoptotic pathway and MAPKs differentially regulate chemotropic responses of retinal growth cones. Neuron 37(6):939–952
126. Spohr TCLS, Dezonne RS, Rehen SK, FCA G (2014) LPA-primed astrocytes induce axonal outgrowth of cortical progenitors by activating PKA signaling pathways and modulating extracellular matrix proteins. Front Cell Neurosci 8:296
127. Spohr TCLS, Dezonne RS, Rehen SK, FCA G (2011) Astrocytes treated by lysophosphatidic acid induce axonal outgrowth of cortical progenitors through extracellular matrix protein and epidermal growth factor signaling pathway. J Neurochem 119(1):113–123
128. Bräuer AU, Savaskan NE, Kühn H, Prehn S, Ninnemann O, Nitsch R (2003) A new phospholipid phosphatase, PRG-1, is involved in axon growth and regenerative sprouting. Nat Neurosci 6(6):572–578
129. Spohr TCS, Choi JW, Gardell SE, Herr DR, Rehen SK et al (2008) Lysophosphatidic acid receptor-dependent secondary effects via astrocytes promote neuronal differentiation. J Biol Chem 283(12):7470–7479
130. Kolevzon A, Gross R, Reichenberg A (2007) Prenatal and perinatal risk factors for autism: a review and integration of findings. Arch Pediatr Adolesc Med 161(4):326–333
131. Durfee SM, Kim FM, Benson CB (2001) Postnatal outcome of fetuses with the prenatal diagnosis of asymmetric hydrocephalus. J Ultrasound Med 20(3):263–268
132. Fukumizu M, Takashima S, Becker LE (1995) Neonatal posthemorrhagic hydrocephalus: neuropathologic and immunohistochemical studies. Pediatr Neurol 13(3):230–234
133. Corcoran A, O'Connor JJ (2013) Hypoxia-inducible factor signalling mechanisms in the central nervous system. Acta Physiol (Oxf) 208(4):298–310
134. Nyakas C, Buwalda B, Luiten PG (1996) Hypoxia and brain development. Prog Neurobiol 49(1):1–51
135. Lin C-I, Chen C-N, Huang M-T, Lee S-J, Lin C-H, Chang C-C et al (2008) Lysophosphatidic acid upregulates vascular endothelial growth factor-C and tube formation in human endothelial cells through LPA(1/3), COX-2, and NF-kappaB activation- and EGFR transactivation-dependent mechanisms. Cell Signal 20(10):1804–1814
136. Boksa P (2008) Maternal infection during pregnancy and schizophrenia. J Psychiatry Neurosci 33(3):183–185
137. Beaulieu J-M, Gainetdinov RR, Caron MG (2009) Akt/GSK3 signaling in the action of psychotropic drugs. Annu Rev Pharmacol Toxicol 49:327–347
138. Beaulieu J-M, Zhang X, Rodriguiz RM, Sotnikova TD, Cools MJ, Wetsel WC et al (2008) Role of GSK3 beta in behavioral abnormalities induced by serotonin deficiency. Proc Natl Acad Sci U S A 105(4):1333–1338
139. Shano S, Moriyama R, Chun J, Fukushima N (2008) Lysophosphatidic acid stimulates astrocyte proliferation through LPA1. Neurochem Int 52(1–2):216–220
140. Roberts C, Winter P, Shilliam CS, Hughes ZA, Langmead C, Maycox PR et al (2005) Neurochemical changes in LPA1 receptor deficient mice—a putative model of schizophrenia. Neurochem Res 30(3):371–377

141. Castilla-Ortega E, Hoyo-Becerra C, Pedraza C, Chun J, Rodríguez De Fonseca F, Estivill-Torrús G et al (2011) Aggravation of chronic stress effects on hippocampal neurogenesis and spatial memory in LPA1 receptor knockout mice. PLoS One 6(9):e25522
142. Hassanein SMA, Moharram H, Monib AM, Ramy ARMRA, Ghany WA (2008) Perinatal ventriculomegaly. J Pediatric Neurol 6:293–307
143. Allenby PA, Gould NS, Thomas C (1985) Congenital posthemorrhagic hydrocephalus: report of a case. Fetal Pediatr Pathol 4(3–4):303–308
144. Morales DM, Holubkov R, Inder TE, Ahn HC, Mercer D, Rao R et al (2015) Cerebrospinal fluid levels of amyloid precursor protein are associated with ventricular size in post-hemorrhagic hydrocephalus of prematurity. PLoS One 10(3):e0115045
145. McAllister JP (2012) Pathophysiology of congenital and neonatal hydrocephalus. Semin Fetal Neonatal Med 17(5):285–294
146. McMullen AB, Baidwan GS, McCarthy KD (2012) Morphological and behavioral changes in the pathogenesis of a novel mouse model of communicating hydrocephalus. PLoS One 7(1):e30159
147. Bolen AL, Naren AP, Yarlagadda S, Beranova-Giorgianni S, Chen L, Norman D et al (2011) The phospholipase A1 activity of lysophospholipase A-I links platelet activation to LPA production during blood coagulation. J Lipid Res 52(5):958–970
148. Lam S-J, Kumar S (2014) Evolution of fetal ventricular dilatation in relation to severity at first presentation. J Clin Ultrasound 42(4):193–198
149. Graham E, Duhl A, Ural S, Allen M, Blakemore K, Witter F (2001) The degree of antenatal ventriculomegaly is related to pediatric neurological morbidity. J Matern Fetal Med 10(4):258–263
150. Gezer C, Ekin A, Ozeren M, Taner CE, Ozer O, Koc A et al (2014) Chromosome abnormality incidence in fetuses with cerebral ventriculomegaly. J Obstet Gynaecol 34(5):387–391
151. Yang AH, Kaushal D, Rehen SK, Kriedt K, Kingsbury MA, McConnell MJ et al (2003) Chromosome segregation defects contribute to aneuploidy in normal neural progenitor cells. J Neurosci 23(32):10454–10462
152. Dudenhausen JW (2014) Practical obstetrics. De Gruyter, Berlin, 511 p
153. Terry M, Calhoun BC, Walker W, Apodaca C, Martin L, Pierce B et al (2000) Aneuploidy and isolated mild ventriculomegaly. Attributable risk for isolated fetal marker. Fetal Diagn Ther 15(6):331–334
154. Sonek J, Croom C (2014) Second trimester ultrasound markers of fetal aneuploidy. Clin Obstet Gynecol 57(1):159–181
155. Zhang J, Williams MA, Rigamonti D (2006) Genetics of human hydrocephalus. J Neurol 253(10):1255–1266
156. Nomura ML, Barini R, De Andrade KC, Milanez H, Simoni RZ, Peralta CFA et al (2010) Congenital hydrocephalus: gestational and neonatal outcomes. Arch Gynecol Obstet. 282(6):607–611
157. Vergani P, Locatelli A, Strobelt N, Cavallone M, Ceruti P, Paterlini G et al (1998) Clinical outcome of mild fetal ventriculomegaly. Am J Obstet Gynecol 178(2):218–222
158. Páez P, Bátiz L-F, Roales-Buján R, Rodríguez-Pérez L-M, Rodríguez S, Jiménez AJ et al (2007) Patterned neuropathologic events occurring in hyh congenital hydrocephalic mutant mice. J Neuropathol Exp Neurol 66(12):1082–1092
159. Miller JM, Kumar R, McAllister JP, Krause GS (2006) Gene expression analysis of the development of congenital hydrocephalus in the H-Tx rat. Brain Res 1075(1):36–47
160. Kaushal D, Contos JJA, Treuner K, Yang AH, Kingsbury MA, Rehen SK et al (2003) Alteration of gene expression by chromosome loss in the postnatal mouse brain. J Neurosci 23(13):5599–5606
161. Rehen SK, McConnell MJ, Kaushal D, Kingsbury MA, Yang AH, Chun J (2001) Chromosomal variation in neurons of the developing and adult mammalian nervous system. Proc Natl Acad Sci U S A 98(23):13361–13366
162. Peterson SE, Yang AH, Bushman DM, Westra JW, Yung YC, Barral S et al (2012) Aneuploid cells are differentially susceptible to caspase-mediated death during embryonic cerebral cortical development. J Neurosci 32(46):16213–16222

163. McConnell MJ, Kaushal D, Yang AH, Kingsbury MA, Rehen SK, Treuner K et al (2004) Failed clearance of aneuploid embryonic neural progenitor cells leads to excess aneuploidy in the Atm-deficient but not the Trp53-deficient adult cerebral cortex. J Neurosci 24(37):8090–8096
164. Li J, Xu M, Zhou H, Ma J, Potter H (1997) Alzheimer presenilins in the nuclear membrane, interphase kinetochores, and centrosomes suggest a role in chromosome segregation. Cell 90(5):917–927
165. Geller LN, Potter H (1999) Chromosome missegregation and trisomy 21 mosaicism in Alzheimer's disease. Neurobiol Dis 6(3):167–179
166. Yurov YB, Vostrikov VM, Vorsanova SG, Monakhov VV, Iourov IY (2001) Multicolor fluorescent in situ hybridization on post-mortem brain in schizophrenia as an approach for identification of low-level chromosomal aneuploidy in neuropsychiatric diseases. Brain Dev 23(Suppl 1):S186–S190
167. Rolig RL, McKinnon PJ (2000) Linking DNA damage and neurodegeneration. Trends Neurosci 23(9):417–424
168. Lewis KE, Lubetsky MJ, Wenger SL, Steele MW (1995) Chromosomal abnormalities in a psychiatric population. Am J Med Genet 60(1):53–54
169. Konstantareas MM, Homatidis S (1999) Chromosomal abnormalities in a series of children with autistic disorder. J Autism Dev Disord 29(4):275–285
170. Oliveira G, Matoso E, Vicente A, Ribeiro P, Marques C, Ataíde A et al (2003) Partial tetrasomy of chromosome 3q and mosaicism in a child with autism. J Autism Dev Disord 33(2):177–185
171. Kim IH, Carlson BR, Heindel CC, Kim H, Soderling SH (2012) Disruption of wave-associated Rac GTPase-activating protein (Wrp) leads to abnormal adult neural progenitor migration associated with hydrocephalus. J Biol Chem 287(46):39263–39274
172. Waltereit R, Leimer U, von Bohlen Und Halbach O, Panke J, Hölter SM, Garrett L et al (2012) Srgap3^{-}/$^{-}$ mice present a neurodevelopmental disorder with schizophrenia-related intermediate phenotypes. FASEB J 26(11):4418–4428
173. Abouhamed M, Grobe K, San IVLC, Thelen S, Honnert U, Balda MS et al (2009) Myosin IXa regulates epithelial differentiation and its deficiency results in hydrocephalus. Mol Biol Cell 20(24):5074–5085
174. Carter CS, Vogel TW, Zhang Q, Seo S, Swiderski RE, Moninger TO et al (2012) Abnormal development of NG2+PDGFR-α+ neural progenitor cells leads to neonatal hydrocephalus in a ciliopathy mouse model. Nat Med 18(12):1797–1804
175. Yokota Y, Eom T-Y, Stanco A, Kim W-Y, Rao S, Snider WD et al (2010) Cdc42 and Gsk3 modulate the dynamics of radial glial growth, inter-radial glial interactions and polarity in the developing cerebral cortex. Development 137(23):4101–4110
176. Hwang M, Peddibhotla S, McHenry P, Chang P, Yochum Z, Park KU et al (2012) P190B RhoGAP regulates chromosome segregation in cancer cells. Cancers (Basel) 4(2):475–489
177. Ye X, Ishii I, Kingsbury MA, Chun J (2002) Lysophosphatidic acid as a novel cell survival/apoptotic factor. Biochim Biophys Acta 1585(2–3):108–113
178. Kingsbury MA, Rehen SK, Ye X, Chun J (2004) Genetics and cell biology of lysophosphatidic acid receptor-mediated signaling during cortical neurogenesis. J Cell Biochem 92(5):1004–1012
179. Potapova TA, Zhu J, Li R (2013) Aneuploidy and chromosomal instability: a vicious cycle driving cellular evolution and cancer genome chaos. Cancer Metastasis Rev 32(3–4):377–389
180. Estivill-Torrús G, Llebrez-Zayas P, Matas-Rico E, Santín L, Pedraza C, De Diego I et al (2008) Absence of LPA1 signaling results in defective cortical development. Cereb Cortex 18(4):938–950
181. Yamane M, Furuta D, Fukushima N (2010) Lysophosphatidic acid influences initial neuronal polarity establishment. Neurosci Lett 480(2):154–157

Chapter 5
Fate Through Fat: Neutral Lipids as Regulators of Neural Stem Cells

Laura K. Hamilton and Karl J.L. Fernandes

Abbreviations

3xTg-AD	Triple transgenic Alzheimer's disease
AA	Arachidonic acid
AD	Alzheimer's disease
Apo	Apolipoprotein
AraC	beta-cytosine arabinoside
ATP	Adenosine triphosphate
aNSC	Active neural stem cell
BBB	Blood–brain barrier
BLBP	Brain lipid binding protein
BrdU	Bromodeoxyuridine
CSF	Cerebral spinal fluid
DCX	Doublecortin
DESI	Desorption electrospray ionization
DG	Dentate gyrus
DHA	Docosahexaenoic acid
DNA	Deoxyribonucleic acid
DPA	Docosapentaenoic acid
EGF	Epidermal growth factors
EPA	Eicosapentaenoic acid

L.K. Hamilton • K.J.L. Fernandes (✉)
Research Center of the University of Montreal Hospital (CRCHUM),
Tour Viger, 900 Rue St. Denis, Room R09-474, Montreal, QC, Canada, H2X 0A9

Department of Neurosciences, Faculty of Medicine, Université de Montréal,
Montréal, QC, Canada, H2X 0A9
e-mail: laura.kerr.hamilton@umontreal.ca; karl.jl.fernandes@umontreal.ca

A. Pébay, R.C.B. Wong (eds.), *Lipidomics of Stem Cells*, Stem Cell Biology and Regenerative Medicine, DOI 10.1007/978-3-319-49343-5_5

FA	Fatty acid
FABP	Fatty acid binding protein
FAO	Fatty acid oxidation
FASN	Fatty acid synthase
GABA	Gamma-aminobutyric acid
GC	Granule cell
GFAP	Glial fibrillary acid protein
GLAST	Glutamate aspartate transporter
ICV	Intracerebroventricular
IMS	Imaging mass spectroscopy
LC	Liquid chromatography
LDLR	Low-density lipoprotein receptor
MALDI	Matrix-assisted laser desorption ionization
MRS	Magnetic resonance spectroscopy
MS	Mass spectroscopy
NeuN	Neuronal nuclei
NMR	Nuclear magnetic resonance
NSC	Neural stem cell
OA	Oleic acid
OB	Olfactory bulb
PA	Palmitic acid
PET	Positron emission tomography
qNSC	Quiescent neural stem cell
RNA	Ribosomal nucleic acid
ROS	Reactive oxygen species
SA	Stearic acid
SCD-1	Stearoyl CoA desaturase-1
SGZ	Subgranular zone
SIMS	Secondary ion mass spectrometry
SVZ	Subventricular zone
TAG	Triacylglycerol
TAP	Transit-amplifying progenitor
TLC	Thin layer chromatography
VLDLR	Very low density lipoprotein receptor
WT	Wild type

5.1 Neural Stem Cells and Adult Neurogenesis

The discovery of dividing cells in the brains of adult mammals during the mid-1960s changed our perception of brain plasticity and the potential for brain repair. It is now accepted that postnatal proliferation persists in two main brain regions, the

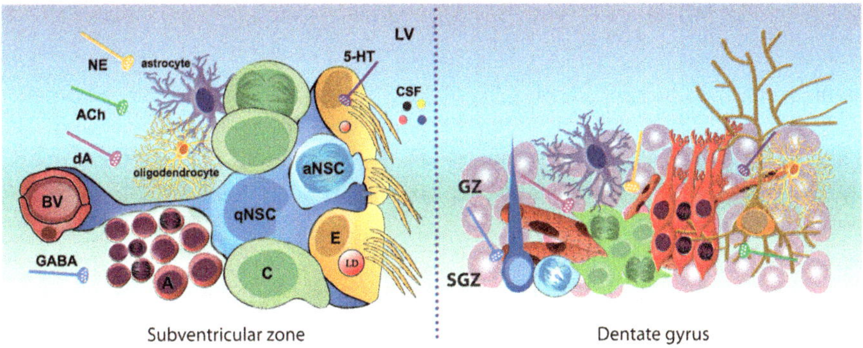

Fig. 5.1 Adult neurogenic niches. Combinatorial signals arising from neurotransmitter innervations (Acetylcholine (ACh), Serotonin (5-HT), Dopamine (dA), Norepinephrine (NE), and GABA), circulating blood factors, and within the cerebral spinal fluid (CSF) regulate the behaviour of neurogenic niches. Quiescent neural stem cells (qNSCs) are found beneath a border of ciliated ependymal cells (Type E) in the SVZ of the lateral ventricle and in the subgranular zone of the dentate gyrus. Upon mitogenic signals, neural stem cells become activated (aNSC), giving rise to transit-amplifying progenitors (TAP, Type C). After several rounds of division, progenitors differentiate into neuroblasts (Type A), oligodendrocytes, and astrocytes. Abbreviations: *BV* blood vessel, *LV* lateral ventricle, *GZ* granular zone, *SGZ* subgranular zone

subventricular zone (SVZ) of the lateral ventricles and the dentate gyrus (DG) of the hippocampus, in virtually all mammals studied, including rodents [1], primates [2, 3], and humans [4, 5]. In these locations, pools of infrequently dividing neural stem cells (NSCs) maintain populations of highly proliferative transit amplifying progenitors (TAPs) that ultimately give rise to fate-committed neuronal or glial progenitors. These TAPs migrate to their final destinations and differentiate into functional postmitotic neurons and glial cells important for various aspects of cerebral function and plasticity [6–10].

Specialized microenvironments, or niches, control NSC activity during adulthood. Within these niches, combinatorial extrinsic signals arising from NSCs, TAPs, neuroblasts, ependymal cells, microglia, extracellular matrix molecules, the cerebrospinal fluid (CSF), and the vasculature interact with cell-intrinsic mechanisms to control proliferation, self-renewal capacity, fate determination, migration, differentiation, and survival (Fig. 5.1) [11–14]. Although NSC pools appear to be conserved throughout life in the SVZ and DG, NSC output declines during adulthood and aging, as well as in the context of multiple neurodegenerative diseases. At the cellular level, it remains unclear whether these alterations involve changes in NSC pool size and/or activity, and at the molecular level, the mechanisms involved in these declines likewise remain poorly defined. Nevertheless, important progress has been made in understanding the control of where, how, and why adult neurogenesis normally occurs.

5.1.1 NSC Niches: Form and Function

In this section, we present an overview of the SVZ and DG niches. While additional sites of neurogenesis may exist within the adult brain, in regions such as the hypothalamus [15, 16] and primate striatum [17], the SVZ and DG remain the primary and best-studied NSC niches, and have provided the clearest insights into the regulation and biological roles of NSCs.

5.1.1.1 Subventricular Zone

NSCs in the SVZ of the forebrain lateral ventricles are ideally positioned to be regulated by environmental signals (Fig. 5.1). Pockets of astrocyte-like NSCs are embedded within the ventricular walls, in intimate association with the ventricle-lining ependymal cells. At their apical surface, these pockets of NSCs contact the ventricular lumen at the centre of ependymal pinwheel structures, allowing them to be regulated by ependymal cell paracrine signals, by factors within the CSF, and by contacts with serotonergic fibres that densely innervate the ventricular surface [18, 19]. At their basal surfaces, NSCs contact the underlying SVZ vasculature, allowing for potential regulation by blood-borne molecules and circulating cells and by vascular endothelial cells [20, 21]. NSCs are also regulated by neighbouring cell types within the SVZ, including by their TAP and neuroblast progeny [22], by the resident immune cells, microglia [23], and by innervation from multiple local and distant neuronal populations [24–28]. Taken together, these studies reveal that NSCs are subject to diverse types and sources of regulation, including local signals from the SVZ itself, brain activity-associated neurotransmitters, and long-range circulating signals.

NSCs in the SVZ have been implicated in the production of both neurons and various glial cell populations. More than 30,000 neuroblasts exit the rodent SVZ each day [29], to begin tangential migration towards the olfactory bulbs (OB) via the rostral migratory stream [30, 31]. About half of these neuroblasts survive, differentiate into periglomerular and granule olfactory interneurons [31, 32], and integrate into the OB circuitry [33, 34]. By genetically blocking the production of newly generated neurons, these have been shown to play a functional role in fine olfactory discrimination, olfactory sensitivity, and more generally, olfactory plasticity [35–37].

SVZ NSCs have also been shown to produce oligodendrocytes, ependymal cells, and astrocytes. A subpopulation of TAPs within the SVZ are olig2-expressing migratory progenitors that enter into the adjacent corpus callosum and striatum to differentiate into oligodendrocytes [38]. Ependymal cells of the lateral ventricle are generated from NSCs during development [39] and show little if any proliferation in the non-injured adult; aging-related ependymal loss is compensated for by SVZ NSCs that proliferate and incorporate within the ependymal layer, eventually differentiating into new ependymal cells [40]. Generation of differentiated (non-neurogenic) astrocytes within the SVZ is less clear under physiological conditions, as NSCs themselves possess an astrocytic phenotype. However, NSC-derived astrocytes are abundantly produced in the SVZ following brain injury, migrate to the lesion site, and participate in protective astrogliosis [41]. Thus, adult SVZ NSCs produce all principal neural cell types of the CNS, and participate in the homeostatic maintenance, regeneration, and repair of the brain.

5.1.1.2 Dentate Gyrus

NSCs in the DG are subject to many of the same local and longer-range regulatory mechanisms as in the SVZ, but with notable differences resulting from the distinct anatomy of the DG niche [42, 43]. Most obviously, the DG niche does not directly border the ventricular system, and thus DG NSCs are not in physical contact with ependymal cells or the CSF. Rather, DG NSCs are located in the subgranular zone (SGZ) of the DG, adjacent to the dense concentration of excitatory granule cells and Gamma-aminobutyric acid (GABA)-ergic interneurons found within the dentate granule layer. Indeed, tonic activation of GABA(A) receptors on NSCs by parvalbumin-expressing GABAergic interneurons has been shown to be a major NSC quiescence signal [44]. In addition to the local circuitry, inputs to the DG from the entorhinal cortex link the overall activation state of the DG to levels of cortical activity [11–14, 45, 46]. One consequence of these DG-specific features is that hippocampal neurogenesis is particularly sensitive to life experience, including factors such as physical activity and stress [47–50].

The physiological roles of adult hippocampal neurogenesis have now been investigated using many complementary and increasingly precise approaches to suppress neurogenesis, including irradiation, pharmacological, and genetic tools. For example, a tag and ablate strategy has been used to selectively eliminate a population of predominantly mature, adult-generated neurons either before or after learning and without affecting ongoing neurogenesis; this revealed that removal of these neurons after learning, but not before, resulted in degradation of existing contextual fear and water maze memories, without affecting non-hippocampal-dependent memory [51]. Suppression of adult neurogenesis impairs population coding of similar contexts [52], and it appears that young DG neurons mediate pattern separation while old DG neurons facilitate pattern completion [53–55]. At present, there are multiple functional roles identified for adult hippocampal neurogenesis including learning [2], spatial memory and pattern separation [54, 56–61], and regulation of stress and emotion [62, 63].

5.1.2 Heterogeneity Within the NSC Continuum

There is now evidence for considerable heterogeneity within the NSC compartment itself. Recent studies indicate that "NSCs" might actually represent a collection or hierarchy of distinguishable cells that have partially overlapping functional characteristics and, importantly, distinct physiological requirements.

NSCs were originally identified in the SVZ, based on the ability of a rare subpopulation of SVZ cells to grow in vitro into free-floating colonies of undifferentiated cells called "neurospheres" in the presence of epidermal growth factor (EGF) [64]. While some groups define EGF-induced formation of self-renewing multipotent neurospheres as a characteristic unique to NSCs, others attribute this characteristic to the early TAP stage of the neurogenic lineage [65]. Moreover, while NSCs are generally considered niche-anchored cells, a subset of cells migrating following brain lesions retain neurosphere-forming characteristics [66], possibly suggesting retention of NSC properties in some migratory progenitors. Recent transcriptomic studies of the NSC population have identified significant genetic differences and similarities during the continuum from quiescent NSC (qNSC) to activated NSC (aNSC) to TAP (discussed further in relation to lipid metabolism below). Layered onto this blurred line between stages of the NSC lineage are dorsoventral, mediolateral, and anteroposterior patterning signals, which program NSCs with a regional specification that delimits the neural cell types they can ultimately differentiate into [67–69]. Interestingly, adult NSCs are derived embryonically from an Oct4-expressing "primitive" NSC, and a recent study has also found evidence for persistence of small numbers of such primitive NSCs in the adult brain [70]. Overall, the heterogeneity within the NSC compartment is likely responsible for the surprising functional differences observed between cells expressing different combinations of classical NSC markers (i.e. nestin, GLAST, GFAP, and BLBP) [71–74].

Thus, while individual cells within the NSC lineage may be distinguishable on the basis of anatomical or genetic criteria under control conditions, their partially overlapping functional characteristics complicate the assignment of strict definitions in different physiological and pathological situations. Notably, the transcriptomic differences between cells within the NSC lineage have helped revealed that regulatory processes act preferentially at particular stages of the NSC lineage; as discussed later, fatty acid (FA) metabolism is an example of one of these processes.

5.1.3 Human Neurogenesis

Creative strategies have been used to study neurogenesis in humans over the past two decades, and these have provided evidence that neurogenesis likewise occurs in the adult human brain. Initial studies analysed hippocampal tissues from deceased cancer patients that had received injections of the thymidine analog, bromodeoxyuridine (BrdU), prior to tumour removal. Co-labelling of post-mortem tissues from these patients using neuronal markers such as doublecortin (DCX) and neuronal nuclei (NeuN) revealed the presence of BrdU-positive newly generated neurons within the DG [4, 75]. Kukekov and colleagues subsequently isolated multipotent neurospheres from surgical biopsy specimens containing SVZ or hippocampus of adult human brains, suggesting the presence of NSCs in these regions [5].

Importantly, while studies based on BrdU or neurosphere formation revealed the presence of adult neurogenesis and NSCs in humans, they could not provide a quantitative measure of their occurrence: neurospheres reflect NSC potential rather than activity, while interpretation of human BrdU incorporation studies is limited by factors such as sub-saturating dosages and aged and/or diseased patients.

Immunohistochemical studies of the primate and human SVZ have revealed important similarities and differences compared to rodents. The human SVZ possesses a hypocellular gap beneath the ventricle-lining ependymal layer, but maintains a sub-ependymal neurogenic niche containing neural precursors and neuroblasts [76–78]. Three distinct subtypes of astrocytes have been identified along the lateral wall of the lateral ventricles. These astrocytes are located at different locations along the anterior-posterior length of the ventricle, and vary in size, ultrastructure, and relationship to the ependymal zone [77]. A subpopulation of these astrocytes proliferate in vivo and behave as multipotent precursor cells in vitro [78], implying that SVZ astrocytes of the adult human brain indeed are NSCs. Consistent with this, cells expressing neuronal markers TuJ1 and DCX have been observed in the SVZ and the majority of these cells have an elongated morphology suggesting they are in migration [17, 77–80]. However, controversial conclusions have been reached concerning the fate of SVZ neurogenesis, with some authors reporting abundant proliferation and neuroblast migration to the human OBs [81–83], and others finding little or no evidence of a rostral migratory pathway [84]. The current consensus from these studies is that SVZ-derived olfactory neurogenesis is abundant during developmental periods and rapidly declines during the first few years postnatally [84].

Recently, the Frisen group has developed an innovative cellular birth-dating technique based on measurement of deoxyribonucleic acid (DNA) concentrations of 14C, whose atmospheric levels spiked during the 1960s as a result of nuclear bomb testing. Using this technique, cells in various human tissues have been carbon-dated, including neurons and glial cells in multiple brain regions. In the human OB, neuronal age corresponded precisely to an individual's age, providing no evidence for ongoing incorporation of new neurons within the adult OB [85]. In contrast, substantial levels of adult hippocampal neurogenesis were detected, estimated at approximately 700 new neurons per day per hippocampus [86]. Modelling of these data suggests that a remarkable 100% of dentate gyrus granule neurons are replaced postnatally (compared to 15% in mice), with roughly one-third of these neurons being replaced regularly. Interestingly, while adult human olfactory neurogenesis is not detected, there are similar levels of proliferation in the human SVZ and the human DG, raising the question of the fate of new SVZ cells. Carbon dating of neurons in the SVZ-adjacent striatum revealed that approximately 25% of striatal interneurons are newly generated postnatally [17, 87], suggesting that the striatum may represent a major target of SVZ neurogenesis in primates. Furthermore, ongoing production of oligodendrocytes was likewise detected.

Together, this work suggests that NSCs and adult neurogenesis are retained in the adult human brain. Approaches to test the physiological functions of adult human neurogenesis remain to be developed. However, given the important roles of the primate hippocampus and striatum in higher cognitive processes, including learning, memory, and emotional regulation, human neurogenesis is likely to play significant roles in higher brain functions. Notably, NSCs can still be isolated from the SVZ of elderly subjects, including those with severe neurodegeneration, such as Alzheimer's disease (AD) [76]. This gives hope that when the precise regulation of NSCs is understood, endogenous regeneration will be feasible.

5.2 Neutral Lipid Metabolism in the Adult Brain

Lipids are one of the fundamental classes of biomolecules, along with nucleic acids [contributing to DNA/ribonucleic acid (RNA)], amino acids (contributing to proteins), and sugars (contributing to carbohydrates). Lipids are molecules constituted of chains of hydrocarbons (CH2) whose wide-ranging cellular and physiological functions include (but are not limited to) membrane structure, energy metabolism and storage, second messengers for growth/proliferation/survival signals, transcriptional regulation, intercellular signalling, inflammation, electrical insulation, and protection [88, 89]. To execute such tasks, the body absorbs, synthesizes, and modifies all major classes of lipids. Here, we have focused on "neutral" or simple lipids, which are non-polar species that will break down into no more than 1–2 types of lipid molecules. Neutral lipids include free FAs, the FA storage form as triglycerides (glycerol + 3 FA), as well as sterols (predominantly cholesterol) (Fig. 5.2). Below, we highlight some of the key players explored in this review (Fig. 5.3).

5.2.1 *Fatty Acids*

FAs are composed of a carboxylic acid with a hydrocarbon chain usually of even number of carbon atoms, ranging from 4 to 28. FAs can be classified according to the length of the hydrocarbon chain: short—(2–5C), medium—(6–12C), long—(13–22C), and very long chain (>23C). They can also be classified according to the presence of C-C double bonds: saturated FA (SFA) have no C-C double bonds; unsaturated FA have either one C-C double bond (monounsaturated, MUFA) or multiple C-C double bonds (polyunsaturated, PUFA). Cytoplasmic FAs are bound to amphipathic transport proteins (fatty acid binding proteins, FABPs) that target them to different cellular compartments. In addition to existing as monomers and triglycerides, FAs can also be incorporated into a wide range of more complex membrane-associated lipids (Fig. 5.2), including glycerophospholipids (glycerol + 2 FA + 1 phosphate group), ceramides (sphingosine + 1 FA), phosphosphingolipids (sphingosine + 1 FA + 1 phosphocholine group), and glycosphingolipids (sphingosine + 1 FA + 1 oligosaccharide).

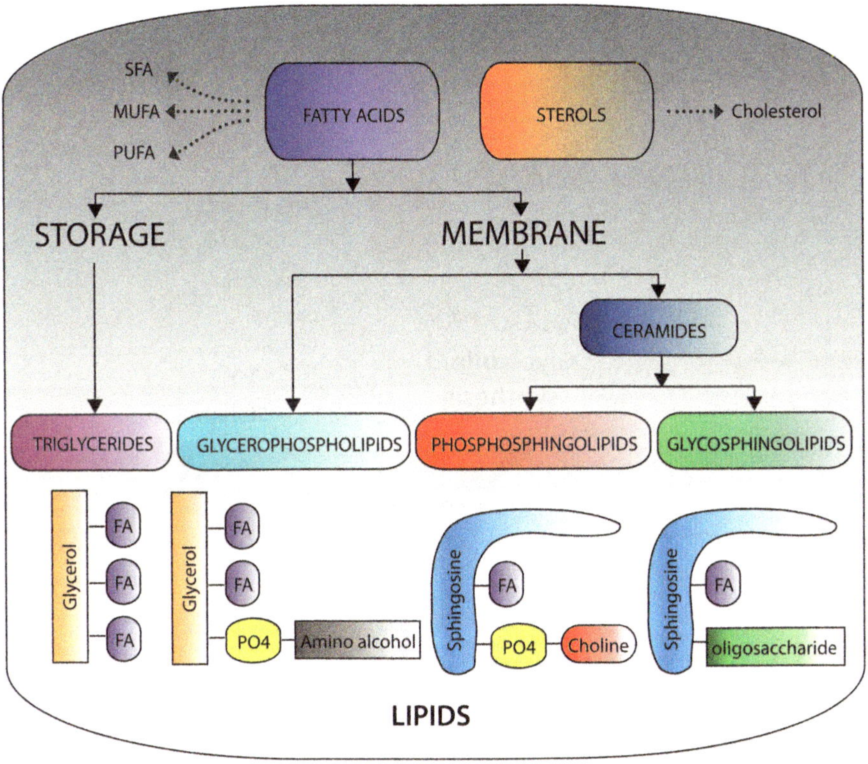

Fig. 5.2 Lipid classification. Fatty acids (FAs) and sterols provide the building blocks for all major lipid classes. Intracellular saturated (SFA), monounsaturated (MUFA), and polyunsaturated (PUFA) can be stored as triglycerides (a glycerol with 3 FA chains) or incorporated into structural membrane lipids. Membrane lipids can be divided into two subgroups, glycerophospholipids (a glycerol with 2 FAs and a phosphate) and ceramides (a sphingosine with one FA). Ceramides can be further modified into phosphosphingolipids (by addition of phosphocholine) or into glycosphingolipids (by addition of an oligosaccharide)

5.2.2 *Triacylglycerides*

Excess intracellular FAs are converted into triacylglycerols (TAGs) (glycerol + 3 FA) within the endoplasmic reticulum, and these TAGs are subsequently packaged within a hydrophilic phospholipid shell called a lipid droplet [90–92]. The mobilization of FAs from TAGs is carried out by lipid droplet-associated lipases, which sequentially release the FAs [93] (Fig. 5.3). FA metabolites formed from the breakdown of TAGs are ultimately used for a variety of purposes, such as signal transduction pathways, membrane biosynthesis, adenosine triphosphate (ATP) production through β-oxidation, or generation of inflammatory eicosanoids.

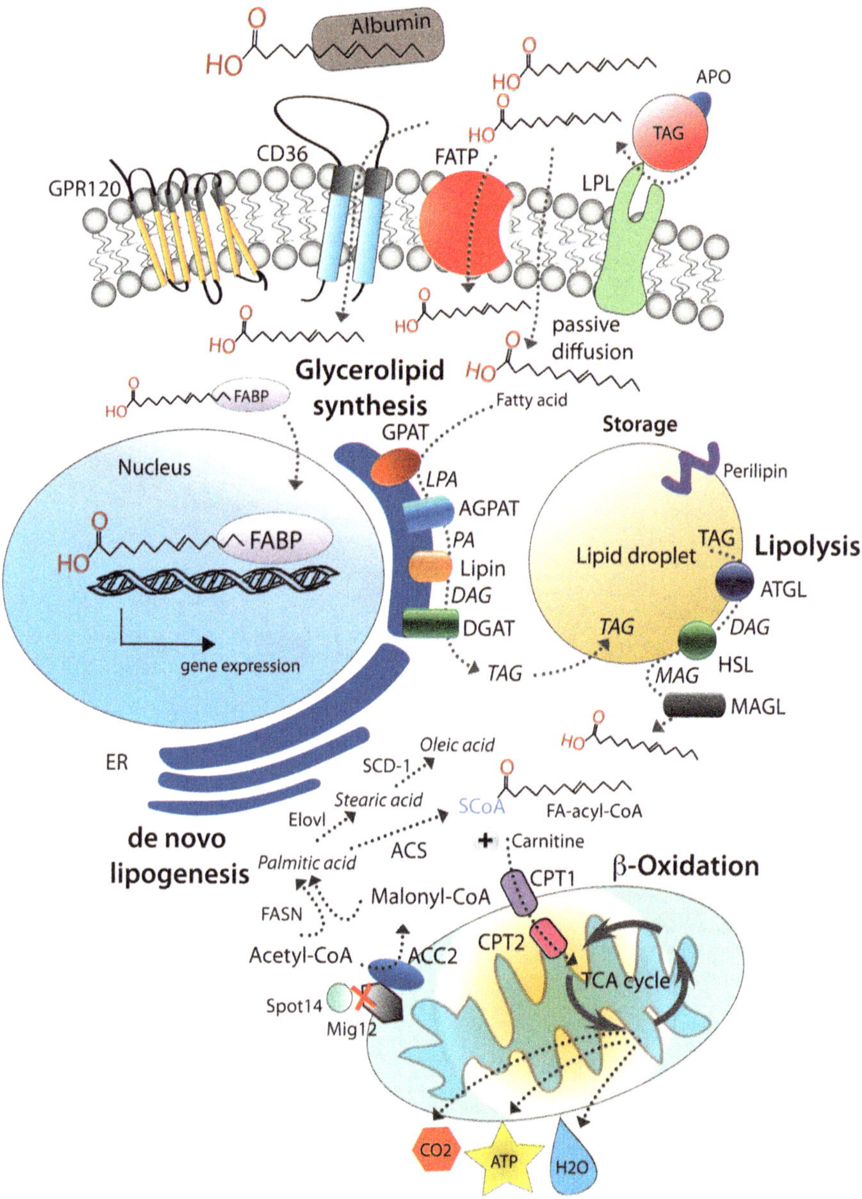

Fig. 5.3 Fatty acid metabolism. Circulating fatty acids (FAs) are bound to albumin or in triacylglyceride (TAG) form within lipoproteins. Lipoproteins dock with various families of plasma membrane receptors, allowing lipoprotein lipase (LPL) to release the FA cargo. These free FAs can subsequently enter the cell by passive diffusion or via transporters such as CD36 or FATP, or alternatively can activate lipid sensing G-protein-coupled receptors such as the GPR120 family. Cytoplasmic FAs are bound by fatty acid binding proteins (FABPs) during intracellular transport. FAs at the endoplasmic reticulum can enter the *glycerolipid synthesis* pathway for *storage* in lipid droplets. This occurs by sequential FA addition by (1) GPAT (glycerol-3-phosphate acid acyltransferase)

5.2.3 Lipid Droplets

Our current understanding of lipid droplet biology comes mainly from studies in peripheral tissues. Under physiological conditions, excess TAGs and cholesterol are stored in lipid droplets in order to buffer their toxicity and to provide a reserve for periods of diminished nutrient availability. In times of lipid deficiency, lipases break down these stores to provide membrane building blocks, such as FAs or sterols, for cellular division and integrity [91]. This protective function is probably the reason for the abundant accumulation of lipid droplets in many disease states characterized by aberrant lipid supply and metabolism, such as obesity, atherosclerosis, and fatty liver disease [94–96], as well as cancer and neurodegeneration [95, 97, 98]. Interestingly, recent studies in Drosophila have shown that reactive oxygen species (ROS) and mitochondrial defects, which are critical factors in many neurodegenerative diseases, trigger accumulation of lipid droplets in glial cells; depending on the context, these glial lipid droplets can be either neuroprotective or a source of damaging peroxidated lipids for adjacent neurons [99, 100].

5.2.4 Apolipoproteins

Cholesterol and triglycerides, as well as cholesterol esters and phospholipids, are shuttled through the body by carrier proteins called apolipoprotein (Apo), which regulates their metabolism and distribution [101, 102]. Lipoproteins dock with plasma membrane receptors that have tissue-specific expression patterns. In the central nervous system, the main lipoprotein receptors are SR-B1 (sterols and phosphatidyl-choline), low-density lipoprotein receptor (LDLR) (cholesterol), very low density lipoprotein receptor (VLDLR) and ApoER2 (triglycerides), and LRP1 (cholesterol and triglycerides), although the lipid cargo specificity of these receptors is only partially known.

Fig. 5.3 (continued) yielding LPA (lysophosphatidic acid), (2) AGPAT (acylglycerol-3-phosphate acyltransferase) and Lipin, yielding DAG (diacylglycerol) via a PA (phosphatidic acid) intermediate, and (3) DGAT (diacylglycerol acyltransferase), yielding TAG. When FAs are needed, lipid droplet TAGs undergo *lipolysis* by (1) Fig. 5.3 (continued) ATGL, adipose triglyceride lipase yielding DAG, diacylglycerol (2) HSL, hormone-sensitive lipase yielding MAG, monoacylglycerol, and (3) MAGL, monoacylglycerol lipase, yielding free FAs at each step. FAs are attached to Acyl-CoA by Acyl-CoA synthase (ACS). FA-acyl-CoA enters the mitochondrion for *β-oxidation* via a carnitine shuttle is used. Acyl-CoA is transferred to the hydroxyl group of carnitine by carnitine palmitoyltransferase I (CPT1), located on the cytosolic face of the outer mitochondrial membrane. Acyl-carnitine is shuttled inside by carnitine-acylcarnitine translocase as carnitine is shuttled out. Acyl-carnitine is converted back to acyl-CoA by carnitine palmitoyltransferase II (CPT2), located on the interior face of the inner mitochondrial membrane. Acyl-CoA is shuttled into the TCA cycle for β-oxidation, producing adenosine triphosphate (ATP), carbon dioxide (CO_2), and water (H_2O). De novo *lipogenesis* produces palmitic acid from acetyl-CoA and Malonyl-CoA by the action of fatty acid synthase (FASN). Palmitic acid can be elongated by elongases (elovl) to produce stearic acid, which can be desaturated to oleic acid by stearoyl-CoA desaturase (SCD-1)

The majority of Apos do not cross the blood–brain barrier (BBB), a selectively permeable interface that comprises astrocytes, endothelial cells, smooth muscle pericytes, and fibroblasts [103]. Peripheral FAs are thought to be able to cross the BBB. However, the mechanism of how they do so is still relatively unknown. It has been proposed that FAs can cross the lipid bilayers by a flip-flop mechanism or by passive diffusion [104]. Longer chain FAs are less soluble, but are able to cross cellular membranes via FA transporters such as CD36 and FATP (Fig. 5.3) [105–107]. On the other hand, about a quarter of the body's cholesterol (the most common sterol in the body) is found in the brain, but cholesterol does not cross the BBB. Brain cholesterol is virtually entirely locally synthesized by astrocytes and oligodendrocytes. Within the brain, lipids are carried principally by ApoE (large lipoproteins), ApoA-1 and ApoD (smaller lipoproteins), and ApoJ (less specific). ApoE, ApoD, and ApoJ are mainly synthesized by astrocytes and other glial cells, while ApoA-1 appears to be principally derived from the periphery.

5.3 Neutral Lipids Are Physiological Regulators of Adult NSCs

5.3.1 Neutral Lipid Carriers Are Required for NSC Maintenance and Neurogenesis

Several lines of evidence implicate neutral lipids as important physiological modulators of NSC activity. Initial evidence for a role of neutral lipids in NSC regulation comes from expression patterns and knockout analyses of FABPs and Apos.

5.3.1.1 FABPs

FABPs-3, -5, and -7 are expressed within the adult brain, and FABP5 and FABP7 in particular are expressed by NSCs and their downstream progenitors [108]. A marked decrease in hippocampal NSCs and proliferating neural progenitors was observed when FABP5, FABP7, or both were knocked out.

5.3.1.2 ApoE

The identification of polymorphism at the APOE locus as the primary genetic risk factor for AD has led to many studies on the effects of ApoE2/3/4 on the brain; however, few studies have focussed on neurogenesis. ApoE elimination resulted in an increase in neural precursor proliferation in the DG niche followed by a premature depletion of GFAP and nestin-expressing NSCs. ApoE directly mediated this effect, as retroviral re-expression of ApoE rescued the phenotype [109]. Li et al.

studied ApoE knockout mice as well as mice with knock-in alleles for human ApoE3 or ApoE4 (ApoE4 being the major AD risk factor), and showed that hippocampal neurogenesis is reduced in both ApoE knockout mice and ApoE4 knock-in mice [110]. Interestingly, when ApoE knock-in mice were exposed to environmental enrichment, which is a well-known stimulator of DG neurogenesis [47, 111], WT and ApoE3 mice showed a significant increase in proliferation and neurogenesis while ApoE4 mice did not increase neurogenesis and instead exhibited increased apoptosis [112]. Together, these studies suggest that lipids transported by ApoE regulate NSC activity, but the precise lipid species involved remain unidentified.

5.3.2 *Fatty Acid Metabolism Is Required for Proliferation of NSCs*

More recent studies have linked the process of FA metabolism to NSC behaviour. Knobloch and colleagues performed a series of innovative experiments to reveal that de novo FA synthesis is required for maintaining NSC proliferation and neurogenesis [113]. They found that mRNA of both fatty acid synthase (FASN, a key enzyme in de novo lipogenesis, Fig. 5.3) and the Spot14 enzyme (a negative regulator of malonyl-CoA synthesis, which is used by FASN during de novo lipogenesis) is expressed within NSCs in the SVZ and DG niches. FASN inhibition using orlistat or cerulenin led to a dose-dependent reduction of DG proliferation. Conversely, Spot14^{+} nestin-GFP cells proliferated slower than the Spot14^{-} subpopulation. These studies revealed that adult NSCs/progenitors have a cell autonomous requirement for FASN-dependent FA synthesis in order to proliferate, and that Spot14, a gene highly enriched in more quiescent cells, limits proliferation by inhibiting FA synthesis. Along these lines, Chorna et al. showed that voluntary exercise up-regulates hippocampal FASN expression, as well as levels of palmitic acid (PA) and stearic acid (SA), and that injection of the FASN inhibitor C75 disrupted exercise-induced increases in hippocampal proliferation and cognitive enhancement [114].

Soon after these pioneering studies, multiple transcriptomic studies were published that identified lipid metabolism genes (and FA metabolism in particular) as among the most differentially expressed gene categories between purified populations of quiescent and activated NSCs. Codega and colleagues used a multistep FACS strategy to separate qNSCs and aNSCs from the SVZ; interestingly, microarray comparison of these two populations showed that the lipid metabolism differences included 17-fold higher expression of stearoyl-CoA-desaturase (SCD-1, rate-limiting enzyme in MUFA synthesis, Fig. 5.3) and fourfold higher expression of ApoE in qNSCs compared to aNSCs [72]. In a second study, Lorens-Bobadilla and colleagues studied the SVZ NSC population at the single-cell level using RNA-Sequencing, and found that genes expressed in quiescent and active NSC subpopulations were enriched for FA metabolism and lipid biosynthesis (e.g. *Fasn*) [115]. Similarly, Shin and colleagues performed single-cell RNA-Seq on hippocampal

NSCs and identified multiple lipid metabolism-related gene categories that were enriched in qNSCs and down-regulated in aNSCs, including FA degradation and sphingolipid metabolism. This included the previously mentioned Spot14, as well as Acyl-CoA synthetases involved in the first step of FA β-oxidation (Acsl3, Acsl6, and Acsbg1), further supporting a novel role for active β-oxidation of FAs in the qNSC subpopulation [116].

5.3.3 Distinct Effects of Fatty Acid Classes on NSCs and Neurogenesis

The previous studies indicate that appropriate regulation of FA metabolism is likely to be necessary for normal transition of NSCs from a quiescent to an active state. Lipidomic studies of these early stages of the NSC lineage have yet to be performed, so the individual lipid species involved in this process have yet to be determined. However, in vitro and (rare) in vivo studies show that different classes of FAs can have distinct effects on survival, proliferation, and differentiation of neural precursors.

5.3.3.1 Polyunsaturated Fatty Acids (PUFAs)

Omega-3 PUFAs such as docosahexaenoic acid (DHA, 22:6), eicosapentaenoic acid (EPA, 20:5), and docosapentaenoic acid (DPA, 22:5) have been studied most frequently because of their association with enhanced learning and memory. Neurospheres derived from the embryonic brain and differentiated in the presence of DHA have fewer apoptotic cells, fewer proliferating cells, and produce significantly more numerous and complex neurons [117, 118]. Building on their previous work, Katakura and colleagues went on to show that DHA, EPA, and DPA, but not the omega-6 PUFA arachidonic acid (AA, 20:4) or MUFA oleic acid (OA, 18:1), arrested the cell cycle of neural precursors and promoted neuronal differentiation [119, 120]. In a similar vein, Sakayori and colleagues found that DHA, but not AA, increased neurosphere numbers and neuronal differentiation, while AA selectively increased the number of astrocytes [121]. In vivo adult neurogenesis studies have been largely limited to in vivo high fat diet paradigms. It was shown that postnatal feeding with diets enriched with AA, but not DHA, increased overall cell proliferation in the DG [122], and that dietary administration of DHA in adult rats significantly increased the number of BrdU+/NeuN+ newborn neurons in the DG [118]. However, such studies are inevitably complicated to interpret due to indirect and systemic effects. Together, these studies demonstrate that omega-3 FAs support survival and neuronal differentiation, at least from embryonic neural precursors.

5.3.3.2 Saturated Fatty Acids (SFAs)

Saturated FAs, such as palmitic acid (PA, 16:0), have been linked to AD [123]. PA is the most abundant SFA in the body and is central to numerous cellular processes: for example, it is the precursor to longer chain SFA, MUFA, and PUFA, it is used for de novo synthesis of apoptosis-associated ceramides, and it is used for protein modification via palmitoylation. Treatment of neural precursors with PA dose-dependently increases c-jun N-terminal kinase phosphorylation, alters bax and bcl-2 levels, and increases caspase-mediated apoptosis [124, 125]. At non-toxic levels, PA increased Stat3 signalling and astrogliogenesis [124].

5.3.3.3 Monounsaturated Fatty Acids (MUFAs)

MUFAs are produced by the enzyme SCD-1 via the desaturation of PA into palmitoleic acid (16:1) and stearic acid (SA, 18:0) into oleic acid (OA, 18:1), the most abundant MUFA in the body. Our group recently identified OA-enriched lipid droplets within the SVZ and we demonstrated that OA can regulate NSC proliferation in the adult brain [126]. We modified the neurosphere assay to determine whether OA regulates NSCs, NSCs and progenitors, or only progenitors. We treated adult neural precursors with 50 μM or 100 μM of OA either on the day of plating (Day 0; measure of NSC activation) or after 4 days of neurosphere growth (Day 4; measure of progenitor cell expansion). Interestingly, we found that when OA was administered on D0, 50 μM increased neurosphere number while 100 μM inhibited it by over 50% without changing neurosphere size. Importantly, this was not accompanied by an increase in TUNEL+ apoptotic cells. In contrast, when OA was administered on day 4, it had no effect at either concentration. These findings suggested that OA was impacting NSC activation specifically. We then performed a self-renewal assay by manually picking neurospheres that had been treated with vehicle or 100 μM of OA and re-plated them under standard neurosphere conditions. When secondary neurospheres were counted, we found that the neurospheres that had been previously treated with OA had significantly fewer NSCs per original neurosphere, demonstrating an inhibition of NSC self-renewing divisions. To study OA in vivo, we intracerebroventricularly (ICV) infused OA in WT mice using mini osmotic pumps for 7 days. Quantification of total proliferation, proliferation NSCs, neuroblasts, and proliferation neuroblasts as well as number of pinwheels and neurospheres, all showed no significant difference. This confirmed our in vitro study, showing that OA does not have a widespread effect on neural precursor proliferation and neurogenesis. However, given that NSCs proliferate rarely and our in vitro data showed that OA selectively altered NSC proliferation, a 7-day paradigm was not appropriate to detect changes in NSC proliferation specifically. Therefore, to selectively study NSC activation, we used a classical SVZ repopulation assay using beta-cytosine arabinoside (AraC) [8]. When OA was co-infused with AraC, it inhibited the ability of GFAP+ NSCs to divide and regenerate the SVZ. To study the signalling mechanism by which OA inhibits NSC proliferation, we performed OA experiments using both in vitro treatment and ICV injections. In both cases, OA increased AKT phosphorylation. When LY294002, an AKT-inhibitor, was combined with OA in vitro, it normalized phosphorylation of AKT and converted the negative effects on neurosphere formation to positive (i.e. an increase in neurosphere number compared to vehicle). We then assessed this in vivo using a combination of fate-mapping of GFAP+ NSCs by in vivo electroporation and OA administration by ICV osmotic pumps. As anticipated, when GFAP-cre was electroporated into flox-YFP mice followed by ICV OA pump implantation, OA selectively inhibited the number of YFP+ NSCs. Importantly, when cells were co-electroporated with a kinase dead AKT plasmid, this prevented the decrease in YFP+ NSCs. These data implicate AKT signalling as an important effector of OA's effects on NSCs. Taken together, our study showed that a single FA, OA, can regulate NSC proliferation in the adult brain.

Given that most studies on FAs and NSCs have been performed in vitro, it is important to consider the limitations of their interpretation. Cells are dramatically affected by the culturing process, culture media are minimal and do not replicate the in vivo milieu, and key molecules and cell types found within the stem cell niche may not even be represented in vitro. Moreover, given the active processing of individual FAs within multiple biological pathways, it is uncertain whether particular in vitro and in vivo effects are due to the applied lipid itself or to one of its many metabolites/derivatives. Nevertheless, the preceding studies show that the FA profile of the stem cell microenvironment impacts neural precursor behaviour in significant and complex ways.

5.4 Aberrant Neutral Lipid Metabolism in Brain Disease

Abnormalities in neutral lipid metabolism are beginning to be linked to brain disorders associated with cognitive impairments or neurodegeneration. Below, we discuss emerging evidence that lipid-mediated alterations can cause NSC dysregulation during diseases of adulthood (AD) or development (autism). Interestingly, these examples highlight that disturbances in NSC behaviour can be caused by lipid metabolism abnormalities within NSCs themselves (cell autonomously, in the case of autism) or within their surrounding niche cells (non-cell autonomously, in the case of AD).

5.4.1 Alzheimer's Disease (AD)

AD is the principal cause of dementia, an aging-related degenerative neurological disease associated with premature deterioration of multiple cognitive modalities including learning, memory, and personality. Overall, brain degeneration and synaptic loss proceed in a region-specific and temporally determined manner, beginning in the entorhinal cortex and advancing to the hippocampus and posterior temporal and parietal cortices [127–129]. NSC activity is also deregulated in a wide range of mouse models of AD [126, 130–132] and in the few superficial assessments of human AD neurogenesis [133, 134]. Symptom onset is largely determined by genetic risk factors that separate AD into two forms: familial/early-onset familial (onset <60 years old) and sporadic/late-onset (onset >60 years old). Intriguingly, in both familial and sporadic forms of AD patients display the same cognitive symptoms and pathological hallmarks.

There is strong evidence to support a role for abnormal lipid metabolism in the pathogenesis of AD. German pathologist Alois Alzheimer was the first to describe the neuropathology of AD, uncovering five neuropathologies including focal deposits (*amyloid plaques*), intraneuronal fibrils (*neurofibrillary tangles*), blood vessel abnormalities (*cerebrovascular amyloidosis*), glial reactivity (*gliosis*), and lipoid deposits within non-neuronal cells (*lipid accumulations*) [135]. In recent years, genetic and genome-wide association studies have solidified the importance of the five initial AD pathologies, showing clusters of genetic mutations related to amyloid processes and storage (PS1, PS2, APP, APOE, SORL1, CLU, CRI, PICALM, BIN1, ABCA7), immunity/inflammation (CLU, CRI, EPHA1, ABCA7, MS4A4A/MS4A6E, CD33, CD2AP) and lipid transport and metabolism (APOE, CLU, ABCA7, SORL1) (reviewed in [136, 137]). Notably, polymorphism at the APOE locus is by far the strongest genetic risk factor for sporadic AD, and there is convincing evidence linking aberrant lipid metabolism to neurodegeneration in AD [138–142]. Furthermore, peripheral metabolic conditions such as insulin resistance, obesity, and dyslipidemia have been identified as major AD comorbidities and risk factors (reviewed in [143]).

In spite of these compelling data, the nature and cellular targets of the pathological lipid species in AD have remained obscure. This has been largely due to the technical complexity involved in localizing, identifying, and determining the biological functions of individual lipid species in the adult brain. Recently, we uncovered evidence for a novel FA-mediated mechanism suppressing endogenous NSC activity in AD [126]. Using the neutral lipid dye Oil Red O, we identified a highly specific accumulation of lipid droplets surrounding the brain's ventricular system in AD. These lipid droplets were present within ependymal cells, the main support cell of the adult forebrain NSC niche, and were in both post-mortem human AD brains and triple-transgenic Alzheimer's disease (3xTg-AD) mice. Imaging mass spectrometry (I-MS) revealed these neutral lipids to be 12 TAG species whose side chains were enriched with OA. I-MS was further used to trace the incorporation of the infused OA with a sensitive in vivo metabolic labelling procedure that uses OA comprised entirely of heavy 13C (13C OA). Uptake of 13C OA into each of the 12 AD-associated triglycerides showed that 11 of the 12 AD-associated triglycerides were replicated in WT mice simply by infusion of OA. For example, 13C OA shifted the TAG 50:1 by exactly 18.060 atomic units (incorporation of one OA side chain). These metabolic labelling experiments also demonstrated that some AD-associated triglycerides contained 13C OA that had been elongated (56:4 and

56:5), reduced (52:2 and 52:3), saturated (52:2 and 54:2), and/or desaturated (54:4, 56:4, and 56:5), revealing that OA at the brain-CSF interface can be used as a precursor to locally generate longer chain PUFAs. To attempt to identify the sources of these AD-associated triglycerides, we perform untargeted LC-MS on plasma samples of the same WT and 3xTg-AD mice. Interestingly, we found no differences in any of the 12 triglycerides species or their associated free FA chains, suggesting that the brain itself might be the source. Indeed, microarray analysis of microdissected SVZs from WT and 3xTg-AD mice revealed enrichments for lipid metabolism and FA biosynthesis genes. Of particular interest was an increased expression of SCD-1, the rate-limiting enzyme in OA synthesis. As mentioned earlier, exogenous OA was sufficient to selectively suppress proliferation of wild-type NSCs both in vitro and in vivo. We therefore used a pharmacological approach to SCD-1 activity in 3xTg-AD mice in vivo, and succeeded in reactivating NSCs and restoring overall levels of neural precursor proliferation in both adult neurogenic niches. These studies support a pathogenic mechanism whereby AD-induced perturbation of niche FA metabolism suppresses the homeostatic and regenerative functions of NSCs.

5.4.2 *Autism*

Dietary or genetic disturbances in lipid metabolism are closely linked with the pathogenesis of autism [144], a spectrum of neurodevelopmental disorders that manifest as abnormalities in social interaction, language, communication, and behaviour. Recent studies have found that genetic mutations associated with autism can cause prominent, cell-autonomous alterations in NSC lipid metabolism that are likely to perturb normal brain development [145, 146]. For example, Xie and colleagues showed that reductions in NSC FA metabolism in embryonic mice resulted in diminished NSC pool size within the developing cortex. This was demonstrated by shRNA-mediated knockdown of TMLHE (an autism risk gene involved in synthesis of carnitine, an acyl carrier necessary for FAs to be imported into the mitochondria). Similar results were obtained by knocking down expression of the mitochondrial enzyme CPT1 (also required for FA import into the mitochondria) or by overexpressing a non-phosphorylatable form of the lipid droplet protein Perilipin-1 (which normally must be phosphorylated in order to allow lipolysis of lipid droplet TAGs). These authors went on to show that TMLHE knockdown causes NSC depletion by promoting symmetric differentiating divisions that generate two progenitor daughter cells, and that this could be countered by supplementation with exogenous carnitine [146].

Links between lipid metabolism and NSC regulation in brain diseases are thus beginning to be uncovered, but we have likely barely scratched the surface. More work is clearly needed to define the roles of individual lipid classes and species that mediate alterations in NSCs under pathological conditions. To accomplish this goal, it will be essential to adopt novel research strategies, harnessing the power of recent technological advances in lipidomics, and adapting them to the brain. Below, we outline tools and strategies for studying lipids in the brain and in NSC biology.

5.5 Techniques to Measure Neutral Lipids in the Brain

The measurement, identification, and localization of lipids has been revolutionized by recent advances in lipid analysis methodologies, including lipid extraction protocols, internal standard availability, instrumentation, and bioinformatic tools and software (reviewed by [147]). These tools have enabled scientists to tackle questions that once were impossible and hold great potential for uncovering novel mechanisms of disease.

Since the equipment and expertise required for more advanced lipid measurement techniques are not readily present in neuroscience labs, an informative place to begin is with simple, cost-effective techniques that enable the detection of general lipid classes (Fig. 5.4). Lipid dyes are cheap, quick and afford the advantage of spatial resolution that thin layer chromatography and most MS techniques do not. The most common dyes for neutral lipids include Nile red, BODIPY, Oil Red O, Sudan black, and Filipin. Generally, the staining procedures are easily adapted for use on brain tissues and take less than an hour from start to finish. Thin layer chromatography (TLC) requires some specialized equipment for extracting lipids from homogenized, whole, or microdissected brain tissues but is otherwise relatively straightforward. TLC separates lipids using a "stationary phase" (normally silica gel) made possible by the differences in polarity of the various lipid classes and can be analysed to gain semi-quantitative measures of all major lipid classes. The major limitation of lipid dyes and TLC is the inability to resolve individual lipid species, which can only be achieved by mass spectrometry (MS) or nuclear magnetic resonance (NMR) techniques.

Determining changes in broad lipid classes is a good start towards understanding lipid composition and distribution. However, since individual lipid species can have vastly different and highly specific biological roles, digging deeper into the individual lipid species may be required. To accomplish this, mass spectrometry (MS) experiments are usually required. The first decision to make when deciding to perform MS is whether a targeted or untargeted approach is desired (reviewed by [148]). For example, is the goal to investigate members of a specific lipid class (targeted approach) or the global lipid profile ("lipidome") of a sample? (untargeted approach). The approach used dramatically affects the conclusions and interpretations that can be drawn. On one hand, a targeted approach can be useful when a specific hypothesis is being tested or when there is a limited quantity of the species of interest within the global sample. Targeted approaches increase sensitivity by making use of a biased lipid extraction method, which enriches specifically for the lipid class of interest. On the other hand, the main advantage of an untargeted

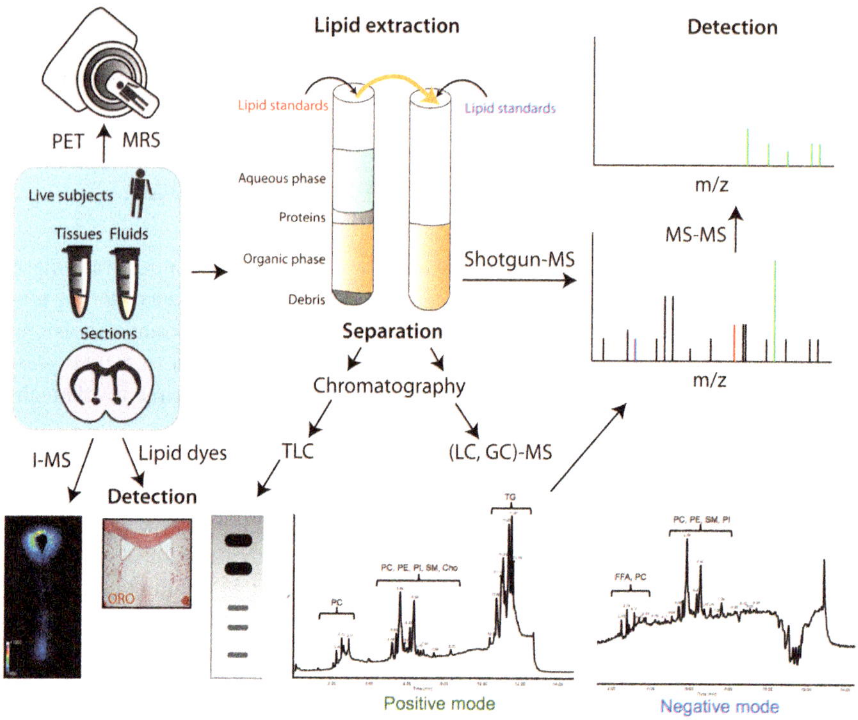

Fig. 5.4 Techniques to study lipids. Lipidomic studies have three main steps: *extraction*, *separation*, and *detection*. Simple techniques such as lipid dyes (e.g. Oil Red O (ORO) for neutral lipid classes) can reveal spatial distribution of lipid accumulations on sections. Thin layer chromatography (TLC) requires lipid extraction and chromatography separation to detect major classes of lipids in biological samples. In order to identify and measure individual lipid species, more complex techniques such as mass spectrometry (MS) are required. Shotgun-MS directly injects isolated lipid extracts into the mass spectrometer for detection. Liquid chromatography (LC) or gas chromatography (GC)-MS separates lipid extracts prior to injection into the MS for detection. Imaging-MS (I-MS) on tissue sections can determine the identity, location, and quantity of lipid species. Lipid changes can be studied in living subjects using positron emission topography (PET) and magnetic resonance spectroscopy (MRS), which can detect and quantify changes in lipid metabolism and lipid species, respectively

approach is that it allows for an unbiased screen of your sample's global lipidome, offering the opportunity for discovery of novel species.

Lipidomic experiments generally consist of three key steps: extraction, separation, and detection (Fig. 5.4). The extraction method chosen and the solvents used to run the samples can greatly affect the classes of lipids that can be analysed. Since lipids are non-polar compounds that are insoluble in water, they are usually enriched by extraction with organic solvents that removes interfering agents, such as proteins, saccharides, or other compounds. Currently, the most widely used extraction method is a derivative of the Folch method [149], which uses chloroform/methanol as the extraction solvent and was later amended by Bligh and Dyer [150]. In addition, many other lipid extraction methods have been developed [151]. For example, for unbiased plasma and CSF lipid extractions, we recently used a methyl-tert-butyl ether extraction protocol [126, 152] that is gaining in popularity due to its lower density than water. This makes the organic phase the upper layer (in contrast to chloroform-based procedures), avoiding the need to cross the other phases when collecting the lipid extract. A key consideration when performing lipid extractions and lipidomic studies is the use of internal standards. It is important to use spiked-in standards during the workflow of the experiment, both before the extraction (to ensure that the lipid classes of interest are successfully extracted) and after the extraction (to confirm that the classes of interest can be detected).

Following lipid extraction from the sample, the next steps are to separate and detect the individual lipid species. MS is the most widely used technique for detection due to its precision in lipid identification with high sensitivity and throughput. There are three main groups of MS approaches for lipidomic studies: shotgun, chromatography-coupled, and imaging (Fig. 5.4). Shotgun-MS [153, 154], in which lipid extracts are directly infused into the MS without pre-separation, is relatively fast and reproducible [154]. However, disadvantages of this approach are increased competition for ionization within the mass spectrometer ("ion suppression") that results in a reduced signal-to-noise ratio, as well as a reduced ability to distinguish between structural isomers. This strategy is generally more appropriate for targeted lipidomic analyses that focus on only one class of lipid. Coupling MS with a prior chromatographic step, such as gas chromatography (GC) or liquid chromatography (LC), allows for the pre-separation of lipids according to their biophysical or structural characteristics; this approach provides more detailed and reliable predictions of lipid identities that is useful in untargeted lipidomic analyses. LC-MS has become one of the most widely used methods for total lipidome analysis, while GC-MS is more suited to profiling FAs and some less polar lipids [155, 156]. During MS, the lipids will then be ionized in either positive or negative ionization mode, depending on ionization tendency of different lipid classes.

The above techniques are increasingly being used in biomarker discovery and disease characterization in the brain [139, 157]. However, a limitation of these more traditional MS techniques is the absence of spatial resolution. This is not an issue when dealing with homogeneous samples, such as plasma or CSF, but can be a significant disadvantage when studying compartmentalized or anatomically complex organs such as the brain. Imaging-MS (I-MS) offers the potential to overcome this limitation by allowing visualization of the spatial distribution of individual species across a thin tissue section. Three main ionization techniques are used with I-MS, including matrix-assisted laser desorption ionization (MALDI) [158], desorption electrospray ionization (DESI) [159], and secondary ion mass spectrometry (SIMS) [160]. MALDI can probe deeper into the tissue samples as compared to other techniques (e.g. SIMS) and provides better spatial resolution in comparison to DESI. Compared to MALDI and SIMS, DESI performs the analysis under ambient conditions, allowing for the recording of spectra in a native tissue environment without sample preparation or pre-separation [161, 162]; this is advantageous in clinical practice for real-time analyses, but yields lower spatial resolution than MALDI or SIMS. Currently, imaging with MALDI-based methods can reach a lateral spatial resolution of 5 μm while SIMS can reach 1 μm, the latter thus being able to resolve lipid alterations in subcellular compartments and membrane domains [163].

The output of MS experiments are mass-over-charge identifications that allow database-aided predictions of species identity. Although MS lipidomic methodology has come a long way, there can still be significant ambiguity in definitive identification of molecules that elute at the same mass-to-charge ratio. MS studies can thus be further strengthened by the addition of a second round of ionization (tandem MS or MS/MS). Tandem MS fractionates parent molecules into their component ions, enabling confirmation of their predicted identities.

The next frontier in studying brain lipid metabolism is the use of non-invasive approaches in living subjects. Functional brain imaging approaches are based on the measurement of metabolic changes that occur rapidly with brain activity. Techniques such as Positron Emission Tomography (PET) and Magnetic Resonance Spectroscopy (MRS) are being developed to track and measure lipid fluxes in vivo. PET measures accumulations of short-lived radio-labelled molecules that are metabolically active, traditionally 18F-fluoro-deoxyglucose (FDG). More recently, however, radio-labelled long-chain FAs such as 11C-arachidonic acid, 11C-palmitic acid, and 18-fluoro-6-thio-heptadecanoic acid (FTHA) have also been used successfully for PET studies. Although only rarely applied to the brain, PET imaging of FA metabolism holds promise for understanding and diagnosing brain lipid metabolism alterations in neurodegenerative diseases. For example, Karmi and colleagues used 11C-palmitate and FTHA to demonstrate that the brains of patients with peripheral metabolic syndrome have increased uptake of circulating FAs [164]. MRS is also widely used in both clinical and preclinical research [165]. Mobile lipids, including cholesterol esters and triglycerides, have enough rotational freedom to generate signal on MRS [166]. Given the findings of defects in FA synthesis and mobilization in neurodegenerative diseases such as AD [126] and autism [146], further development and optimization of brain imaging approaches may provide a means of early identification of presymptomatic populations, diagnostics, and tracking of clinical outcomes.

5.6 Conclusion

To enhance and sustain appropriate levels of neurogenesis throughout life and following damage or degeneration, it will be critical to reveal the cocktail of NSC regulators present under normal and pathological conditions. The arrival of techniques allowing scientists to better identify, measure, and localize lipids has led to a deeper understanding of global lipid metabolism. Even though these techniques are only beginning to be adapted to studies on the brain, current data has shown that NSCs appear to have enhanced sensitivity to lipid regulation. Further studies are needed to extend our knowledge of lipid metabolism in the brain in general and in NSC regulation.

References

1. Altman J, Das GD (1965) Autoradiographic and histological evidence of postnatal hippocampal neurogenesis in rats. J Comp Neurol 124(3):319–335
2. Gould E et al (1999) Hippocampal neurogenesis in adult Old World primates. Proc Natl Acad Sci U S A 96(9):5263–5267
3. Kornack DR, Rakic P (1999) Continuation of neurogenesis in the hippocampus of the adult macaque monkey. Proc Natl Acad Sci U S A 96(10):5768–5773
4. Eriksson PS et al (1998) Neurogenesis in the adult human hippocampus. Nat Med 4(11):1313–1317
5. Kukekov VG et al (1999) Multipotent stem/progenitor cells with similar properties arise from two neurogenic regions of adult human brain. Exp Neurol 156(2):333–344
6. Alvarez-Buylla A, Lim DA (2004) For the long run: maintaining germinal niches in the adult brain. Neuron 41(5):683–686
7. Doetsch F, Caille I, Lim DA, Garcia-Verdugo JM, Alvarez-Buylla A (1999a) Subventricular zone astrocytes are neural stem cells in the adult mammalian brain. Cell 97(6):703–716
8. Doetsch F, Garcia-Verdugo JM, Alvarez-Buylla A (1999b) Regeneration of a germinal layer in the adult mammalian brain. Proc Natl Acad Sci U S A 96(20):11619–11624
9. Imura T, Kornblum HI, Sofroniew MV (2003) The predominant neural stem cell isolated from postnatal and adult forebrain but not early embryonic forebrain expresses GFAP. J Neurosci 23(7):2824–2832
10. Seri B, Garcia-Verdugo JM, McEwen BS, Alvarez-Buylla A (2001) Astrocytes give rise to new neurons in the adult mammalian hippocampus. J Neurosci 21(18):7153–7160
11. Fuchs E, Tumbar T, Guasch G (2004) Socializing with the neighbors: stem cells and their niche. Cell 116(6):769–778
12. Morrison SJ, Spradling AC (2008) Stem cells and niches: mechanisms that promote stem cell maintenance throughout life. Cell 132(4):598–611
13. Suh H, Deng W, Gage FH (2009) Signaling in adult neurogenesis. Annu Rev Cell Dev Biol 25:253–275
14. Walker MR, Patel KK, Stappenbeck TS (2009) The stem cell niche. J Pathol 217(2):169–180
15. Kokoeva MV, Yin H, Flier JS (2005) Neurogenesis in the hypothalamus of adult mice: potential role in energy balance. Science 310(5748):679–683
16. Li J, Tang Y, Cai D (2012) IKKbeta/NF-kappaB disrupts adult hypothalamic neural stem cells to mediate a neurodegenerative mechanism of dietary obesity and pre-diabetes. Nat Cell Biol 14(10):999–1012
17. Ernst A et al (2014) Neurogenesis in the striatum of the adult human brain. Cell 156(5):1072–1083
18. Mirzadeh Z, Merkle FT, Soriano-Navarro M, Garcia-Verdugo JM, Alvarez-Buylla A (2008) Neural stem cells confer unique pinwheel architecture to the ventricular surface in neurogenic regions of the adult brain. Cell Stem Cell 3(3):265–278
19. Tong CK, Alvarez-Buylla A (2014) SnapShot: adult neurogenesis in the V-SVZ. Neuron 81(1):220–220 e221
20. Shen Q et al (2008) Adult SVZ stem cells lie in a vascular niche: a quantitative analysis of niche cell-cell interactions. Cell Stem Cell 3(3):289–300
21. Tavazoie M et al (2008) A specialized vascular niche for adult neural stem cells. Cell Stem Cell 3(3):279–288
22. Lim DA, Alvarez-Buylla A (2014) Adult neural stem cells stake their ground. Trends Neurosci 37(10):563–571
23. Apple DM, Fonseca RS, Kokovay E (2016) The role of adult neurogenesis in psychiatric and cognitive disorders. Brain Res.
24. Bordey A (2011) Adult-born neuron development is controlled by GABAA receptor subtypes (Commentary on Duveau et al.). Eur J Neurosci 34(3):361

25. Hoglinger GU, Arias-Carrion O, Ipach B, Oertel WH (2014) Origin of the dopaminergic innervation of adult neurogenic areas. J Comp Neurol 522(10):2336–2348
26. Lennington JB et al (2011) Midbrain dopamine neurons associated with reward processing innervate the neurogenic subventricular zone. J Neurosci 31(37):13078–13087
27. Paez-Gonzalez P, Asrican B, Rodriguez E, Kuo CT (2014) Identification of distinct ChAT(+) neurons and activity-dependent control of postnatal SVZ neurogenesis. Nat Neurosci 17(7):934–942
28. Young SZ, Taylor MM, Bordey A (2011) Neurotransmitters couple brain activity to subventricular zone neurogenesis. Eur J Neurosci 33(6):1123–1132
29. Alvarez-Buylla A, Garcia-Verdugo JM, Tramontin AD (2001) A unified hypothesis on the lineage of neural stem cells. Nat Rev Neurosci 2(4):287–293
30. Jankovski A, Sotelo C (1996) Subventricular zone-olfactory bulb migratory pathway in the adult mouse: cellular composition and specificity as determined by heterochronic and heterotopic transplantation. J Comp Neurol 371(3):376–396
31. Lois C, Alvarez-Buylla A (1994) Long-distance neuronal migration in the adult mammalian brain. Science 264(5162):1145–1148
32. Yamaguchi M, Mori K (2005) Critical period for sensory experience-dependent survival of newly generated granule cells in the adult mouse olfactory bulb. Proc Natl Acad Sci U S A 102(27):9697–9702
33. Petreanu L, Alvarez-Buylla A (2002) Maturation and death of adult-born olfactory bulb granule neurons: role of olfaction. J Neurosci 22(14):6106–6113
34. Winner B, Cooper-Kuhn CM, Aigner R, Winkler J, Kuhn HG (2002) Long-term survival and cell death of newly generated neurons in the adult rat olfactory bulb. Eur J Neurosci 16(9):1681–1689
35. Bath KG et al (2008) Variant brain-derived neurotrophic factor (Val66Met) alters adult olfactory bulb neurogenesis and spontaneous olfactory discrimination. J Neurosci 28(10): 2383–2393
36. Breton-Provencher V, Lemasson M, Peralta MR 3rd, Saghatelyan A (2009) Interneurons produced in adulthood are required for the normal functioning of the olfactory bulb network and for the execution of selected olfactory behaviors. J Neurosci 29(48):15245–15257
37. Gheusi G et al (2000) Importance of newly generated neurons in the adult olfactory bulb for odor discrimination. Proc Natl Acad Sci U S A 97(4):1823–1828
38. Menn B et al (2006) Origin of oligodendrocytes in the subventricular zone of the adult brain. J Neurosci 26(30):7907–7918
39. Spassky N et al (2005) Adult ependymal cells are postmitotic and are derived from radial glial cells during embryogenesis. J Neurosci 25(1):10–18
40. Luo J, Shook BA, Daniels SB, Conover JC (2008) Subventricular zone-mediated ependyma repair in the adult mammalian brain. J Neurosci 28(14):3804–3813
41. Benner EJ et al (2013) Protective astrogenesis from the SVZ niche after injury is controlled by Notch modulator Thbs4. Nature 497(7449):369–373
42. Kempermann G, Song H, Gage FH (2015) Neurogenesis in the adult hippocampus. Cold Spring Harb Perspect Biol 7(9):a018812
43. Ming GL, Song H (2011) Adult neurogenesis in the mammalian brain: significant answers and significant questions. Neuron 70(4):687–702
44. Song J et al (2012) Neuronal circuitry mechanism regulating adult quiescent neural stem-cell fate decision. Nature 489(7414):150–154
45. Aimone JB et al (2014) Regulation and function of adult neurogenesis: from genes to cognition. Physiol Rev 94(4):991–1026
46. Hsieh J (2012) Orchestrating transcriptional control of adult neurogenesis. Genes Dev 26(10):1010–1021
47. Gregoire CA, Bonenfant D, Le Nguyen A, Aumont A, Fernandes KJ (2014) Untangling the influences of voluntary running, environmental complexity, social housing and stress on adult hippocampal neurogenesis. PLoS One 9(1):e86237

48. Kobilo T, Yuan C, van Praag H (2011) Endurance factors improve hippocampal neurogenesis and spatial memory in mice. Learn Mem 18(2):103–107
49. Schoenfeld TJ, Gould E (2012) Stress, stress hormones, and adult neurogenesis. Exp Neurol 233(1):12–21
50. van Praag H, Kempermann G, Gage FH (1999) Running increases cell proliferation and neurogenesis in the adult mouse dentate gyrus. Nat Neurosci 2(3):266–270
51. Arruda-Carvalho M, Sakaguchi M, Akers KG, Josselyn SA, Frankland PW (2011) Posttraining ablation of adult-generated neurons degrades previously acquired memories. J Neurosci 31(42):15113–15127
52. Niibori Y et al (2012) Suppression of adult neurogenesis impairs population coding of similar contexts in hippocampal CA3 region. Nat Commun 3:1253
53. Aimone JB, Deng W, Gage FH (2010) Adult neurogenesis: integrating theories and separating functions. Trends Cogn Sci 14(7):325–337
54. Aimone JB, Deng W, Gage FH (2011) Resolving new memories: a critical look at the dentate gyrus, adult neurogenesis, and pattern separation. Neuron 70(4):589–596
55. Nakashiba T et al (2012) Young dentate granule cells mediate pattern separation, whereas old granule cells facilitate pattern completion. Cell 149(1):188–201
56. Clelland CD et al (2009) A functional role for adult hippocampal neurogenesis in spatial pattern separation. Science 325(5937):210–213
57. Deng W, Saxe MD, Gallina IS, Gage FH (2009) Adult-born hippocampal dentate granule cells undergoing maturation modulate learning and memory in the brain. J Neurosci 29(43):13532–13542
58. Imayoshi I et al (2008) Roles of continuous neurogenesis in the structural and functional integrity of the adult forebrain. Nat Neurosci 11(10):1153–1161
59. Kitamura T et al (2009) Adult neurogenesis modulates the hippocampus-dependent period of associative fear memory. Cell 139(4):814–827
60. Sahay A, Wilson DA, Hen R (2011) Pattern separation: a common function for new neurons in hippocampus and olfactory bulb. Neuron 70(4):582–588
61. Shors TJ et al (2001) Neurogenesis in the adult is involved in the formation of trace memories. Nature 410(6826):372–376
62. Cameron HA, Glover LR (2015) Adult neurogenesis: beyond learning and memory. Annu Rev Psychol 66:53–81
63. Snyder JS, Glover LR, Sanzone KM, Kamhi JF, Cameron HA (2009) The effects of exercise and stress on the survival and maturation of adult-generated granule cells. Hippocampus 19(10):898–906
64. Reynolds BA, Weiss S (1992) Generation of neurons and astrocytes from isolated cells of the adult mammalian central nervous system. Science 255(5052):1707–1710
65. Doetsch F, Petreanu L, Caille I, Garcia-Verdugo JM, Alvarez-Buylla A (2002) EGF converts transit-amplifying neurogenic precursors in the adult brain into multipotent stem cells. Neuron 36(6):1021–1034
66. Faiz M et al (2008) Substantial migration of SVZ cells to the cortex results in the generation of new neurons in the excitotoxically damaged immature rat brain. Mol Cell Neurosci 38(2):170–182
67. Fuentealba LC et al (2015) Embryonic origin of postnatal neural stem cells. Cell 161(7):1644–1655
68. Merkle FT et al (2014) Adult neural stem cells in distinct microdomains generate previously unknown interneuron types. Nat Neurosci 17(2):207–214
69. Merkle FT, Mirzadeh Z, Alvarez-Buylla A (2007) Mosaic organization of neural stem cells in the adult brain. Science 317(5836):381–384
70. Sachewsky N et al (2014) Primitive neural stem cells in the adult mammalian brain give rise to GFAP-expressing neural stem cells. Stem Cell Reports 2(6):810–824
71. Calzolari F et al (2015) Fast clonal expansion and limited neural stem cell self-renewal in the adult subependymal zone. Nat Neurosci 18(4):490–492

72. Codega P et al (2014) Prospective identification and purification of quiescent adult neural stem cells from their in vivo niche. Neuron 82(3):545–559
73. Giachino C et al (2014) Molecular diversity subdivides the adult forebrain neural stem cell population. Stem Cells 32(1):70–84
74. Mich JK et al (2014) Prospective identification of functionally distinct stem cells and neurosphere-initiating cells in adult mouse forebrain. Elife 3:e02669
75. Knoth R et al (2010) Murine features of neurogenesis in the human hippocampus across the lifespan from 0 to 100 years. PLoS One 5(1):e8809
76. Leonard BW et al (2009) Subventricular zone neural progenitors from rapid brain autopsies of elderly subjects with and without neurodegenerative disease. J Comp Neurol 515(3): 269–294
77. Quinones-Hinojosa A et al (2006) Cellular composition and cytoarchitecture of the adult human subventricular zone: a niche of neural stem cells. J Comp Neurol 494(3):415–434
78. Sanai N et al (2004) Unique astrocyte ribbon in adult human brain contains neural stem cells but lacks chain migration. Nature 427(6976):740–744
79. Bernier PJ, Vinet J, Cossette M, Parent A (2000) Characterization of the subventricular zone of the adult human brain: evidence for the involvement of Bcl-2. Neurosci Res 37(1):67–78
80. Weickert CS et al (2000) Localization of epidermal growth factor receptors and putative neuroblasts in human subependymal zone. J Comp Neurol 423(3):359–372
81. Bedard A, Parent A (2004) Evidence of newly generated neurons in the human olfactory bulb. Brain Res Dev Brain Res 151(1–2):159–168
82. Curtis MA et al (2007) Human neuroblasts migrate to the olfactory bulb via a lateral ventricular extension. Science 315(5816):1243–1249
83. Maresh A, Rodriguez Gil D, Whitman MC, Greer CA (2008) Principles of glomerular organization in the human olfactory bulb--implications for odor processing. PLoS One 3(7):e2640
84. Lim DA, Alvarez-Buylla A (2016) The adult ventricular-subventricular zone (V-SVZ) and olfactory bulb (OB) neurogenesis. Cold Spring Harb Perspect Biol 8(5)
85. Bergmann O et al (2012) The age of olfactory bulb neurons in humans. Neuron 74(4):634–639
86. Spalding KL et al (2013) Dynamics of hippocampal neurogenesis in adult humans. Cell 153(6):1219–1227
87. Ernst A, Frisen J (2015) Adult neurogenesis in humans-common and unique traits in mammals. PLoS Biol 13(1):e1002045
88. Hotamisligil GS (2006) Inflammation and metabolic disorders. Nature 444(7121):860–867
89. Saltiel AR, Kahn CR (2001) Insulin signalling and the regulation of glucose and lipid metabolism. Nature 414(6865):799–806
90. Digel M, Ehehalt R, Fullekrug J (2010) Lipid droplets lighting up: insights from live microscopy. FEBS Lett 584(11):2168–2175
91. Martin S, Parton RG (2006) Lipid droplets: a unified view of a dynamic organelle. Nature reviews. Mol Cell Biol 7(5):373–378
92. Ohsaki Y et al (2009) Biogenesis of cytoplasmic lipid droplets: from the lipid ester globule in the membrane to the visible structure. Biochim Biophys Acta 1791(6):399–407
93. Zechner R et al (2012) Fat signals—lipases and lipolysis in lipid metabolism and signaling. Cell Metab 15(3):279–291
94. Cohen JD (2011) Rationale for aggressive lipid lowering in high-risk patients. J Am Osteopath Assoc 111(4 Suppl 3):eS7–e12
95. Greenberg AS et al (2011) The role of lipid droplets in metabolic disease in rodents and humans. J Clin Invest 121(6):2102–2110
96. Walther TC, Farese RV Jr (2012) Lipid droplets and cellular lipid metabolism. Annu Rev Biochem 81:687–714
97. Bozza PT, Viola JP (2010) Lipid droplets in inflammation and cancer. Prostaglandins Leukot Essent Fatty Acids 82(4–6):243–250
98. Greenberg AS, Coleman RA (2011) Expanding roles for lipid droplets. Trends Endocrinol Metab 22(6):195–196

99. Bailey AP et al (2015) Antioxidant role for lipid droplets in a stem cell niche of drosophila. Cell 163(2):340–353
100. Liu L et al (2015) Glial lipid droplets and ROS induced by mitochondrial defects promote neurodegeneration. Cell 160(1–2):177–190
101. Bu G (2009) Apolipoprotein E and its receptors in Alzheimer's disease: pathways, pathogenesis and therapy. Nat Rev Neurosci 10(5):333–344
102. Mahley RW, Nathan BP, Pitas RE, Apolipoprotein E (1996) Structure, function, and possible roles in Alzheimer's disease. Ann N Y Acad Sci 777:139–145
103. Zlokovic BV (2008) The blood-brain barrier in health and chronic neurodegenerative disorders. Neuron 57(2):178–201
104. Kamp F et al (2003) Rapid flip-flop of oleic acid across the plasma membrane of adipocytes. J Biol Chem 278(10):7988–7995
105. Hamilton JA (1999) Transport of fatty acids across membranes by the diffusion mechanism. Prostaglandins Leukot Essent Fatty Acids 60(5–6):291–297
106. Mitchell RW, On NH, Del Bigio MR, Miller DW, Hatch GM (2011) Fatty acid transport protein expression in human brain and potential role in fatty acid transport across human brain microvessel endothelial cells. J Neurochem 117(4):735–746
107. Pan Y et al (2015) Fatty acid-binding protein 5 facilitates the blood-brain barrier transport of docosahexaenoic acid. Mol Pharm 12(12):4375–4385
108. Matsumata M et al (2012) The effects of Fabp7 and Fabp5 on postnatal hippocampal neurogenesis in the mouse. Stem Cells 30(7):1532–1543
109. Yang CP, Gilley JA, Zhang G, Kernie SG (2011) ApoE is required for maintenance of the dentate gyrus neural progenitor pool. Development 138(20):4351–4362
110. Li G et al (2009) GABAergic interneuron dysfunction impairs hippocampal neurogenesis in adult apolipoprotein E4 knockin mice. Cell Stem Cell 5(6):634–645
111. Mustroph ML et al (2012) Aerobic exercise is the critical variable in an enriched environment that increases hippocampal neurogenesis and water maze learning in male C57BL/6 J mice. Neuroscience 219:62–71
112. Levi O, Michaelson DM (2007) Environmental enrichment stimulates neurogenesis in apolipoprotein E3 and neuronal apoptosis in apolipoprotein E4 transgenic mice. J Neurochem 100(1):202–210
113. Knobloch M et al (2013) Metabolic control of adult neural stem cell activity by Fasn-dependent lipogenesis. Nature 493(7431):226–230
114. Chorna NE et al (2013) Fatty acid synthase as a factor required for exercise-induced cognitive enhancement and dentate gyrus cellular proliferation. PLoS One 8(11):e77845
115. Llorens-Bobadilla E et al (2015) Single-cell transcriptomics reveals a population of dormant neural stem cells that become activated upon brain injury. Cell Stem Cell 17(3):329–340
116. Shin J et al (2015) Single-cell RNA-Seq with waterfall reveals molecular cascades underlying adult neurogenesis. Cell Stem Cell 17(3):360–372
117. Katakura M et al (2009) Docosahexaenoic acid promotes neuronal differentiation by regulating basic helix-loop-helix transcription factors and cell cycle in neural stem cells. Neuroscience 160(3):651–660
118. Kawakita E, Hashimoto M, Shido O (2006) Docosahexaenoic acid promotes neurogenesis in vitro and in vivo. Neuroscience 139(3):991–997
119. Katakura M et al (2013) Omega-3 polyunsaturated Fatty acids enhance neuronal differentiation in cultured rat neural stem cells. Stem Cells Int 2013:490476
120. Rashid MA, Katakura M, Kharebava G, Kevala K, Kim HY (2013) N-Docosahexaenoylethanolamine is a potent neurogenic factor for neural stem cell differentiation. J Neurochem 125(6):869–884
121. Sakayori N et al (2011) Distinctive effects of arachidonic acid and docosahexaenoic acid on neural stem/progenitor cells. Genes Cells 16(7):778–790
122. Maekawa M et al (2009) Arachidonic acid drives postnatal neurogenesis and elicits a beneficial effect on prepulse inhibition, a biological trait of psychiatric illnesses. PLoS One 4(4):e5085

123. Patil S, Chan C (2005) Palmitic and stearic fatty acids induce Alzheimer-like hyperphosphorylation of tau in primary rat cortical neurons. Neurosci Lett 384(3):288–293
124. Wang Z et al (2014) Palmitic acid affects proliferation and differentiation of neural stem cells in vitro. J Neurosci Res 92(5):574–586
125. Yuan Q et al (2013) Palmitic acid increases apoptosis of neural stem cells via activating c-Jun N-terminal kinase. Stem Cell Res 10(2):257–266
126. Hamilton LK et al (2015) Aberrant lipid metabolism in the forebrain niche suppresses adult neural stem cell proliferation in an animal model of Alzheimer's disease. Cell Stem Cell 17(4):397–411
127. Braak H, Braak E (1991a) Demonstration of amyloid deposits and neurofibrillary changes in whole brain sections. Brain Pathol 1(3):213–216
128. Braak H, Braak E (1991b) Neuropathological stageing of Alzheimer-related changes. Acta Neuropathol 82(4):239–259
129. Braak H, Braak E (1995) Staging of Alzheimer's disease-related neurofibrillary changes. Neurobiol Aging 16(3):271–278 discussion 278-284
130. Chuang TT (2010) Neurogenesis in mouse models of Alzheimer's disease. Biochim Biophys Acta 1802(10):872–880
131. Lazarov O, Marr RA (2010) Neurogenesis and Alzheimer's disease: at the crossroads. Exp Neurol 223(2):267–281
132. Lazarov O, Mattson MP, Peterson DA, Pimplikar SW, van Praag H (2010) When neurogenesis encounters aging and disease. Trends Neurosci 33(12):569–579
133. Perry EK et al (2012) Neurogenic abnormalities in Alzheimer's disease differ between stages of neurogenesis and are partly related to cholinergic pathology. Neurobiol Dis 47(2): 155–162
134. Ziabreva I et al (2006) Altered neurogenesis in Alzheimer's disease. J Psychosom Res 61(3):311–316
135. Alzheimer A (1907) Über eine eigenartige Erkrankung der Hirnrinde. Allg Z Psychiatrie Psychisch-Gerichtlich Med 64:146–148
136. Karch CM, Cruchaga C, Goate AM (2014) Alzheimer's disease genetics: from the bench to the clinic. Neuron 83(1):11–26
137. Tosto G, Reitz C (2013) Genome-wide association studies in Alzheimer's disease: a review. Curr Neurol Neurosci Rep 13(10):381
138. Astarita G et al (2011) Elevated stearoyl-CoA desaturase in brains of patients with Alzheimer's disease. PLoS One 6(10):e24777
139. Fraser T, Tayler H, Love S (2010) Fatty acid composition of frontal, temporal and parietal neocortex in the normal human brain and in Alzheimer's disease. Neurochem Res 35(3): 503–513
140. Hussain G, Schmitt F, Loeffler JP, Gonzalez de Aguilar JL (2013) Fatting the brain: a brief of recent research. Front Cell Neurosci 7:144
141. Podtelezhnikov AA et al (2011) Molecular insights into the pathogenesis of Alzheimer's disease and its relationship to normal aging. PLoS One 6(12):e29610
142. Tanzi RE (2012) The genetics of Alzheimer disease. Cold Spring Harb Perspect Med 2(10)
143. Pasinetti GM, Eberstein JA (2008) Metabolic syndrome and the role of dietary lifestyles in Alzheimer's disease. J Neurochem 106(4):1503–1514
144. Tamiji J, Crawford DA (2010) The neurobiology of lipid metabolism in autism spectrum disorders. Neurosignals 18(2):98–112
145. Mariani J et al (2015) FOXG1-dependent dysregulation of GABA/glutamate neuron differentiation in autism spectrum disorders. Cell 162(2):375–390
146. Xie Z, Jones A, Deeney JT, Hur SK, Bankaitis VA (2016) Inborn errors of long-chain fatty acid beta-oxidation link neural stem cell self-renewal to autism. Cell Rep 14(5):991–999
147. Rolim AE, Henrique-Araujo R, Ferraz EG, de Araujo Alves Dultra FK, Fernandez LG (2015) Lipidomics in the study of lipid metabolism: current perspectives in the omic sciences. Gene 554(2):131–139

148. Patti GJ, Yanes O, Siuzdak G (2012) Innovation: metabolomics: the apogee of the omics trilogy. Nat Rev Mol Cell Biol 13(4):263–269
149. Folch J, Lees M, Sloane Stanley GH (1957) A simple method for the isolation and purification of total lipides from animal tissues. J Biol Chem 226(1):497–509
150. Bligh EG, Dyer WJ (1959) A rapid method of total lipid extraction and purification. Can J Biochem Physiol 37(8):911–917
151. Reis A et al (2013) A comparison of five lipid extraction solvent systems for lipidomic studies of human LDL. J Lipid Res 54(7):1812–1824
152. Matyash V, Liebisch G, Kurzchalia TV, Shevchenko A, Schwudke D (2008) Lipid extraction by methyl-tert-butyl ether for high-throughput lipidomics. J Lipid Res 49(5):1137–1146
153. Han X, Gross RW (2005) Shotgun lipidomics: electrospray ionization mass spectrometric analysis and quantitation of cellular lipidomes directly from crude extracts of biological samples. Mass Spectrom Rev 24(3):367–412
154. Han X, Yang K, Gross RW (2012) Multi-dimensional mass spectrometry-based shotgun lipidomics and novel strategies for lipidomic analyses. Mass Spectrom Rev 31(1):134–178
155. Bogusz S Jr et al (2012) Solid-phase microextraction combined with comprehensive two-dimensional gas chromatography for fatty acid profiling of cell wall phospholipids. J Sep Sci 35(18):2438–2444
156. Li M, Zhou Z, Nie H, Bai Y, Liu H (2011) Recent advances of chromatography and mass spectrometry in lipidomics. Anal Bioanal Chem 399(1):243–249
157. Cunnane SC et al (2012) Plasma and brain fatty acid profiles in mild cognitive impairment and Alzheimer's disease. J Alzheimers Dis 29(3):691–697
158. Cornett DS, Reyzer ML, Chaurand P, Caprioli RM (2007) MALDI imaging mass spectrometry: molecular snapshots of biochemical systems. Nat Methods 4(10):828–833
159. Wiseman JM, Ifa DR, Song Q, Cooks RG (2006) Tissue imaging at atmospheric pressure using desorption electrospray ionization (DESI) mass spectrometry. Angew Chem Int Ed Engl 45(43):7188–7192
160. Todd PJ, McMahon JM, Short RT, McCandlish CA (1997) Organic SIMS of biologic tissue. Anal Chem 69(17):529A–535A
161. Cooks RG, Ouyang Z, Takats Z, Wiseman JM (2006) Detection technologies. Ambient mass spectrometry. Science 311(5767):1566–1570
162. Dill AL, Ifa DR, Manicke NE, Ouyang Z, Cooks RG (2009) Mass spectrometric imaging of lipids using desorption electrospray ionization. J Chromatogr B Analyt Technol Biomed Life Sci 877(26):2883–2889
163. Gross RW, Holcapek M (2014) Lipidomics. Anal Chem 86(17):8505
164. Karmi A et al (2010) Increased brain fatty acid uptake in metabolic syndrome. Diabetes 59(9):2171–2177
165. Zhu H, Barker PB (2011) MR spectroscopy and spectroscopic imaging of the brain. Methods Mol Biol 711:203–226
166. Ferreira CR, Silber MH, Chang T, Murnick JG, Kirmse B (2015) Cerebral lipid accumulation detected by MRS in a child with carnitine palmitoyltransferase 2 deficiency: a case report and review of the literature on genetic etiologies of lipid peaks on MRS. JIMD Rep 28:69–74

Chapter 6
Cannabinoids as Regulators of Neural Development and Adult Neurogenesis

Alline C. Campos, Juan Paraíso-Luna, Manoela V. Fogaça, Francisco S. Guimarães, and Ismael Galve-Roperh

Abbreviations

2-AG	2-arachidonoylglycerol
AEA	Anandamide
Ca^{2+}	Calcium
CB_1	Cannabinoid receptor type 1
CB_2	Cannabinoid receptor type 2
CBD	Cannabidiol
DAGL	Diacylglycerol lipase
DG	Dentate gyrus
E/I	Excitation/inhibition
ECB	Endocannabinoids
MAGL	Monoacylglycerol lipase

A.C. Campos (✉) • M.V. Fogaça • F.S. Guimarães
Department of Pharmacology, Medical School of Ribeirão Preto, University of São Paulo, 3900 Bandeirantes Avenue, Ribeirão Preto, SP, Brazil
e-mail: allinecampos@usp.br; manoelafogaca@usp.br; fsguimar@fmrp.usp.br

J. Paraíso-Luna • I. Galve-Roperh
Department of Biochemistry and Molecular Biology I, School of Biology, Complutense University, and Neurochemistry Universitary Research Institute, C/Jose Antonio Novais 12, Madrid 28040, Spain

CIBERNED, Center for Networked Biomedical Research in Neurodegenerative Diseases, C/Valderebollo 5, Madrid 28031, Spain
e-mail: juanparaiso@ucm.es; igr@quim.ucm.es

A. Pébay, R.C.B. Wong (eds.), *Lipidomics of Stem Cells*, Stem Cell Biology and Regenerative Medicine, DOI 10.1007/978-3-319-49343-5_6

NP	Neural progenitor
SGZ	Subgranular zone
SVZ	Subventricular zone
THC	Δ^9-tetrahydrocannabinol

6.1 Cannabinoids

The term cannabinoid was first used to describe a class of substances with similar chemical structures extracted from the plant *Cannabis sativa*. More than 100 cannabinoids have been identified in this plant, including Δ^9-tetrahydrocannabinol (THC), the one responsible for its main psychological effects, and cannabidiol (CBD), the major non-psychotomimetic compound [1]. The observation that the activity of psychoactive cannabinoids was intrinsically related to its chemical structure [2] raised the hypothesis that cannabinoid receptors would be present in the organism. In the late 1980s, the endocannabinoid (ECB) system started to be described with the identification of a specific receptor for THC in the central nervous system (CNS, [3]) that was subsequently cloned and named cannabinoid CB_1 receptor [4].

CB_1 receptors are now considered the most abundant metabotropic receptor in the mammals' CNS and are also present in peripheral tissues. The CB_1 receptors are widely expressed in presynaptic terminals, where they regulate the release of several neurotransmitters (e.g., GABAe, glutamate, serotonin, acetylcholine, dopamine) [5, 6]. A second cannabinoid receptor, named CB_2, was described in 1993 by Munro and colleagues [7]. Although initially thought to be expressed mainly in cells of the hematopoietic and immune systems, more recent studies have challenged this notion demonstrating that CB_2 receptors may be expressed in neurons and is present in microglia and neural stem cells [8–10]. Of note, despite their different localization and, apparently, functions, both CB_1 and CB_2 receptors are coupled to a $G_{i/o}$ protein [11].

In addition to CB_1 and CB_2 receptors, their endogenous ligands (termed endocannabinoids) were also isolated in mammals. The most extensively investigated are those derived from arachidonic acid, arachidonoyl ethanolamide (anandamide-AEA), and 2-arachidonoyl glycerol (2-AG), which are degraded by specific enzymes (Fig. 6.1, [12, 13]). AEA and 2-AG can also interact with other receptors such as proliferator-activated receptors (PPAR- α and γ). Moreover, AEA interacts with GPR55 and the Transient Receptor Potential Vanilloid Type 1 (TRPV1) [14].

Cannabinoids decrease neurotransmitter release by inhibiting calcium (Ca^{2+}) and activating potassium channels [15]. They also affect short-term neuronal activity by reducing the depolarization-induced suppression of inhibition (DSI), mainly in GABAergic synapses, and the depolarization-induced suppression of excitation (DSE), in synapses that release glutamate and the neuropeptide cholecystokinin [16–18]. Moreover, cannabinoids display neuroprotective actions, being involved in the control of glutamate-induced excitotoxicity [19], and are critical regulators of neurodevelopment and adult neurogenesis [20].

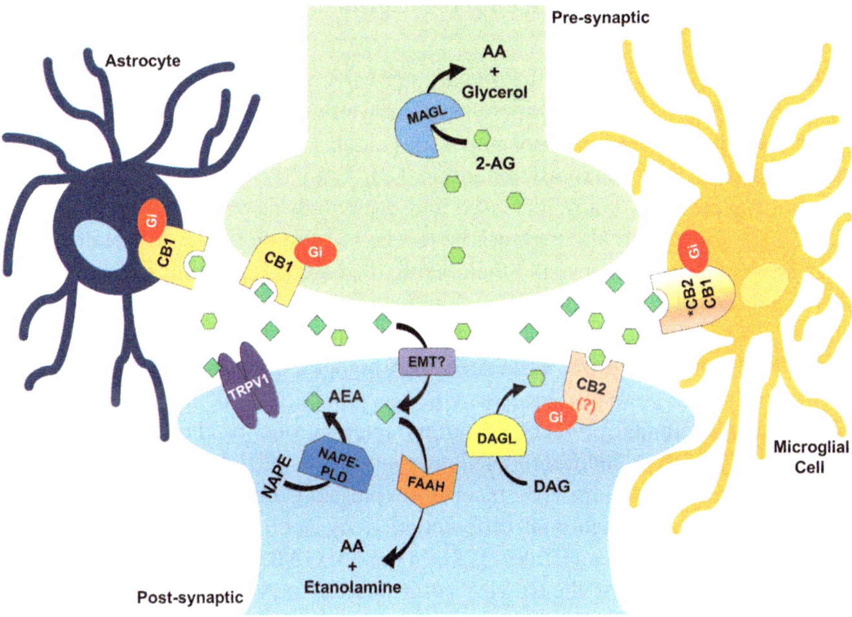

Fig. 6.1. Schematic representation of the endocannabinoid system in the brain. (*?*) Putative expression of CB2 receptor in neurons. *Microglial cells express CB1(constitutive) and CB2 (activated state) receptors. Endocannabinoids are produced in astrocytes, microglia, and neurons

In this chapter, we summarize the main pieces of evidence indicating that cannabinoid signaling on neural stem/progenitor cells affects their proliferation, maturation, and survival. These effects can modify CNS functions, being a potential new avenue for the development of novel therapeutic strategies for neurodegenerative and psychiatric disorders.

6.2 The Neurodevelopmental Role of the Endocannabinoid System

An extensive literature has addressed the consequences of developmental exposure to phytocannabinoids, mostly THC, and also to potent synthetic cannabinoid agonists. These studies have demonstrated that exposure of the immature nervous system to THC, in perinatal stages and/or the adolescence, is associated to numerous behavioral alterations [21]. Experimental evidence indicates that the developing brain is more sensitive to exogenous cannabinoid-induced plastic adaptations. These findings prompted the search of the neurobiological substrate of phytocannabinoid actions.

6.2.1 Expression of the Endocannabinoid System

The ECB system is present and functional since early stages of development, including the primordium of the nervous system, as well as in the restricted neurogenic areas of the adult brain (the hippocampal subgranular zone-SGZ and subventricular zone-SVZ). Along neuronal differentiation, CB_1 and CB_2 receptors show opposite patterns of expression, being increased and decreased, respectively [10, 22]. CB_1 receptors are expressed, although at low levels, in neuroepithelial progenitor cells from early embryonic stages, and their levels increase along neural differentiation [20]. In addition, CB_1 is enriched in white matter areas in embryonic stages, until the acquisition of its final expression pattern in the adult nervous system [23]. In vivo, CB_1 receptor levels are associated with higher expression of differentiation markers of various neuronal lineages. CB_1 receptor activity is more prominent in differentiated pyramidal projection neurons, interneurons, or cholinergic neurons than in their respective undifferentiated progenitor cells [20]. Little is known about the mechanisms controlling CB_1 receptor expression during neurodevelopment. CB_1 is induced during neuronal differentiation by neurotrophins such as brain-derived neurotrophic factor (BDNF) [24]. In mature GABAergic interneurons, CB_1 expression is controlled by the 67-kDa isoform GABA-synthesizing enzyme glutamate decarboxylase [25] and in striatal neurons is regulated by the transcription factor REST via RE1 sites [26].

The CB_1 receptor regulation by ECBS during development is poorly understood. 2-AG and AEA can be synthesized on-demand by surrounding differentiated neurons in response to neuronal activity. In addition, ECB can be produced in a paracrine/autocrine manner by neural progenitors (NPs) [27, 28]. The extracellular or intrinsic mechanisms responsible for ECB production in active neurogenic niches are not entirely understood. NPs produce and release the two major ECB compounds, namely, AEA and 2-AG, in response to increased intracellular Ca^{2+} concentration, and the ECB tone contributes to basal and stimulus-induced NP proliferation via CB_1 receptors [27, 29, 30]. 2-AG levels in neurogenic niches are precisely regulated by diacylglycerol lipase (DAGL) and monoacylglycerol lipase (MAGL) activity. Ablation of DAGLα, but not of the β isoform, interferes with hippocampal and SVZ-derived neurogenesis [31] and pharmacological inhibition of DAGL activity in NP cultures reduces cell proliferation [32]. NPs express FAAH, the primary enzyme involved in AEA degradation, and its genetic ablation or pharmacological inhibition promote NP proliferation [27, 33].

The role of extracellular signaling cues promoting ECB production is solely known for 2-AG generation, whereas signals driving AEA levels remain elusive, as the expression pattern of NAPE-PLD (N-acyl phosphatidylethanolamine phospholipase D) and FAAH (fatty acid amide hydrolase) enzymes responsible for AEA synthesis and degradation, respectively, during brain development remains unknown. Fibroblast growth factor (FGF) in coordination with neural cell adhesion molecule increases 2-AG levels via DAGL coupled with PLCγ activation. Alternatively, NGF via TrkA enhances 2-AG production during neurite outgrowth of cholinergic neurons by controlling the levels of MAGL [14]. In NPs, the high expression levels of DAGLα have been shown to rapidly decrease along their differentiation into GABAergic neuronal cells [34], through a mechanism that relies on the regulation of the transcriptional regulator specificity protein 1. On the contrary, retinoic acid-induced neuronal-like differentiation of neuroblastoma cells increases first DAGLα expression and later DAGLβ [35].

A variety of neuroactive molecules acting via ionotropic and metabotropic receptors have the potential to engage ECB generation via increased Ca^{2+} levels or G_q-PLC activation. These responses may occur after neurotransmitter-mediated neuronal activity and are also associated with spontaneous neuronal activity during cortical development. However, the contribution of spontaneous neuronal activity (during brain development) or neuronal synaptic activity (in adult neurogenic niches) in NP cell fate regulation, via ECB production, remains unknown. In addition to CB_1 receptors, CB_2 receptor activity regulates NP cell proliferation, cell cycle maintenance, and neural differentiation [10, 32, 36]. Whereas CB_2 receptor regulation clearly regulates stem/progenitor cell responses, its expression levels and the identity of neural cells expressing it remain obscure.

6.2.2 *Cannabinoid Signaling Consequences in the Developing Brain*

6.2.2.1 Proliferation

The first pieces of evidence for an active role of cannabinoid signaling in NP cells came from studies on the regulation of adult neurogenesis by pharmacological cannabinoid manipulation or genetic ablation of the CB_1 receptor [20, 37]. These studies evidenced that ablation of CB_1 receptor expression reduced hippocampal and SVZ NP cell proliferation in vivo. Likewise, CB_1 receptor absence in vitro inhibits self-renewal and NP proliferation [27]. Recent findings suggest that the positive role of CB_1 receptor signaling in adult neurogenesis is reminiscent of its role in NP proliferation and identity during cortical development Fig. 6.2a [38].

CB_1 receptor signaling controls neural cell fate decisions during CNS development by regulating the expression of genes responsible for neural identity [39]. In differentiating neuroblasts, CB_1 activation regulates the homeodomain containing transcription factor Pax6 post-translationally via PI3K/Akt-dependent phosphorylation, and this is in turn responsible for its positive actions in neurite outgrowth [40]. In addition, CB_1 activation increases Pax6 expression in cortical progenitors, driving the expansion towards basal intermediate progenitors by inducing the expression of the transcription factor Tbr2/eomes [38].

6.2.2.2 Neuronal Differentiation and Morphogenesis

CB_1 receptor signaling also affects neuronal differentiation acting in post-mitotic cells and, in an independent manner of its regulatory role, in undifferentiated NPs. CB_1 signaling activates NP cell proliferation and pro-survival signaling pathways

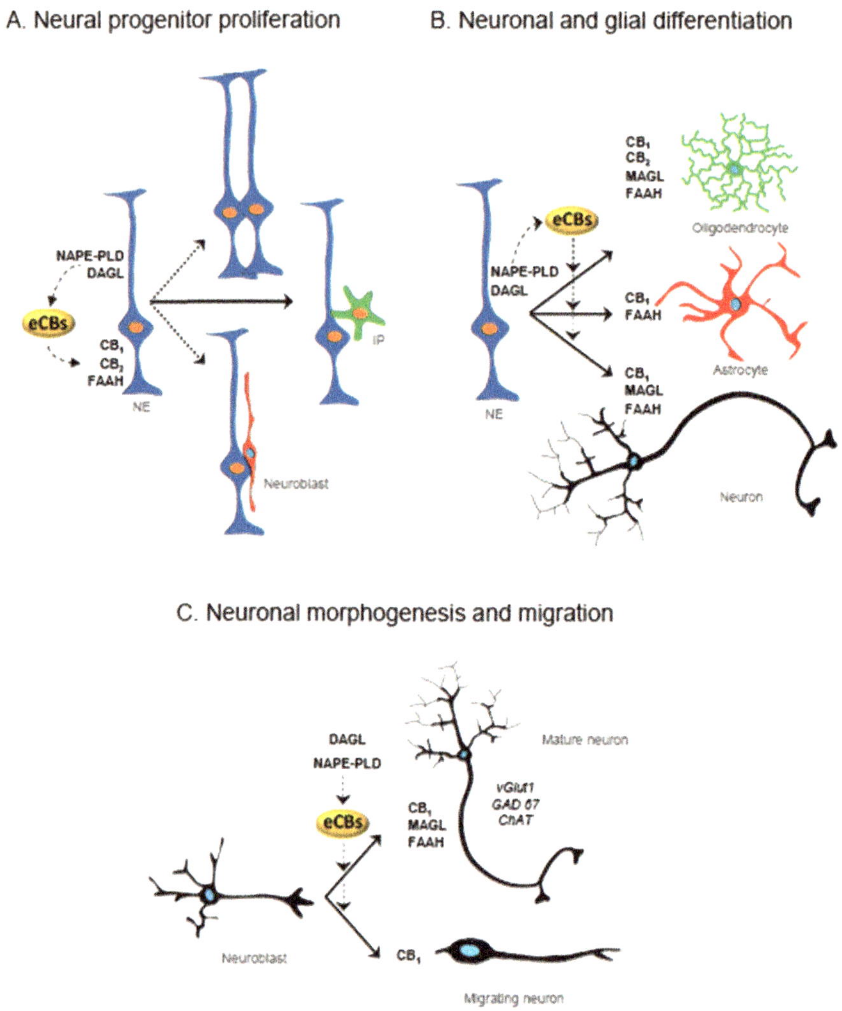

Fig. 6.2. The endocannabinoid system exerts a regulatory role on neural cell fate at different levels. Cannabinoid signaling regulates (**a**) NP proliferation and identity of progenitor cells, (**b**) neuronal and glial differentiation, and (**c**) neuronal morphogenesis and migration

that contribute to the regulation of cell cycle maintenance and the switch between cell proliferation and differentiation/migration. On the other hand, post-mitotic conditional CB_1 receptor ablation does not affect cortical progenitor expansion but only neuronal differentiation (Fig. 6.2b) [41]. CB_1 regulates the balance between the expression of Ctip2 and Satb2, two transcriptional regulators that are involved in the decision switch of deep- versus upper-layer cortical neurons. Ctip2 drives deep-layer cortical neuronal identity and corticospinal connectivity, whereas Satb2 is involved in intracortical projection neurons selectively arising from upper cortical layers [42]. Deletion of CB_1 during mouse cortical development lowered Ctip2 expression and generation of deep-layer V neurons, and this is reflected in the reduced ability for skilled motor activity of CB_1-deficient mice [39].

Cannabinoid signaling also exerts a crucial regulatory role in axon guidance and morphogenesis (Fig. 6.2c) [14]. CB_1 receptor located in axon growth cones of differentiating neurons induces its collapse in response to DAGL-derived 2-AG, [43]. A tight spatiotemporal regulation of 2-AG availability has been suggested accordingly to the differential subcellular localization of 2-AG metabolizing enzymes [44]. MAGL is enriched in tubulin-consolidating axon shafts while DAGL accumulates in actin-rich motile axon tips, thus generating a 2-AG gradient that triggers axonal growth cone collapse. In cortical and retinal neurons, CB_1 regulates axonal growth cone by controlling the plasma membrane localization of the Dcc (deleted in colorectal cancer) receptor [45], whereas in GABAergic interneurons the monomeric G protein RhoA is involved [43]. CB_1 receptor regulation of growth cone collapse and neurite retraction relies on its ability to regulate actomyosin cytoskeleton via RhoA/ROCK signaling and Rac1/WAVE complex [46, 47].

CB_1 receptor regulation of growth cone dynamics is responsible for its role in the establishment of long-range subcortical projections. Ablation or pharmacological blockade of CB_1 receptors in utero alters corticothalamic projections and induces axon fasciculation deficits [48]. The complementary expression pattern of DAGL in thalamocortical axons and of MAGL in corticothalamic and thalamocortical developing axons contribute to the generation of spatially restricted 2-AG pools. It has therefore been suggested a potential role for 2-AG as one of the molecules responsible for the timely developmental coordination between corticothalamic and thalamocortical projection "hand-shaking" [49]. The CB_1 receptor thus exerts an acute/short-term regulation of growth cone signaling in neurite tips, as well as long-lasting changes in neurogenic gene expression that affect neuronal wiring and connectivity.

In postnatal stages, cannabinoid receptor activity regulates astroglial and oligodendroglial differentiation (Fig. 6.2b). CB_1 receptor activity increases astroglial differentiation and GFAP expression in the developing cortex [50]. In oligodendrocyte progenitor cells CB_1 and CB_2 activation promotes the expression of Olig-2 in a PI3K/Akt/mTORC1-dependent manner [51], and their activation by 2AG or WIN55,212-2 administration favors white matter recovery and oligodendrocyte differentiation [52, 53].

Noteworthy, ECB signaling in oligodendrocytes via CB_2 receptors can contribute to neuron axon pathfinding by modulating Slit/Robo signaling in corticothalamic neurons expressing CB_1 receptor [54].

6.3 Pathological Implications of Cannabinoid Signaling in the Developing Brain

The neurodevelopmental role of the ECB system and its ability to regulate neural cell fate has important implications in regard to its potential contribution to neurodevelopmental disorders. Likewise, exposure to plant-derived cannabinoids, cannabinergic drugs interacting with the ECB system (i.e., modulators of ECB synthesis and degradation), or pollutants interfering with the ECB system can induce functional alterations in the adult progeny. Extensive literature exists regarding the consequences of cannabinoid-exposure during adolescence indicating that this is a critical period of susceptibility to deleterious actions produced by these compounds [21]. Less is known about the consequences of prenatal cannabinoid administration or embryonic manipulation of cannabinoid signaling [54, 55]. Cannabinoid-induced alterations of the nervous system development have been demonstrated in different experimental models. In early embryonic chick development, administration of a THC analogue disrupts neurogenesis and affects brain, somite and spinal cord primordium development, indicating that the ECB system is active in early cell fate decisions of neural tube progenitor cells [56]. In pregnant rats, administration of WIN-55,212-2 during the gestational period induces changes in dorsal pallial migrating neuroblasts and marginal zone interneurons [57]. Unfortunately, the impact of WIN-55,212-2 treatment in the progeny's brain was not investigated.

6.3.1 *Neuronal Hyperexcitability and Epileptogenesis*

Constitutive absence of CB_1 receptors in null mice results in increased seizure susceptibility that is mostly attributed to the lack of the neuromodulatory role of presynaptic CB_1 receptors [58]. In addition, the neurodevelopmental alterations associated with the loss of CB_1 receptors in early stages, i.e., during embryonic development when synaptic activity is still absent or emerging, can shed new light on the cellular mechanisms responsible for epileptogenesis and the appropriate balance of excitation/inhibition (E/I). Alterations of neurogenesis and changes of excitatory and inhibitory neuronal cell populations are, therefore, essential for coordinated activity. Considering the evidence that the ECB via CB_1 receptors regulates both excitatory projection neuron specification and GABAergic interneuron morphogenesis and local microcircuits, these alterations can contribute to the higher susceptibility and severity to seizures as a consequence of CB_1 signaling manipulation. In agreement, embryonic THC administration exerts a deleterious impact in deep-cortical layer projection neurons and increases seizure susceptibility via CB_1 receptors [59]. In this study, the impact of THC in interneurons and particularly in CCK basket cells was not investigated, but selective neuronal lineage rescue of CB_1 receptor expression [60] revealed that CB_1 receptors expressed in projection neurons and the GABAergic lineage contribute to seizure susceptibility. Likewise, prenatal THC administration, by interfering with cytoskeleton stability via c-Jun N-terminal kinase and Superior Cervical Ganglion 10/stathmin-2 protein levels, decreases Schaffer collateral-induced long-term depression and perisomatic basket cell surrounding pyramidal cell somata [61]. Interference with the correct generation of different neuronal subpopulations can be responsible for embryonic THC-induced E/I unbalance. In addition to CB_1

receptor regulation of neuronal differentiation, cannabinoid signaling actions in neuronal migration can contribute to developmental epileptogenesis. Genetic ablation of CB_1 receptors during cortical development exerts a radial migration blockade that results in ectopic projection neurons resembling subcortical band heterotopias (Díaz-Alonso, de Salas-Quiroga, Galve-Roperh, personal communication). Noteworthy, transient CB_1 receptor knockdown restricted to embryonic stages exerts long-lasting migration blockade that persists in the adulthood and induces increased seizure susceptibility. The promigratory role of CB1 receptors during brain development (Fig. 6.2c) is in agreement with the described role of the ECB system regulating neuroblasts migration in the adult rostral migratory stream [62]. These findings support the notion that cannabinoid signaling controls the appropriate E/I balance by additional mechanisms to the canonical CB_1 receptor neuromodulation.

6.3.2 *Neuropsychiatric Disorders*

Experimental evidence described herein reveals that defective ECB signaling or developmental exposure to phytocannabinoids can induce alterations in neuronal number, specification and functional properties, or morphological changes that may be responsible not only for seizure susceptibility but also for neuropsychiatric actions of cannabinoid signaling. The neurobiological substrate responsible for the emotional, social interaction, and cognitive changes induced by phytocannabinoid consumption or by an unbalanced ECB signaling during brain development remains largely unknown [54, 55]. In agreement with previous evidence of CB_1 regulation of CCK development, a recent study showed that embryonic THC administration correlated with selective changes in the development of CCK basket cells, but not other interneuron populations. Embryonic THC administration compromised feedforward and feedback inhibition in the progeny [63]. The persistent inhibitory deficits in the adult progeny was associated with deficient social interaction, but not increased anxiety, as reported in many studies where THC was administered in the adolescent period [21]. The impact of THC in CCK development raises the hypothesis of a potential interaction between cannabinoid signaling and autism. Noteworthy, autism-related mutations of neuroligin 3 are associated with changes in CB_1 constitutive activity [64]. THC administration during adolescence, but not later, interferes with GABA maturation and functionality in the prefrontal cortex, highlighting the importance of developmental actions in cannabinoid effects [65]. On the other hand, CB_1 receptor blockade in the adult can counteract several phenotypic markers of the Fragile X model (based on the loss of fragile X mental retardation protein FRMP) [66]. The consequences of manipulating CB_1 receptor signaling during brain development in autism models remain to be investigated. Furthermore, the role of CB_1 in interneuron developmental changes underlying the pathogenesis of schizophrenia constitutes an expanding field of research [67].

6.4 Adult Neurogenesis

At the beginning of the twentieth century, independent researchers reported what they believed to be the first description of mitotic figures in the adult nervous system of mammals [68]. However, this finding was not recognized because of the accepted dogma based on Santiago Ramon y Cajal's view that, reflecting the limitations of the techniques available at that time, it was impossible to identify dividing neurons in the adult brain [69].

For more than 100 years, evidence of adult neurogenesis was denied, as the accepted view was that this process could only happen during embryonic periods, stopping just after birth. In the early 1960s, Joseph Altman, a scientist of the Massachusetts Institute of Technology, using tritiated thymidine administered intraperitoneally in adult rats, reported that "a proliferative region of granule cells was identified in the dentate gyrus of the hippocampus" [70, 71]. Almost 15 years later, Dr. Michael Kaplan presented additional evidence that new neurons are added in specific regions of the young and adult rat brain, including the neocortex, hippocampal formation, and olfactory bulb [72–74]. However, it was the work of [75], which reported that new neurons are indeed generated in the hippocampus of adult humans that established one of the most exciting recent fields in neuroscience: adult neurogenesis.

Adult neurogenesis is a complex process that evolves from the initial division of precursor cells until the effective differentiation and generation of a new functional and integrated neuron. In the words of Dr. G. Kempermann: "Neurogenesis is a process, not an event.". It can be more precisely defined as an in vivo process that involves cell division, survival (not all cells that divide will survive), migration, differentiation, and maturation [76–78]. Neural proliferative capacity has been reported in different brain regions, such as the hypothalamus and the cell layers surrounding the third ventricle [79]. However, the best characterized neurogenic areas in the adult brain are the SVZ of the lateral walls of the lateral ventricle and SGZ of the dentate gyrus (DG) of the hippocampal formation [80]. Both regions have a resident population of neural stem/progenitor cells that can originate neurons, astrocytes, and oligodendrocytes [81].

Despite the half-century of research separating the initial findings of Altman from our current knowledge, the particular function/physiological role of adult neurogenesis, as well as the key regulators of this process, remain under debate. So far, it seems to be a consensus that experience modulates neurogenesis in the adult brain either positively or negatively. Voluntary exercise or enrichment environment enhances proliferation in neurogenic niches [82]. Conversely, chronic stress exposure decreases neurogenesis. However, due the different neurobiological nature of the two main neurogenic niches, it is reasonable to infer that neurogenesis in SVZ and SGZ might be recruited differently and consequently exerts distinct or complementary roles on brain functions [77].

In the SVZ, neurogenesis is regulated by the olfactory experience of the animals [83, 84]. Odor exposure can increase the survival of newborn neurons and improve memory in a learned odor discrimination task, suggesting that neurogenesis in the olfactory bulb is recruited during learning and memory processes related to olfactory stimulation [85]. However, due to the relevance of the hippocampus for several brain functions and its implication on the genesis of neuropsychiatric disorders, much closer attention has been paid to SGZ than SVZ neurogenesis [86, 87].

Hippocampal neurogenesis is proposed to be important for at least some forms of learning and memory. Positive associations between them have been replicated by independent groups in rodents and humans [88–90]. For example, voluntary running and exposure to enriched environments improve learning and memory process with a concomitant increase in cell proliferation and survival of new DG generated neurons [82, 91, 92].

In addition, decreased adult hippocampal neurogenesis has been associated with psychiatric disorders such as anxiety, schizophrenia, and mood disorders. Stressful experiences that can precipitate symptoms of anxiety and mood disorders downregulate hippocampal neurogenesis [33, 93, 94]. Snyder et al. [95] showed that impaired SGZ, but not SVZ, neurogenic capacity facilitates stress-induced depressive-like symptoms and disrupt the essential negative feedback of hippocampus in hypothalamic-pituitary-adrenal (HPA) axis [95]. Adult hippocampal neurogenesis has also been implicated in the mechanism of pattern separation [96, 97]). Pattern separation is a complex concept that involves CA3 region as an associative network between a spatial location and a situation or an object that allows completion of memory during recall [98]. It has been hypothesized that this event is highly regulated by new neurons formed in the DG. In addition, several authors have demonstrated that neurogenesis is relevant for the perception of an event as stressful or not [99]. In the light of psychiatric conditions that involve an initial exposure to a traumatic event, such as posttraumatic stress disorder, the intact capacity of DG to produce new neurons has been associated with a poor ability of fear discrimination and overgeneralization (Besnard and Sahay 2015).

Of note, drugs used in the clinical practice for the treatment of psychiatric disorders, such as antidepressants or lithium, normalize or even facilitate hippocampal neurogenesis [94, 100]. Moreover, compounds with therapeutic potential for psychiatric conditions, such as cannabinoids, also impacts positively in adult hippocampal neurogenesis [33, 101].

6.4.1 Cannabinoids and Adult Neurogenesis

Several independent groups around the world have demonstrated the importance of the ECB system in the modulation of different steps required for neurogenesis: cell proliferation, differentiation, maturation, and survival (Fig. 6.3, [37, 86]). Indeed, activation of CB receptors regulates intracellular pathways involved in cell proliferation, differentiation, survival, and the integration of new cells in already established circuitries, such as the MEK/ERK/CREB and PI3K/Akt/mTOR and BDNF production [14, 37]. Also, voluntary exercise seems to increase adult hippocampal

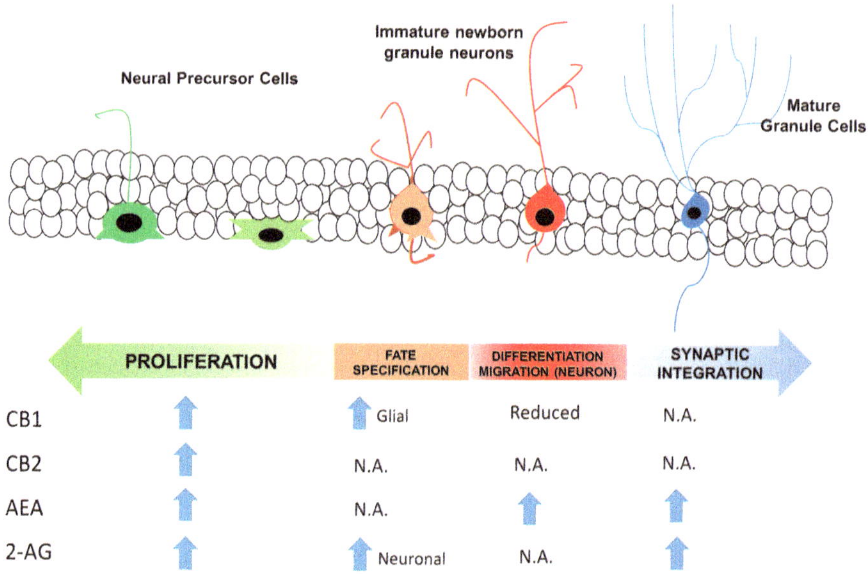

Fig. 6.3. Complex modulation of the endocannabinoid system during the process of adult hippocampal neurogenesis. *Blue arrows* facilitation of the formation of new cells/new neurons, *N.A* data not available or inconclusive. Based on in vivo studies

neurogenesis through a facilitation of CB_1-mediated neurotransmission. Finally, a positive association between cannabinoid-induced neurogenesis and behavioral improvement has been observed in animal models of anxiety, psychosis, depression, and memory impairment (as further discussed in item 1.5 of this chapter). Chronic (10 days), but not the acute administration of HU-210, a synthetic cannabinoid, induces neurogenesis in mice. A very similar picture is found after repeated administration of WIN55,212-2, a CB_1/CB_2 agonist [27, 32, 101, 102].

The two main compounds of the plant *Cannabis sativa*, THC and CBD, also affect adult hippocampal neurogenesis. Repeated treatment with CBD for 15-days prevented β-amyloid-induced neurotoxicity via activation of the proliferator-activated receptor-γ (PAAR-γ), suggesting a mechanism for CBD neuroprotective effects [103]. Wolf et al. [30] suggested that chronic treatment with CBD (42 days) decreases cell proliferation but stimulates cell survival. These responses were mediated by CB_1 receptors, as CBD effects were absent in CB_1 receptor knockout mice. Also, repeated CBD (30 mg/kg) treatment for 14 days prevented a stress-induced decrease in cell survival and differentiation in mice. In non-stressed mice, CBD increased the number of double-labeled BrdU/NeuN cells in the dentate gyrus [33]. These results were associated with increased levels of AEA, but not 2-AG, in the hippocampus of mice treated with CBD [33]. On the other hand, THC, a partial CB_1 receptor agonist, decreased proliferation and, at the same time, spatial memory [30].

The participation of ECB in the modulation of neurogenesis has also been investigated. For example, hippocampal cell proliferation is increased in FAAH deficient mice and in animals treated with URB597, an FAAH inhibitor [27]. On the other hand, the ECB uptake inhibitor, AM404, reversed the trimethylthiazoline (TMT)-induced decrease of neurogenesis [104]. Finally, the genetic ablation of the enzyme responsible for 2-AG synthesis reduced cell proliferation, the number of doublecortin (a neuroblast marker) positive cells, and decreased the survival of newborn cells in the DG [31, 105].

The facilitation of CB_2 signaling also influences adult neurogenesis. Repeated administration of HU-308, a CB_2 receptor agonist, during 5 days, induces neural precursor cells proliferation in the DG. This effect seems to recruit Akt/mTORC1 pathway [36]. In the opposite way, the administration of CB_2 inverse agonist (JTE907) or antagonists (SR144528 or AM630) reduces cell proliferation and the number of BrdU labeled cells in the SVZ and SGZ [10, 32, 36]. The involvement of CB_2 receptors in these results was confirmed by the failure of a CB_2 agonist to induce any change in neurogenesis in animals deficient for this receptor [10, 36].

In the case of studies using pharmacological and genetic regulation of CB_1 receptors, the results are controversial. CB_1-deficient mice exhibit low rates of proliferation, astrogliogenesis, and neurogenesis in the DG and SVZ [27]. Also, repeated administration of the CB_1 antagonists/inverse agonists, SR141716A, and AM251, decreased neurogenesis in some studies [106]. Other groups, nevertheless, suggested that these drugs facilitate neurogenesis [30, 104, 107]. Interestingly, the effects of some of these cannabinergic drugs were preserved in CB_1 but not in $TRPV_1$-deficient mice [107]. These discrepancies may be related to the use of different animal species, strain or gender, cannabinergic drugs, and doses employed. In addition, contradictory results may be the consequence of different BrdU-administration schedule, and time-point of analysis, which may induce alternative interpretations. For example, Wolf et al. [30] found increased cell proliferation 1 and 24 h after treatment with AM251, but a decrease in cell maturation 48 h and 7 days later.

6.4.2 *Neurogenesis, Cannabinoids, and Neuropsychiatric/Neurodegenerative Disorders: What's the Correlation?*

Considering that the ECB system modulates adult neurogenesis and that this process is impaired in neuropsychiatric and neurodegenerative disorders, it is plausible that cannabinoids may induce beneficial or detrimental effects in the brain and influence behavior by controlling newly generated neuron-induced plasticity. Cannabinoids are effective in modulating neurogenesis in various animal models of depression, anxiety disorders, Alzheimer's disease, and cerebral ischemia. Some of these studies are not only based on associative results, but suggest causality, once the direct ablation of hippocampal neurogenesis by different methods prevented the therapeutic effects induced by distinct cannabinoids tested.

Acute treatment with AM404, an ECB uptake inhibitor, reversed the trimethylthiazoline-induced decrease of hippocampal cell proliferation and promoted anxiolytic-like effect [104]. In the same sense, sub-chronic treatment with the CB_1/CB_2 agonist HU210 induced anxiolytic- and antidepressant-like effects accompanied by an increase in neurogenesis [101]. Although a controversial finding, authors suggested that neurogenesis ablation through hippocampal X-ray irradiation prevented HU210-induced behavioral responses [101]. In agreement, repeated injections of CBD reversed the anxiogenic-like responses and the neurogenesis impairment produced by chronic stress in a CB1-dependent manner [33]. These effects were completely lost after ganciclovir administration to transgenic mice that express thymidine kinase under the control of the GFAP promoter, a method used to ablate only adult dividing precursor cells. In accordance, a recent study showed that the enhancement of 2-AG-induced neurotransmission by the MAGL inhibitor, JZL184, also prevented the anxiogenic- and pro-depressive-like effects, as well as the decrease in neurogenesis, induced by chronic stress [108]. Strengthening this hypothesis, the antidepressant-like effect produced by a single injection of the CB_1 antagonist SR141716A was lost after sub-chronic administration of the drug, probably due to the reduction in neurogenesis observed in these animals [106].

Several studies in the literature show that (1) neurogenesis is altered in some neurodegenerative diseases, and (2) cannabinoids can improve behavioral responses, as memory impairment, and brain damage, in animal models of these disorders. For example, Esposito et al. [103] showed that chronic administration of CBD in rats that previously received β-amyloid injection in the hippocampus, an animal model of Alzheimer's disease, decreases reactive gliosis, neuronal damage and facilitates adult hippocampal neurogenesis through PPAR-γ receptors. Also, cannabinoids can ameliorate age-related reduction in neurogenesis, suggesting that these compounds could replenish damaged/death neurons during neurodegeneration [32, 102]. In the middle cerebral artery occlusion rat model, widely used to evaluate cerebral ischemic injury, daily injections of oleoylethanolamide, a monounsaturated analog of anandamide, improved the spatial cognitive impairment concomitant to an increase in BDNF and hippocampal neurogenesis [109]. Also, CB_2 receptor regulation counteracts alcohol-induced decline in neurogenesis [110].

Taken together, these pieces of evidence suggest that cannabinoids could exert anxiolytic- and antidepressant-like effects as well as neuroprotection through an enhancement of adult neurogenesis. New studies using cannabinergic drugs that modulate the ECB tone in long-term studies of animal models of mood, cognitive, or neurodegenerative disorders are urgently needed to clarify these important aspects.

6.5 Conclusions and Perspectives

In this chapter, we have presented evidence indicating that cannabinoids exert an important neurodevelopmental regulatory role on and mediate plastic events in the adult brain (Figs. 6.2 and 6.3). Important unanswered questions, however, remain.

For example, is the modulation of neurogenesis by endocannabinoid signaling always positive, or can it be deleterious in some pathological conditions? What are the precise mechanisms by which cannabinoid regulate neurogenesis, neurodevelopment, and cell fate? What is the role of non-cannabinoid mediated mechanisms (e.g., TRPV1, GPR55, PPAR-γ receptors) in cannabinoid modulation of neurogenesis? What intracellular pathways are involved? These open questions indicate that we are only at the beginning of our journey. However, the results so far clearly support the perspective that new knowledge in this area could bring important contributions to the therapy of neuropsychiatric and neurodevelopmental disorders.

Acknowledgements We would like to thank Franciele Scarante for her technical support on design and graphic art of Figs. 6.1 and 6.3, and members of our research groups for an inspiring scientific environment. ACC and FSG are recipients of FAPESP grants. IGR research is funded by PI15-00310, RTC-2015-3364-1, and S2011-BMD-2336 supported by the Instituto de Salud Carlos III, Mineco (Plan Estatal de I+D+i 2013-2016) and Comunidad de Madrid. Research was cofinanced by the European Development Regional Fund "A way to achieve Europe" (ERDF). JPL is a recipient of FPU (Ministerio de Educación) fellowship. MVF is recipient of a CAPES fellowship.

References

1. Pertwee RG (2005) Pharmacological actions of cannabinoids. Handb Exp Pharmacol 168:1–51
2. Mechoulam R, Lander N, Varkony TH, Kimmel I, Becker O, Ben-Zvi Z et al (1980) Stereochemical requirements for cannabinoid activity. J Med Chem 23(10):1068–1072
3. Devane WA, Dysarz FA 3rd, Johnson MR, Melvin LS, Howlett AC (1988) Determination and characterization of a cannabinoid receptor in rat brain. Mol Pharmacol 34(5):605–613
4. Matsuda LA, Lolait SJ, Brownstein MJ, Young AC, Bonner TI (1990) Structure of a cannabinoid receptor and functional expression of the cloned cDNA. Nature 346(6284):561–564
5. Freund TF, Katona I, Piomelli D (2003) Role of endogenous cannabinoids in synaptic signaling. Physiol Rev 83:1017–1066
6. Katona I, Freund TF (2008) Endocannabinoid signaling as a synaptic circuit breaker in neurological disease. Nat Med 14(9):923–930
7. Munro S, Thomas KL, Abu-Shaar M (1993) Molecular characterization of a peripheral receptor for cannabinoids. Nature (London) 365:61–65
8. Lisboa SF, Gomes FV, Guimaraes FS, Campos AC (2016) Microglial cells as a link between cannabinoids and the immune hypothesis of psychiatric disorders. Front Neurol 7:5
9. Onaivi ES, Ishiguro H, Gong JP, Patel S, Perchuk A, Meozzi PA et al (2006) Discovery of the presence and functional expression of cannabinoid CB2 receptors in brain. Ann N Y Acad Sci 1074:514–536
10. Palazuelos J, Aguado T, Egia A, Mechoulam R, Guzman M, Galve-Roperh I (2006) Non-psychoactive CB2 cannabinoid agonists stimulate neural progenitor proliferation. FASEB J 20(13):2405–2407
11. Howlett AC (2002) The cannabinoid receptors. Prostaglandins Other Lipid Mediat 68–69:619–631
12. Devane WA, Hanus L, Breuer A, Pertwee RG, Stevenson LA, Griffin G et al (1992) Isolation and structure of a brain constituent that binds to the cannabinoid receptor. Science 258(5090):1946–1949

13. Mechoulam R, Ben-Shabat S, Hanus L, Ligumsky M, Kaminski NE, Schatz AR et al (1995) Identification of an endogenous 2-monoglyceride, present in canine gut, that binds to cannabinoid receptors. Biochem Pharmacol 50(1):83–90
14. Maccarrone M, Guzman M, Mackie K, Doherty P, Harkany T (2014) Programming of neural cells by (endo)cannabinoids: from physiological rules to emerging therapies. Nat Rev Neurosci 15(12):786–801
15. Szabo B, Schlicker E (2005) Effects of cannabinoids on neurotransmission. Handb Exp Pharmacol 168:327–365
16. Diana MA, Marty A (2004) Endocannabinoid-mediated short-term synaptic plasticity: depolarization-induced suppression of inhibition (DSI) and depolarization-induced suppression of excitation (DSE). Br J Pharmacol 142(1):9–19
17. Wilson RI, Nicoll RA (2001) Endogenous cannabinoids mediate retrograde signalling at hippocampal synapses. Nature 410(6828):588–592
18. Yoshida T, Hashimoto K, Zimmer A, Maejima T, Araishi K, Kano M (2002) The cannabinoid CB1 receptor mediates retrograde signals for depolarization-induced suppression of inhibition in cerebellar Purkinje cells. J Neurosci 22(5):1690–1697
19. Marsicano G, Goodenough S, Monory K, Hermann H, Eder M, Cannich A et al (2003) CB1 cannabinoid receptors and on-demand defense against excitotoxicity. Science 302(5642):84–88
20. Galve-Roperh I, Chiurchiu V, Diaz-Alonso J, Bari M, Guzman M, Maccarrone M (2013) Cannabinoid receptor signaling in progenitor/stem cell proliferation and differentiation. Prog Lipid Res 52(4):633–650
21. Rubino T, Parolaro D (2016) The impact of exposure to cannabinoids in adolescence: insights from animal models. Biol Psychiatry 79(7):578–585
22. Begbie J, Doherty P, Graham A (2004) Cannabinoid receptor, CB1, expression follows neuronal differentiation in the early chick embryo. J Anat 205(3):213–218
23. Romero J, Garcia-Palomero E, Berrendero F, Garcia-Gil L, Hernandez ML, Ramos JA et al (1997) Atypical location of cannabinoid receptors in white matter areas during rat brain development. Synapse 26(3):317–323
24. Maison P, Walker DJ, Walsh FS, Williams G, Doherty P (2009) BDNF regulates neuronal sensitivity to endocannabinoids. Neurosci Lett 467(2):90–94
25. Eggan SM, Lazarus MS, Stoyak SR, Volk DW, Glausier JR, Huang ZJ et al (2012) Cortical glutamic acid decarboxylase 67 deficiency results in lower cannabinoid 1 receptor messenger RNA expression: implications for schizophrenia. Biol Psychiatry 71(2):114–119
26. Blazquez C, Chiarlone A, Sagredo O, Aguado T, Pazos MR, Resel E et al (2011) Loss of striatal type 1 cannabinoid receptors is a key pathogenic factor in Huntington's disease. Brain 134(Pt 1):119–136
27. Aguado T, Monory K, Palazuelos J, Stella N, Cravatt B, Lutz B et al (2005) The endocannabinoid system drives neural progenitor proliferation. FASEB J 19(12):1704–1706
28. Butti E, Bacigaluppi M, Rossi S, Cambiaghi M, Bari M, Cebrian Silla A et al (2012) Subventricular zone neural progenitors protect striatal neurons from glutamatergic excitotoxicity. Brain 135(Pt 11):3320–3335
29. Rubio-Araiz A, Arevalo-Martin A, Gomez-Torres O, Navarro-Galve B, Garcia-Ovejero D, Suetterlin P et al (2008) The endocannabinoid system modulates a transient TNF pathway that induces neural stem cell proliferation. Mol Cell Neurosci 38(3):374–380
30. Wolf SA, Bick-Sander A, Fabel K, Leal-Galicia P, Tauber S, Ramirez-Rodriguez G et al (2010) Cannabinoid receptor CB1 mediates baseline and activity-induced survival of new neurons in adult hippocampal neurogenesis. Cell Commun Signal 8:12
31. Gao Y, Vasilyev DV, Goncalves MB, Howell FV, Hobbs C, Reisenberg M et al (2010) Loss of retrograde endocannabinoid signaling and reduced adult neurogenesis in diacylglycerol lipase knock-out mice. J Neurosci 30(6):2017–2024
32. Goncalves MB, Suetterlin P, Yip P, Molina-Holgado F, Walker DJ, Oudin MJ et al (2008) A diacylglycerol lipase-CB2 cannabinoid pathway regulates adult subventricular zone neurogenesis in an age-dependent manner. Mol Cell Neurosci 38(4):526–536

33. Campos AC, Ortega Z, Palazuelos J, Fogaca MV, Aguiar DC, Diaz-Alonso J et al (2013) The anxiolytic effect of cannabidiol on chronically stressed mice depends on hippocampal neurogenesis: involvement of the endocannabinoid system. Int J Neuropsychopharmacol 16(6):1407–1419
34. Walker DJ, Suetterlin P, Reisenberg M, Williams G, Doherty P (2010) Down-regulation of diacylglycerol lipase-alpha during neural stem cell differentiation: identification of elements that regulate transcription. J Neurosci Res 88(4):735–745
35. Jung KM, Astarita G, Thongkham D, Piomelli D (2011) Diacylglycerol lipase-alpha and -beta control neurite outgrowth in neuro-2a cells through distinct molecular mechanisms. Mol Pharmacol 80(1):60–67
36. Palazuelos J, Ortega Z, Diaz-Alonso J, Guzman M, Galve-Roperh I (2012) CB2 cannabinoid receptors promote neural progenitor cell proliferation via mTORC1 signaling. J Biol Chem 287(2):1198–1209
37. Prenderville JA, Kelly AM, Downer EJ (2015) The role of cannabinoids in adult neurogenesis. Br J Pharmacol 172(16):3950–3963
38. Díaz-Alonso J, Aguado T, de Salas-Quiroga A, Ortega Z, Guzman M, Galve-Roperh I (2015) CB1 cannabinoid receptor-dependent activation of mTORC1/Pax6 signaling drives Tbr2 expression and basal progenitor expansion in the developing mouse cortex. Cereb Cortex 25(9):2395–2408
39. Díaz-Alonso J, Aguado T, Wu CS, Palazuelos J, Hofmann C, Garcez P, et al. The CB(1) cannabinoid receptor drives corticospinal motor neuron differentiation through the Ctip2/Satb2 transcriptional regulation axis. J Neurosci 2012a;32(47):16651–65
40. Bromberg KD, Ma'ayan A, Neves SR, Iyengar R (2008) Design logic of a cannabinoid receptor signaling network that triggers neurite outgrowth. Science 320(5878):903–909
41. Díaz-Alonso J, Guzman M, Galve-Roperh I. Endocannabinoids via CB(1) receptors act as neurogenic niche cues during cortical development. Philos Trans R Soc Lond B Biol Sci 2012b;367(1607):3229-41
42. Greig LC, Woodworth MB, Galazo MJ, Padmanabhan H, Macklis JD (2013) Molecular logic of neocortical projection neuron specification, development and diversity. Nat Rev Neurosci 14(11):755–769
43. Berghuis P, Rajnicek AM, Morozov YM, Ross RA, Mulder J, Urban GM et al (2007) Hardwiring the brain: endocannabinoids shape neuronal connectivity. Science 316(5828):1212–1216
44. Keimpema E, Barabas K, Morozov YM, Tortoriello G, Torii M, Cameron G et al (2010) Differential subcellular recruitment of monoacylglycerol lipase generates spatial specificity of 2-arachidonoyl glycerol signaling during axonal pathfinding. J Neurosci 30(42): 13992–14007
45. Argaw A, Duff G, Zabouri N, Cecyre B, Chaine N, Cherif H et al (2011) Concerted action of CB1 cannabinoid receptor and deleted in colorectal cancer in axon guidance. J Neurosci 31(4):1489–1499
46. Njoo C, Agarwal N, Lutz B, Kuner R (2015) The cannabinoid receptor CB1 interacts with the WAVE1 complex and plays a role in actin dynamics and structural plasticity in neurons. PLoS Biol 13(10):e1002286
47. Roland AB, Ricobaraza A, Carrel D, Jordan BM, Rico F, Simon A et al (2014) Cannabinoid-induced actomyosin contractility shapes neuronal morphology and growth. Elife 3:e03159
48. Mulder J, Aguado T, Keimpema E, Barabas K, Ballester Rosado CJ, Nguyen L et al (2008) Endocannabinoid signaling controls pyramidal cell specification and long-range axon patterning. Proc Natl Acad Sci U S A 105(25):8760–8765
49. Wu CS, Zhu J, Wager-Miller J, Wang S, O'Leary D, Monory K et al (2010) Requirement of cannabinoid CB(1) receptors in cortical pyramidal neurons for appropriate development of corticothalamic and thalamocortical projections. Eur J Neurosci 32(5):693–706
50. Aguado T, Palazuelos J, Monory K, Stella N, Cravatt B, Lutz B et al (2006) The endocannabinoid system promotes astroglial differentiation by acting on neural progenitor cells. J Neurosci 26(5):1551–1561

51. Gomez O, Sanchez-Rodriguez A, Le M, Sanchez-Caro C, Molina-Holgado F, Molina-Holgado E (2011) Cannabinoid receptor agonists modulate oligodendrocyte differentiation by activating PI3K/Akt and the mammalian target of rapamycin (mTOR) pathways. Br J Pharmacol 163(7):1520–1532
52. Arevalo-Martin A, Garcia-Ovejero D, Molina-Holgado E (2010) The endocannabinoid 2-arachidonoylglycerol reduces lesion expansion and white matter damage after spinal cord injury. Neurobiol Dis 38(2):304–312
53. Tomas-Roig J, Wirths O, Salinas-Riester G, Havemann-Reinecke U (2016) The cannabinoid CB1/CB2 agonist WIN55212.2 promotes oligodendrocyte differentiation in vitro and neuroprotection during the cuprizone-induced central nervous system demyelination. CNS Neurosci Ther 22(5):387–395
54. Alpár A, Tortoriello G, Calvigioni D, Niphakis MJ, Milenkovic I, Bakker J et al (2014) Endocannabinoids modulate cortical development by configuring Slit2/Robo1 signalling. Nat Commun 5:4421
55. Di Marzo V, Stella N, Zimmer A (2015) Endocannabinoid signalling and the deteriorating brain. Nat Rev Neurosci 16(1):30–42
56. Psychoyos D, Hungund B, Cooper T, Finnell RH (2008) A cannabinoid analogue of Delta9-tetrahydrocannabinol disrupts neural development in chick. Birth Defects Res B Dev Reprod Toxicol 83(5):477–488
57. Saez TM, Aronne MP, Caltana L, Brusco AH (2014) Prenatal exposure to the CB1 and CB2 cannabinoid receptor agonist WIN 55,212-2 alters migration of early-born glutamatergic neurons and GABAergic interneurons in the rat cerebral cortex. J Neurochem 129(4):637–648
58. Soltesz I, Alger BE, Kano M, Lee SH, Lovinger DM, Ohno-Shosaku T et al (2015) Weeding out bad waves: towards selective cannabinoid circuit control in epilepsy. Nat Rev Neurosci 16(5):264–277
59. de Salas-Quiroga A, Diaz-Alonso J, Garcia-Rincon D, Remmers F, Vega D, Gomez-Canas M et al (2015) Prenatal exposure to cannabinoids evokes long-lasting functional alterations by targeting CB1 receptors on developing cortical neurons. Proc Natl Acad Sci U S A 112(44):13693–13698
60. Ruehle S, Remmers F, Romo-Parra H, Massa F, Wickert M, Wortge S et al (2013) Cannabinoid CB1 receptor in dorsal telencephalic glutamatergic neurons: distinctive sufficiency for hippocampus-dependent and amygdala-dependent synaptic and behavioral functions. J Neurosci 33(25):10264–10277
61. Tortoriello G, Morris CV, Alpar A, Fuzik J, Shirran SL, Calvigioni D et al (2014) Miswiring the brain: Delta9-tetrahydrocannabinol disrupts cortical development by inducing an SCG10/stathmin-2 degradation pathway. EMBO J 33(7):668–685
62. Oudin MJ, Gajendra S, Williams G, Hobbs C, Lalli G, Doherty P (2011) Endocannabinoids regulate the migration of subventricular zone-derived neuroblasts in the postnatal brain. J Neurosci 31(11):4000–4011
63. Vargish GA, Pelkey KA, Yuan X, Chittajallu R, Collins D, Fang C et al (2016) Persistent inhibitory circuit defects and disrupted social behaviour following in utero exogenous cannabinoid exposure. Mol Psychiatry
64. Földy C, Malenka RC, Sudhof TC (2013) Autism-associated neuroligin-3 mutations commonly disrupt tonic endocannabinoid signaling. Neuron 78(3):498–509
65. Cass DK, Flores-Barrera E, Thomases DR, Vital WF, Caballero A, Tseng KY (2014) CB1 cannabinoid receptor stimulation during adolescence impairs the maturation of GABA function in the adult rat prefrontal cortex. Mol Psychiatry 19(5):536–543
66. Busquets-Garcia A, Gomis-Gonzalez M, Guegan T, Agustin-Pavon C, Pastor A, Mato S et al (2013) Targeting the endocannabinoid system in the treatment of fragile X syndrome. Nat Med 19(5):603–607
67. Volk DW, Lewis DA (2016) The role of endocannabinoid signaling in cortical inhibitory neuron dysfunction in schizophrenia. Biol Psychiatry 79(7):595–603
68. Allen E (1912) The cessation of mitosis in the central nervous system of the albino rat. J Comp Neurol 19:547–568

69. Gross CG (2000) Neurogenesis in the adult brain: death of a dogma. Nat Rev Neurosci 1:67–73
70. Altman J (1963) Autoradiographic investigation of cell proliferation in the brains of rats and cats. Anat Rec 145:573–591
71. Altman J, Das GD (1965) Autoradiographic and histological evidence of postnatal hippocampal neurogenesis in rats. J Comp Neurol 124(3):319–335
72. Kaplan MS (1981) Neurogenesis in the 3-month-old rat visual cortex. J Comp Neurol 195(2):323–338
73. Kaplan MS (1983) Proliferation of subependymal cells in the adult primate CNS: differential uptake of DNA labelled precursors. J Hirnforsch 24(1):23–33
74. Kaplan MS (2001) Environment complexity stimulates visual cortex neurogenesis: death of a dogma and a research career. Trends Neurosci 24(10):617–620
75. Eriksson PS, Perfilieva E, BjörkEriksson T, Alborn AM, Nordborg C, Peterson DA, Gage FH (1998) Neurogenesis in the adult human hippocampus. Nat Med 4(11):1313–1317
76. Deng W, Aimone JB, Gage FH (2010) New neurons and new memories: how does adult hippocampal neurogenesis affect learning and memory? Nat Rev Neurosci 11(5):339–350
77. Kempermann G (2008) The neurogenic reserve hypothesis: what is adult hippocampal neurogenesis good for? Trends Neurosci 31(4):163–169
78. Suh H, Deng W, Gage FH (2009) Signaling in adult neurogenesis. Annu Rev Cell Dev Biol 25:253–275
79. Chaker Z, George C, Petrovska M, Caron JB, Lacube P, Caille I et al (2016) Hypothalamic neurogenesis persists in the aging brain and is controlled by energy-sensing IGF-I pathway. Neurobiol Aging 41:64–72
80. Kempermann G, Jessberger S, Steiner B, Kronenberg G (2004) Milestones of neuronal development in the adult hippocampus. Trends Neurosci 27(8):447–452
81. Gage FH (2000) Mammalian neural stem cells. Science 287:1433–1438
82. Van Praag H, Kempermann G, Gage FH (1999) Running increases cell proliferation and neurogenesis in the adult mouse dentate gyrus. Nat Neurosci 2:266–270
83. Lledo PM, Alonso M, Grubb MS (2006) Adult neurogenesis and functional plasticity in neuronal circuits. Nat Rev Neurosci 7(3):179–193
84. Lledo PM, Saghatelyan A (2005) Integrating new neurons into the adult olfactory bulb: joining the network, life-death decisions, and the effects of sensory experience. Trends Neurosci 28(5):248–254
85. Alonso M, Viollet C, Gabellec MM, Meas-Yedid V, Olivo-Marin JC, Lledo PM (2006) Olfactory discrimination learning increases the survival of adult-born neurons in the olfactory bulb. J Neurosci 26(41):10508–10513
86. Fogaça MV, Galve-Roperh I, Guimaraes FS, Campos AC (2013) Cannabinoids, neurogenesis and antidepressant drugs: is there a link? Curr Neuropharmacol 11(3):263–275
87. Opendak M, Gould E (2015) Adult neurogenesis: a substrate for experience-dependent change. Trends Cogn Sci 19(3):151–161
88. Coras R, Siebzehnrubl FA, Pauli E, Huttner HB, Njunting M, Kobow K et al (2010) Low proliferation and differentiation capacities of adult hippocampal stem cells correlate with memory dysfunction in humans. Brain 133(11):3359–3372
89. Gould E, Beylin A, Tanapat P, Reeves A, Shors TJ (1999) Learning enhances adult neurogenesis in the hippocampal formation. Nat Neurosci 2(3):260–265
90. Kempermann G, Gage FH (2002) Genetic determinants of adult hippocampal neurogenesis correlate with acquisition, but not probe trial performance, in the water maze task. Eur J Neurosci 16(1):129–136
91. Kee NJ, Preston E, Wojtowicz JM (2001) Enhanced neurogenesis after transient global ischemia in the dentate gyrus of the rat. Exp Brain Res 136(3):313–320
92. Tashiro A, Makino H, Gage FH (2007) Experience-specific functional modification of the dentate gyrus through adult neurogenesis: a critical period during an immature stage. J Neurosci 27(12):3252–3259
93. Czéh B, Michaelis T, Watanabe T, Frahm J, de Biurrun G, van Kampen M et al (2001) Stress-induced changes in cerebral metabolites, hippocampal volume, and cell proliferation

are prevented by antidepressant treatment with tianeptine. Proc Natl Acad Sci U S A 98(22):12796–12801
94. Santarelli L, Saxe M, Gross C, Surget A, Battaglia F, Dulawa S et al (2003) Requirement of hippocampal neurogenesis for the behavioral effects of antidepressants. Science 301(5634):805–809
95. Snyder JS, Soumier A, Brewer M, Pickel J, Cameron HA (2011) Adult hippocampal neurogenesis buffers stress responses and depressive behaviour. Nature 476(7361):458–461
96. Besnard A, Sahay A (2016) Adult hippocampal neurogenesis, fear generalization, and stress. Neuropsychopharmacology 41(1):24–44
97. Sahay A, Wilson DA, Hen R (2011) Pattern separation: a common function for new neurons in hippocampus and olfactory bulb. Neuron 70(4):582–588
98. Rolls ET (2013) The mechanisms for pattern completion and pattern separation in the hippocampus. Front Syst Neurosci 7:74
99. Egeland M, Zunszain PA, Pariante CM (2015) Molecular mechanisms in the regulation of adult neurogenesis during stress. Nat Rev Neurosci 16(4):189–200
100. David DJ, Samuels BA, Rainer Q, Wang JW, Marsteller D, Mendez I et al (2009) Neurogenesis-dependent and -independent effects of fluoxetine in an animal model of anxiety/depression. Neuron 62(4):479–493
101. Jiang W, Zhang Y, Xiao L, Van Cleemput J, Ji SP, Bai G et al (2005) Cannabinoids promote embryonic and adult hippocampus neurogenesis and produce anxiolytic- and antidepressant-like effects. J Clin Invest 115(11):3104–3116
102. Marchalant Y, Brothers HM, Wenk GL (2009) Cannabinoid agonist WIN-55,212-2 partially restores neurogenesis in the aged rat brain. Mol Psychiatry 14(12):1068–1069
103. Esposito G, Scuderi C, Valenza M, Togna GI, Latina V, De Filippis D et al (2011) Cannabidiol reduces Abeta-induced neuroinflammation and promotes hippocampal neurogenesis through PPARgamma involvement. PLoS One 6(12):e28668
104. Hill MN, Kambo JS, Sun JC, Gorzalka BB, Galea LA (2006) Endocannabinoids modulate stress-induced suppression of hippocampal cell proliferation and activation of defensive behaviours. Eur J Neurosci 24(7):1845–1849
105. Jenniches I, Ternes S, Albayram O, Otte DM, Bach K, Bindila L et al (2016) Anxiety, stress, and fear response in mice with reduced endocannabinoid levels. Biol Psychiatry 79(10): 858–868
106. Lee S, Kim DH, Yoon SH, Ryu JH (2009) Sub-chronic administration of rimonabant causes loss of antidepressive activity and decreases doublecortin immunoreactivity in the mouse hippocampus. Neurosci Lett 467(2):111–116
107. Jin K, Xie L, Kim SH, Parmentier-Batteur S, Sun Y, Mao XO et al (2004) Defective adult neurogenesis in CB1 cannabinoid receptor knockout mice. Mol Pharmacol 66(2):204–208
108. Zhang Z, Wang W, Zhong P, Liu SJ, Long JZ, Zhao L et al (2015) Blockade of 2-arachidonoylglycerol hydrolysis produces antidepressant-like effects and enhances adult hippocampal neurogenesis and synaptic plasticity. Hippocampus 25(1):16–26
109. Yang LC, Guo H, Zhou H, Suo DQ, Li WJ, Zhou Y et al (2015) Chronic oleoylethanolamide treatment improves spatial cognitive deficits through enhancing hippocampal neurogenesis after transient focal cerebral ischemia. Biochem Pharmacol 94(4):270–281
110. Rivera P, Blanco E, Bindila L, Alen F, Vargas A, Rubio L et al (2015) Pharmacological activation of CB2 receptors counteracts the deleterious effect of ethanol on cell proliferation in the main neurogenic zones of the adult rat brain. Front Cell Neurosci 9:379

Chapter 7
Ceramide-1-Phosphate and Its Role in Trafficking of Normal Stem Cells and Cancer Metastasis

Gabriela Schneider and Mariusz Z. Ratajczak

Abbreviations

BM	Bone marrow
C1P	Ceramide-1-phosphate
CERK	Ceramide kinase
EPC	Endothelial progenitor cells
FGF-2	Fibroblast growth factor-2
HSPC	Hematopoietic stem/progenitor cell
LPA	Lysophosphatidic acid
LPC	Lysophosphatidylocholine
LPP	Lipid phosphate phosphatase
MAPK	Mitogen activated protein kinase
MCP-1	Macrophage chemoattractant protein-1
MMP	Metalloproteinase
MSC	Mesenchymal stem cell
mTOR1	Mammalian target of rapamycin 1
NF-κB	Nuclear factor kappa B
PA	Phosphatidic acid
PAH	Pulmonary artery hypertension

G. Schneider (✉)
Stem Cell Institute at the James Graham Brown Cancer Center, University of Louisville, 500 S. Floyd Street, Rm. 103, Louisville, KY 40202, USA
e-mail: gabriela.schneider@louisville.edu

M.Z. Ratajczak
Stem Cell Institute at the James Graham Brown Cancer Center, University of Louisville, 500 S. Floyd Street, Rm. 103, Louisville, KY 40202, USA

Department of Regenerative Medicine, Warsaw Medical University, Warsaw, Poland

A. Pébay, R.C.B. Wong (eds.), *Lipidomics of Stem Cells*, Stem Cell Biology and Regenerative Medicine, DOI 10.1007/978-3-319-49343-5_7

PGE_2	Prostaglandin E2
PI3-K	Phosphatidylinositol 3-kinase
PLD	Phospholipase D
S1P	Sphingosine-1-phosphate
SDF-1	α-Chemokine stromal-derived factor 1
SMase	Sphingomyelinase
VSEL	Very small embryonic-like stem cell

7.1 Introduction

Ceramide-1-phosphate (C1P) belongs to the sphingolipids which are important components of cell membrane, and some of them are playing also critical function in regulation of key cell processes. In particular, sphingosine-1-phosphate (S1P) and ceramide-1-phosphate (C1P) were shown to be involved in regulation of cell proliferation [1–5], apoptosis [6, 7], survival [8, 9], cell migration [10–12], embryonic development [13], or inflammation [14]. Unlike S1P, which has been shown to be secreted from normal activated cells as an extracellular signaling molecule, C1P is mainly released from damaged cells and tissues due to irradiation, toxic effect of chemotherapy, myocardial infarction, or ischemia [11, 12, 15, 16].

C1P is a direct metabolite of ceramide, and its biosynthesis occurs in the Golgi apparatus from which C1P can be transported to plasma membrane and probably other organelles mainly by specific ceramide phosphate transfer protein [17]. So far only one enzyme, known as a ceramide kinase (CERK), has been identified to directly phosphorylate ceramide resulting in formation of C1P [18, 19]. Interestingly, mice with CERK knockout have only slightly decreased level of C1P in comparison to wild-type littermates, thus suggesting the presence of other pathways that leads to C1P synthesis [20, 21]. Transfer of fatty acyl chain to S1P or degradation of sphingomyelin by phospholipase D (PLD) could also result in C1P formation and thus become additional source of this bioactive lipid in cell [22]. However, so far no S1P acyl transferase has been identified in living organism [1, 21]. In contrast, sphingomyelinase D (SMase D) activity has been detected in the toxins of some bacteria and the venom of a variety of arthropods including spiders of the gender *Loxosceles* but so far there is no evidence that SMase D exist in mammalian cells [22].

Although in vitro studies indicate that SMase D can generate C1P from sphingomyelin [23], it can also hydrolyze lysophospholipids such as lysophosphatidylcholine (LPC), lysophosphatidylinositol, or lysophosphatidylglycerol resulting in generation of lysophosphatidic acid (LPA) [24, 25]. Moreover, recent studies indicate that SMase D from venom of spiders catalyze exclusively transphosphatidylation rather than hydrolic reaction thereby forming cyclic ceramide phosphate (1,2) instead of C1P or LPA [26]. Alternatively, similarly to S1P for which two sphingosine kinases were identified, we cannot exclude the presence of additional isoforms of ceramide kinases with overlapping activity that could lead to phosphorylation of ceramide. This hypothesis can be supported by some studies indicating the involvement of CERK-like kinase in murine retina development [27]. Recently, human CERK-like enzyme was also identified in retina [28]; however, in standard in vitro studies this enzyme was unable to phosphorylate ceramide, and therefore its role in cell biology remains unclear [29]. Specific C1P phosphatases or promiscuous lipid phosphate phosphatases (LPP1-3) dephosphorylate C1P to ceramide and thus participate in its degradation.

Unlike ceramide which is often pro-apoptotic, C1P has been reported to promote cell growth, survival, glucose uptake, and cell migration through unknown plasma membrane receptor/s that does not bind other sphingolipids including ceramides, S1P, or sphingomyelin [10]. Interestingly, however there are not identified C1P receptor/s yet, putative C1P receptor is pertussis toxin sensitive therefore most likely belongs to $G_{\alpha i}$ protein-coupled receptor family [10]. In macrophages, the C1P receptor seems to be of low affinity since relatively high concentration (5–20 μM) of C1P are needed to induce its activation [10, 30]. However, increase in intracellular calcium concentration in pulmonary endothelial cells was observed at much lower concentration of C1P (0.6 nM) [31], and since these changes were observed within few seconds post C1P stimulation of cells, it suggests that this effect is rather receptor mediated. Relatively lower concentration of C1P (0.5 μM) was also needed for induction of calcium mobilization and elevation in inositol (1,4,5)-triphosphate level in Jurkat cells [32]. Interestingly, such concentrations seem to be still within physiological range since in murine serum concentration of C1P up to 20 μM were observed [33]. In contrast, in serum of human individuals 0.5 μM or lower concentrations of C1P were detected and obtained values varied from ~0.2–~0.6 μM depending on methodology of sample processing [34]. Interestingly, the same authors observed a decrease in long chain C1P level (C_{26}-C1P) in response to fasting [34].

Unlike serum, in which mostly C_{18}-C1P and C_{26}-C1P are detected [34], the major intracellular form of C1P is C_{16}-C1P [35]. The intracellular level of C1P was reported to be ~ 2 pmol/10^6 cells [33, 36] but it can reach concentration of up to 45 pmol/10^6 in macrophages (C_{16}-C1P) [20]. Moreover, the level of intracellular C1P can increase after stimulation with different factors, e.g., after exposure of resting macrophages to macrophage colony-stimulating factor [2]. Intracellular C1P concentration can also be increased upon treatment of cells with pro-inflammatory agonists such as ATP or A23197 [35]. Extracellular increase in C1P level is rather the effect of organ/tissue injury and was observed in response to ischemia [11], myocardial infarction [16], irradiation [12, 15], or as a result of chemotherapy [12].

The first reported biological effect of C1P stimulation was induction of rat fibroblast proliferation [1] which was later confirmed in other cell types [2–4]. Further studies revealed that CERK activity and thus exogenous C1P concentration might affect proliferation of some cancer cells [36, 37] as well as normal mesangial cells and fibroblasts [38]. Molecular studies indicate that C1P potently stimulates intracellular calcium mobilization [31, 39] and glucose uptake [39]. Subsequent studies demonstrated C1P involvement in cell survival [8] and inhibition of apoptosis mainly due to blockage of ceramide synthesis [6, 40]. The list of biological effects of C1P also includes stimulation of phagocytosis [41], degranulation [42], and regulation of inflammation [14] as well as regulation of cell migration and invasion,

Table 7.1 List of cell types for which C1P-induced migration was observed with corresponding effective C1P concentration and possible signaling pathways involved in this process

	Cell type	C1P concentration	Signaling pathways involved in migration	References
Immune cells	Macrophages (Raw264.7 and J774A.1)	5–50 μM	pMAPK 44/42, MAPKp38, PI3K/AKT, NF-κB	[10, 30]
Stem cells	HSCs	10–100 μM	pMAPK 44/42, MAPKp38, AKT, Stat-3, Stat-5	[15]
	MSCs	0.5–50 μM priming: 100–200 μM	pMAPK 44/42, AKT	[11, 4]
	EPCs	0.1–50 μM	pMAPK 44/42, AKT	[11]
	VSELs	50 μM	pMAPK 44/42, AKT	[11]
Cancer cells	Rhabdomyosarcoma (ARMS and ERMS)	0.5–10 μM	pMAPK 44/42, PI3K/AKT	[12]
	Pancreatic cancer	5–30 μM	PI3K, PI3K/AKT, mTOR, RhoA	[65]
	THP-1 monocytes	20–30 μM	pMAPK 44/42?, MAPKp38?, PI3/AKT?, NF-κB?	[30]

which in more details will be discussed below in this chapter. Table 7.1 provides working concentration of C1P for different cell types as well as downstream-activated pathways that play a role in migration of normal and malignant cells that will be discussed in this chapter.

7.2 C1P Stimulates Migration of Macrophages

One of the first studies indicating the potential role of exogenous C1P in stimulation of migration of different cells was performed studying its effect on macrophages using Transwells [10]. Increased migration of these cells was observed when C1P was added to the lower chamber in concentration of 30–50 μM and was dependent on activation of pMAPK42/44 and PI3-K/Akt pathways, as inhibition of these pathways completely abolished C1P-induced macrophage chemotaxis [10]. Further studies revealed that stimulation of macrophages with C1P and phosphorylation of pMAPK42/44 and Akt led to release from cells of macrophage chemoattractant protein-1 (MCP-1) [30]. In addition, C1P also stimulate the binding of nuclear factor kappa B (NF-κB) to DNA and blockage of this transcription factor resulted in complete inhibition of MCP-1 release and macrophage migration [10]. MCP-1 was also shown to be involved in C1P-induced migration of THP-1 cells and in addition to already known signaling pathways, MAPKp38 also had been shown to be important downstream effector of C1P stimulation both in macrophages and THP-1 cell line [30].

Sequestration of MCP-1 with a neutralizing antibody or treatment with MCP-1 siRNA abolished C1P-stimulated cell migration of macrophages and THP-1 cells. Similarly, inhibition of the pathways involved in C1P-stimulated MCP-1 release completely blocked C1P-directed cell migration. This suggests that C1P promotes MCP-1 release in different cell types and that this chemokine might be a major mediator of C1P-stimulated cell migration [30]. Moreover, migration of macrophages in response to C1P was also associated with increase in expression of metalloproteinases (MMP) -2 and -9 as well as induction of actin polymerization and increased phosphorylation of focal adhesion protein—paxillin [43]. Blockage of any of these proteins either by pharmacological tools or with specific siRNA reduced C1P-induced migration of macrophages [43].

Another interesting finding was identification of phosphatidic acid (PA), which is structurally related to C1P, as a potential natural antagonist of the C1P receptor [44]. Although PA alone did not affect macrophage migration significantly, it was able to displace radiolabeled C1P from its membrane-binding site and thus to inhibit C1P-stimulated macrophage migration. Moreover, treatment of macrophages with exogenous PLD, an enzyme that produces PA from membrane phospholipids, also inhibited C1P-stimulated cell migration [44].

7.3 C1P Regulates Hematopoietic Stem/Progenitor Cells Trafficking

For many years, it was believed that the α-chemokine stromal-derived factor 1 (SDF-1) was the major chemoattractant in peripheral blood to regulate trafficking of hematopoietic stem/progenitor cells (HSPCs). SDF-1 binds to its $G_{\alpha i}$ protein- coupled receptor CXCR4 which is present at the surface of HSPCs [45, 46]. However, this explanation has been challenged by several observations supporting SDF-1–CXCR4-independent homing and mobilization mechanisms [15, 47–49]. Additionally, SDF-1 was shown to be a potent chemoattractant for HSPCs when used in supraphysiological doses. Moreover, myeloablative conditioning for transplantation induces a highly proteolytic microenvironment in the bone marrow (BM) that leads to degradation of SDF-1; therefore, its level is not optimal to stimulate homing of HSPCs [28]. In support of this notion, new potent chemoattractants for HSPCs, such as proteolytic enzyme-resistant sphingophospholipids S1P and C1P have been identified [15, 50] (Fig. 7.1). By employing liquid chromatography electrospray ionization tandem mass spectrometry, it has been shown that the level of C1P measured in BM microenvironment increases after conditioning for hematopoietic transplant by lethal irradiation [15]. Moreover, C1P, in a similar way as S1P, induces several signaling pathways in murine Sca-1$^+$ cells that are enriched for HSPCs, including pMAPK 42/44, MAPKp38, AKT, and several Stat proteins, as well as strongly chemoattracts murine HSPCs [15].

Interestingly, at the same time C1P does not affect the clonogenecity of murine progenitors from all major hematopoietic lineages [15]. C1P stimulation also

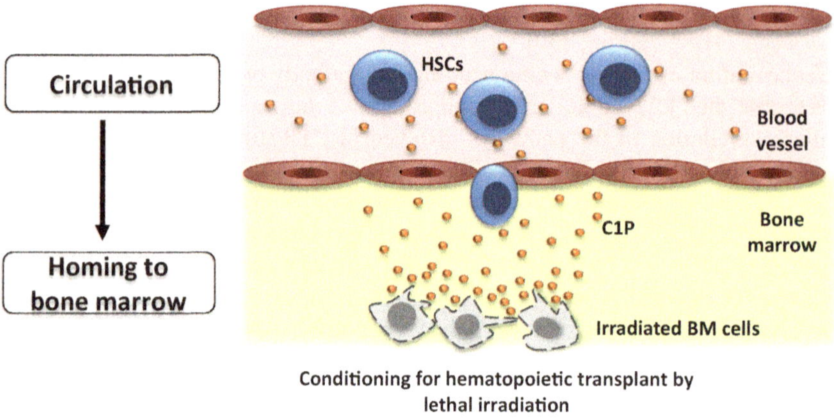

Fig. 7.1 Role of C1P in homing of HSCs. As a result of conditioning for hematopoietic transplant by lethal irradiation, there is an increase in C1P level in BM environment. C1P chemoattracts HSCs, which has been given to recipients and increases their homing to their final destination

increased adhesion of HSPCs to BM-derived fibroblast [15]. However, despite the fact that the overall SDF-1 level decreases in murine BM after lethal irradiation, the responsiveness of HSPCs to an SDF-1 gradient can be significantly enhanced by some factors such as prostaglandin E2 (PGE_2) [51] and an additional effect of C1P on engraftment of HSPCs may be related to an increase of PGE_2 level in BM and an increase of PGE_2-mediated pro-homing activities [51, 52]. In support of this, C1P induces activity of cytosolic phospholipases A_2, which regulates production of arachidonic acid, a substrate for PGE_2 synthesis [53].

7.4 C1P Regulates Migration of Mesenchymal Stem/Stromal Cells (MSCs), Endothelial Progenitor Cells (EPCs), and Very Small Embryonic-Like Stem Cells (VSELs)

It has been demonstrated in several animal and clinical models that stem cells are mobilized into peripheral blood after organ or tissue injury and supposedly play a role in regeneration of damaged organ/tissues [54–60]. These circulating stem cells could potentially contribute to tissue repair directly like EPCs that support formation of new blood vessels [54, 56] or VSELs that differentiate and replace damaged cells [60]. It is also possible that some stem cells could act indirectly as a source of several growth factors and microvesicles/exosomes (e.g., MSCs) that provide trophic signals that inhibit cell apoptosis and stimulate vascularization of damaged tissues [61, 62].

Interestingly, S1P and C1P levels increase in response to tissue/organ injuries in biological fluids [11, 12, 15, 16], which led to the question of whether these

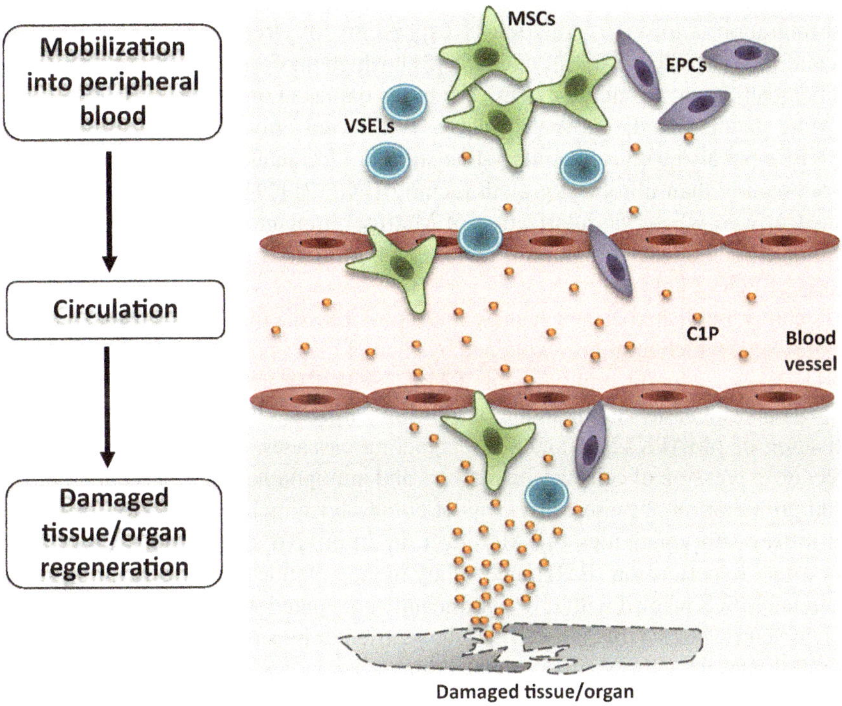

Fig. 7.2 Role of C1P in chemoattracting MSCs, EPCs, and VSELs to damaged tissue/organ. C1P is released from damaged tissues and attracts several stem cell population such as MSCs, EPCs, or VSELs. These stem cells can participate in regeneration directly (angiogenesis, differentiation into cells of damaged tissue) or indirectly (paracrine effect, release of microvesicles). Since organ or tissue injury may be the result of ischemia or hypoxia in growing tumor, C1P effect can be both positive and negative. On the one hand, it will play a role in physiological process of regeneration, on the other hand, MSCs and EPCs recruited to growing tumor may support cancer growth and expansion

bioactive lipids can play a role in trafficking of non-hematopoietic stem cells to damaged tissues. Indeed, C1P similarly to S1P is a potent chemoattractant for MSCs, EPCs, [16] endothelial cells (HUVEC), and VSELs [11] (Fig. 7.2). Moreover, response of cells to biologically relevant doses of C1P was much stronger than to other bioactive lipids such as LPA and LPC or to already known chemoattractants such as SDF-1. Importantly, SDF-1 induces migration of cells in experimental models only when used at supraphysiological doses, whereas lower, more physiological doses do not induce cell migration [11].

As observed in macrophages and HSPCs, stimulation with C1P induces activation of AKT and pMAPK42/44 signaling pathways both in MSCs and HUVEC [11]. Moreover, in MSCs, C1P induces expression of SDF-1 at the level of mRNA [11]. This has been further confirmed at the protein level in cell extracts as well as in conditioned media obtained from MSCs [11]. Moreover, C1P also enhances expression of cyclooxygenase-2 both at mRNA and protein levels in MSCs [11].

This suggests that similarly to the observations in HSCPs, C1P additional effect on cell migration is through induction of expressions of already identified chemoattractant factors such as SDF-1 and PGE_2 [51], which play together an important role in chemoattracting circulating stem cells to the damaged organs [63, 64].

More detailed studies revealed that C1P does not only stimulate migration of HUVECs but also triggers capillary-like structure formation in 3D-Matrigel assay, more potently than fibroblast growth factor-2 (FGF-2) [11]. This was further confirmed in vivo where vascularization of Matrigel implants was analyzed, demonstrating a potent vascularization effect of C1P in vivo at least comparable to the effects of FGF-2 [11].

Recently, published data have indicated a possible role of C1P as a priming agent for MSCs [4] which might be explained by chemokinetic properties of this bioactive lipid [12]. Accordingly, pretreatment of MSCs with C1P improved migration activity in Transwell assays compared to non-primed MSCs. This effect correlated with activation of pMAPK42/44 and AKT signaling cascades. C1P priming had little effect on expression of cell surface markers and multipotency of MSCs but it potentiated proliferation of these cells, grow of colony-forming unit-fibroblast, and their anti-inflammatory activities [4]. Moreover, in an in vivo animal model of pulmonary artery hypertension (PAH) induced by monocrotaline, a single administration of human MSCs primed with C1P significantly attenuated the PAH-related increase in right ventricular systolic pressure, right ventricular hypertrophy, and thickness of α-smooth muscle actin-positive cells residing around the vessel wall [4]. Thus, this study showed that C1P priming increases the effects of MSC therapy by enhancing the migratory, self-renewal, and anti-inflammatory activities of these cells. To summarize this exciting data, priming of MSCs by C1P could be implemented in the clinic as a novel promising option for the treatment of PAH patients [4] and most likely patients with other disorders.

7.5 C1P Modulates Migration and Invasion of Cancer Cells

Bioactive lipids such as LPA, LPC, or S1P have already been reported to stimulate migration of cancer cells in wide variety of experimental studies. They were also found to stimulate cell resistance to chemotheraphy, stimulate cell proliferation, and promote angiogenesis so crucial for malignant growth [65–67]. However, for a long time there was no evidence on whether C1P could also modulate migration of cancer cells.

Recently, it was shown that C1P, similarly to S1P, is a potent chemoattractant for rhabdomyosarcoma and pancreatic cancer cell lines [12, 68] (Fig. 7.3). This effect is $G_{\alpha i}$ protein-coupled receptor dependent since pretreatment of both cell types with pertussis toxin completely abolished migratory responses to a C1P gradient [12, 68]. C1P, as already discussed above, was also shown also stimulate migration of immortalized THP-1 cell line, and this effect was associated with the release of MCP-1 [30]. It was also reported that C1P-induced migration of cancer cells depends on activation of AKT and pMAK42/44 pathways, as addition of appropriate inhibitors efficiently

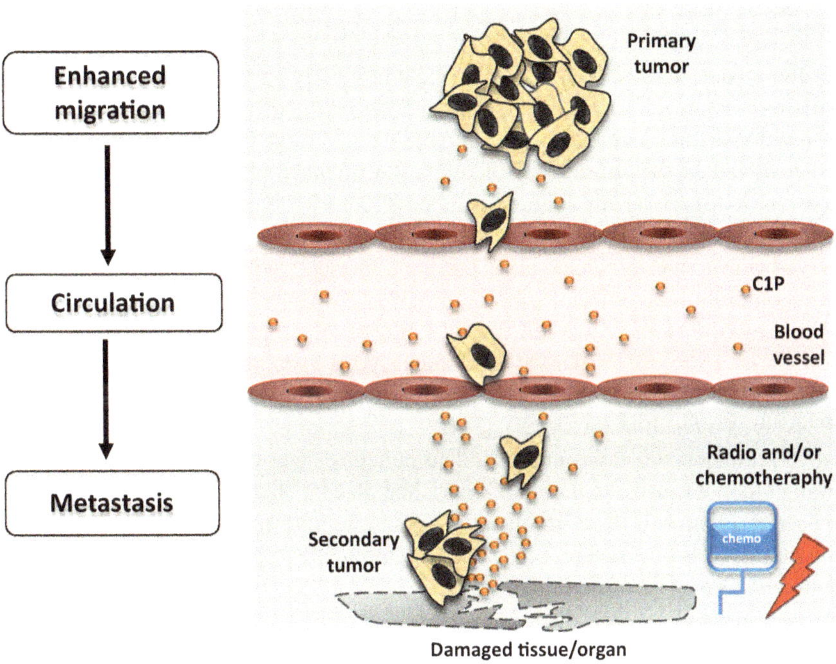

Fig. 7.3 C1P plays a role in formation of a prometastatic environment. With damages due to radio/chemotheraphy, cells release C1P, which chemoattracts cancer cells that survive the initial treatment. Cancer cells migrate (metastasize) to distant locations where they can form secondary tumors

inhibit C1P-induced migration of rhabdomyosarcoma and pancreatic cancer cells [12, 68]. In the case of pancreatic cancer, the mammalian target of rapamycin 1 (mTOR1) signaling as well as RhoA small GTPase were found to be involved in regulation of migration of these cells [68]. Interestingly, C1P and S1P are upregulated in response to radio-chemotheraphy in different tissues and C1P contributes to the induction of unwanted prometastatic environment as a side effect of this therapy [12]. Therefore, there is a need to identify C1P receptor/s and to develop antagonists for these receptors as well as to employ C1P binding/inactivating agents which could be used in clinic to ameliorate the metastatic effects of C1P.

It is also worth mentioning that the activity of CERK which plays a role in C1P synthesis also correlates with cell migration since overexpression of CERK in pancreatic cancer cells enhances spontaneous migration of these cells [65, 68]. This activity was potently blocked in control experiments with selective CERK inhibitors or specific siRNA [65, 68]. Similar studies showed that CERK is also required for migration of mouse fibroblasts, which show elevated level of C1P during the early stages of wound healing [66, 69]. This suggests that by appropriate control of CERK activity in cells we could on the one hand decrease dissemination of cancer cells (CERK inhibition) and on the other we could improve wound healing after injury (CERK activation).

7.6 Conclusions

Over the recent years, bioactive lipids, including C1P, have emerged as important regulators of stem cell trafficking. C1P stimulates egress of HSPCs as well as MSCs, EPCs, and VSELs from bone marrow into peripheral blood. C1P has also been shown to be involved in the stimulation of migration of macrophages and trafficking of cancer cells. Moreover, in contrast to peptide-based chemoattractants that induce migration in supraphysiological concentrations, the pro-migratory effect of C1P is observed within physiological values present in peripheral blood, lymph, or interstitial tissue fluid.

The level of C1P increases due to tissue or organ damage in response to different injuries. Therefore, C1P released from damaged tissues might chemoattract stem cells which could then participate in tissue regeneration. However, the increase of C1P levels as a result of the cytotoxic effect of chemotherapy and radiotherapy can create a prometastatic environment and stimulates cancer cells that survived initial treatment to metastasize to these C1P-enriched places. Therefore, there is an important need to identify C1P receptors and to develop small-molecule compounds that could inhibit C1P-C1P receptor axis and thus modulate migratory properties of normal stem cells as well as malignant cells.

References

1. Gomez-Munoz A, Duffy PA, Martin A, O'Brien L, Byun HS, Bittman R et al (1995) Short-chain ceramide-1-phosphates are novel stimulators of DNA synthesis and cell division: antagonism by cell-permeable ceramides. Mol Pharmacol 47:833–839
2. Gangoiti P, Granado MH, Wang SW, Kong JY, Steinbrecher UP, Gomez-Munoz A (2008) Ceramide 1-phosphate stimulates macrophage proliferation through activation of the PI3-kinase/PKB, JNK and ERK1/2 pathways. Cell Signal 20:726–736. doi:10.1016/j.cellsig.2007.12.008
3. Gangoiti P, Bernacchioni C, Donati C, Cencetti F, Ouro A, Gomez-Munoz A et al (2012) Ceramide 1-phosphate stimulates proliferation of C2C12 myoblasts. Biochimie 94:597–607
4. Lim J, Kim Y, Heo J, Kim KH, Lee S, Lee SW et al (2016) Priming with ceramide-1 phosphate promotes the therapeutic effect of mesenchymal stem/stromal cells on pulmonary artery hypertension. Biochem Biophys Res Commun 473:35–41
5. Gao Z, Wang H, Xiao FJ, Shi XF, Zhang YK, Xu QQ et al (2016) SIRT1 mediates Sphk1/S1P-induced proliferation and migration of endothelial cells. Int J Biochem Cell Biol 74:152–160
6. Gomez-Munoz A, Kong JY, Salh B, Steinbrecher UP (2004) Ceramide-1-phosphate blocks apoptosis through inhibition of acid sphingomyelinase in macrophages. J Lipid Res 45:99–105
7. Gomez-Munoz A, Kong J, Salh B, Steinbrecher UP (2003) Sphingosine-1-phosphate inhibits acid sphingomyelinase and blocks apoptosis in macrophages. FEBS Lett 539:56–60
8. Gomez-Munoz A, Kong JY, Parhar K, Wang SW, Gangoiti P, Gonzalez M et al (2005) Ceramide-1-phosphate promotes cell survival through activation of the phosphatidylinositol 3-kinase/protein kinase B pathway. FEBS Lett 579:3744–3750
9. Rodriguez AM, Graef AJ, LeVine DN, Cohen IR, Modiano JF, Kim JH (2015) Association of sphingosine-1-phosphate (S1P)/S1P receptor-1 pathway with cell proliferation and survival in canine hemangiosarcoma. J Vet Intern Med 29:1088–1097
10. Granado MH, Gangoiti P, Ouro A, Arana L, Gonzalez M, Trueba M et al (2009) Ceramide 1-phosphate (C1P) promotes cell migration Involvement of a specific C1P receptor. Cell Signal 21:405–412. doi:10.1016/j.cellsig.2008.11.003

11. Kim C, Schneider G, Abdel-Latif A, Mierzejewska K, Sunkara M, Borkowska S et al (2013) Ceramide-1-phosphate regulates migration of multipotent stromal cells and endothelial progenitor cells—implications for tissue regeneration. Stem Cells 31:500–510
12. Schneider G, Bryndza E, Abdel-Latif A, Ratajczak J, Maj M, Tarnowski M et al (2013) Bioactive lipids S1P and C1P are prometastatic factors in human rhabdomyosarcoma, and their tissue levels increase in response to radio/chemotherapy. Mol Cancer Res 11:793–807
13. Mizugishi K, Yamashita T, Olivera A, Miller GF, Spiegel S, Proia RL (2005) Essential role for sphingosine kinases in neural and vascular development. Mol Cell Biol 25:11113–11121
14. Lamour NF, Subramanian P, Wijesinghe DS, Stahelin RV, Bonventre JV, Chalfant CE (2009) Ceramide 1-phosphate is required for the translocation of group IVA cytosolic phospholipase A2 and prostaglandin synthesis. J Biol Chem 284:26897–26907
15. Kim CH, Wu W, Wysoczynski M, Abdel-Latif A, Sunkara M, Morris A et al (2012) Conditioning for hematopoietic transplantation activates the complement cascade and induces a proteolytic environment in bone marrow: a novel role for bioactive lipids and soluble C5b-C9 as homing factors. Leukemia 26:106–116
16. Karapetyan AV, Klyachkin YM, Selim S, Sunkara M, Ziada KM, Cohen DA et al (2013) Bioactive lipids and cationic antimicrobial peptides as new potential regulators for trafficking of bone marrow-derived stem cells in patients with acute myocardial infarction. Stem Cells Dev 22:1645–1656
17. Simanshu DK, Kamlekar RK, Wijesinghe DS, Zou X, Zhai X, Mishra SK et al (2013) Non-vesicular trafficking by a ceramide-1-phosphate transfer protein regulates eicosanoids. Nature 500:463–467
18. Bajjalieh SM, Martin TF, Floor E (1989) Synaptic vesicle ceramide kinase. A calcium-stimulated lipid kinase that co-purifies with brain synaptic vesicles. J Biol Chem 264:14354–14360
19. Kolesnick RN, Hemer MR (1990) Characterization of a ceramide kinase activity from human leukemia (HL-60) cells. Separation from diacylglycerol kinase activity. J Biol Chem 265:18803–18808
20. Graf C, Zemann B, Rovina P, Urtz N, Schanzer A, Reuschel R et al (2008) Neutropenia with impaired immune response to Streptococcus pneumoniae in ceramide kinase-deficient mice. J Immunol 180:3457–3466
21. Boath A, Graf C, Lidome E, Ullrich T, Nussbaumer P, Bornancin F (2008) Regulation and traffic of ceramide 1-phosphate produced by ceramide kinase: comparative analysis to glucosylceramide and sphingomyelin. J Biol Chem 283:8517–8526
22. Rivera IG, Ordonez M, Presa N, Gomez-Larrauri A, Simon J, Trueba M et al (2015) Sphingomyelinase D/ceramide 1-phosphate in cell survival and inflammation. Toxins (Basel) 7:1457–1466
23. Pettus BJ, Bielawska A, Spiegel S, Roddy P, Hannun YA, Chalfant CE (2003) Ceramide kinase mediates cytokine- and calcium ionophore-induced arachidonic acid release. J Biol Chem 278:38206–38213
24. van Meeteren LA, Frederiks F, Giepmans BN, Pedrosa MF, Billington SJ, Jost BH et al (2004) Spider and bacterial sphingomyelinases D target cellular lysophosphatidic acid receptors by hydrolyzing lysophosphatidylcholine. J Biol Chem 279:10833–10836
25. Lee S, Lynch KR (2005) Brown recluse spider (*Loxosceles reclusa*) venom phospholipase D (PLD) generates lysophosphatidic acid (LPA). Biochem J 391:317–323
26. Lajoie DM, Zobel-Thropp PA, Kumirov VK, Bandarian V, Binford GJ, Cordes MH (2013) Phospholipase D toxins of brown spider venom convert lysophosphatidylcholine and sphingomyelin to cyclic phosphates. PLoS One 8:e72372
27. Graf C, Niwa S, Muller M, Kinzel B, Bornancin F (2008) Wild-type levels of ceramide and ceramide-1-phosphate in the retina of ceramide kinase-like-deficient mice. Biochem Biophys Res Commun 373:159–163
28. Tuson M, Marfany G, Gonzalez-Duarte R (2004) Mutation of CERKL, a novel human ceramide kinase gene, causes autosomal recessive retinitis pigmentosa (RP26). Am J Hum Genet 74:128–138

29. Bornancin F, Mechtcheriakova D, Stora S, Graf C, Wlachos A, Devay P et al (2005) Characterization of a ceramide kinase-like protein. Biochim Biophys Acta 1687:31–43
30. Arana L, Ordonez M, Ouro A, Rivera IG, Gangoiti P, Trueba M et al (2013) Ceramide 1-phosphate induces macrophage chemoattractant protein-1 release: involvement in ceramide 1-phosphate-stimulated cell migration. Am J Physiol Endocrinol Metab 304:E1213–E1226
31. Gijsbers S, Mannaerts GP, Himpens B, Van Veldhoven PP (1999) N-acetyl-sphingenine-1-phosphate is a potent calcium mobilizing agent. FEBS Lett 453:269–272
32. Colina CM, Gubbins KE (2005) Vapor-liquid and vapor-liquid-liquid equilibria of carbon dioxide/n-perfluoroalkane/n-alkane ternary mixtures. J Phys Chem B 109:2899–2910. doi:10.1021/jp046612d
33. Mietla JA, Wijesinghe DS, Hoeferlin LA, Shultz MD, Natarajan R, Fowler AA 3rd et al (2013) Characterization of eicosanoid synthesis in a genetic ablation model of ceramide kinase. J Lipid Res 54:1834–1847. doi:10.1194/jlr.M035683
34. Hammad SM, Pierce JS, Soodavar F, Smith KJ, Al Gadban MM, Rembiesa B et al (2010) Blood sphingolipidomics in healthy humans: impact of sample collection methodology. J Lipid Res 51:3074–3087
35. Lamour NF, Stahelin RV, Wijesinghe DS, Maceyka M, Wang E, Allegood JC et al (2007) Ceramide kinase uses ceramide provided by ceramide transport protein: localization to organelles of eicosanoid synthesis. J Lipid Res 48:1293–1304. doi:10.1194/jlr.M700083-JLR200
36. Pastukhov O, Schwalm S, Zangemeister-Wittke U, Fabbro D, Bornancin F, Japtok L et al (2014) The ceramide kinase inhibitor NVP-231 inhibits breast and lung cancer cell proliferation by inducing M phase arrest and subsequent cell death. Br J Pharmacol 171:5829–5844. doi:10.1111/bph.12886
37. Mitra P, Maceyka M, Payne SG, Lamour N, Milstien S, Chalfant CE et al (2007) Ceramide kinase regulates growth and survival of A549 human lung adenocarcinoma cells. FEBS Lett 581:735–740
38. Pastukhov O, Schwalm S, Romer I, Zangemeister-Wittke U, Pfeilschifter J, Huwiler A (2014) Ceramide kinase contributes to proliferation but not to prostaglandin E2 formation in renal mesangial cells and fibroblasts. Cell Physiol Biochem 34:119–133
39. Ouro A, Arana L, Gangoiti P, Rivera IG, Ordonez M, Trueba M et al (2013) Ceramide 1-phosphate stimulates glucose uptake in macrophages. Cell Signal 25:786–795
40. Granado MH, Gangoiti P, Ouro A, Arana L, Gomez-Munoz A (2009) Ceramide 1-phosphate inhibits serine palmitoyltransferase and blocks apoptosis in alveolar macrophages. Biochim Biophys Acta 1791:263–272
41. Hinkovska-Galcheva V, Boxer LA, Kindzelskii A, Hiraoka M, Abe A, Goparju S et al (2005) Ceramide 1-phosphate, a mediator of phagocytosis. J Biol Chem 280:26612–26621
42. Mitsutake S, Kim TJ, Inagaki Y, Kato M, Yamashita T, Igarashi Y (2004) Ceramide kinase is a mediator of calcium-dependent degranulation in mast cells. J Biol Chem 279:17570–17577
43. Ordonez M, Rivera IG, Presa N, Gomez-Munoz A (2016) Implication of matrix metalloproteinases 2 and 9 in ceramide 1-phosphate-stimulated macrophage migration. Cell Signal 28:1066–1074
44. Ouro A, Arana L, Rivera IG, Ordonez M, Gomez-Larrauri A, Presa N et al (2014) Phosphatidic acid inhibits ceramide 1-phosphate-stimulated macrophage migration. Biochem Pharmacol 92:642–650. doi:10.1016/j.bcp.2014.10.005
45. Aiuti A, Webb IJ, Bleul C, Springer T, Gutierrez-Ramos JC (1997) The chemokine SDF-1 is a chemoattractant for human CD34+ hematopoietic progenitor cells and provides a new mechanism to explain the mobilization of CD34+ progenitors to peripheral blood. J Exp Med 185:111–120
46. Ratajczak MZ, Serwin K, Schneider G (2013) Innate immunity derived factors as external modulators of the CXCL12-CXCR4 axis and their role in stem cell homing and mobilization. Theranostics 3:3–10. doi:10.7150/thno.4621
47. Ma Q, Jones D, Springer TA (1999) The chemokine receptor CXCR4 is required for the retention of B lineage and granulocytic precursors within the bone marrow microenvironment. Immunity 10:463–471

48. Christopherson KW 2nd, Hangoc G, Mantel CR, Broxmeyer HE (2004) Modulation of hematopoietic stem cell homing and engraftment by CD26. Science 305:1000–1003. doi:10.1126/science.1097071
49. Onai N, Zhang Y, Yoneyama H, Kitamura T, Ishikawa S, Matsushima K (2000) Impairment of lymphopoiesis and myelopoiesis in mice reconstituted with bone marrow-hematopoietic progenitor cells expressing SDF-1-intrakine. Blood 96:2074–2080
50. Ratajczak MZ, Lee H, Wysoczynski M, Wan W, Marlicz W, Laughlin MJ et al (2010) Novel insight into stem cell mobilization-plasma sphingosine-1-phosphate is a major chemoattractant that directs the egress of hematopoietic stem progenitor cells from the bone marrow and its level in peripheral blood increases during mobilization due to activation of complement cascade/membrane attack complex. Leukemia 24:976–985. doi:10.1038/leu.2010.53
51. Hoggatt J, Singh P, Sampath J, Pelus LM (2009) Prostaglandin E2 enhances hematopoietic stem cell homing, survival, and proliferation. Blood 113:5444–5455. doi:10.1182/blood-2009-01-201335
52. Pelus LM, Hoggatt J, Singh P (2011) Pulse exposure of haematopoietic grafts to prostaglandin E2 in vitro facilitates engraftment and recovery. Cell Prolif 44(Suppl. 1):22–29. doi:10.1111/j.1365-2184.2010.00726.x
53. Corallini F, Bossi F, Gonelli A, Tripodo C, Castellino G, Mollnes TE et al (2009) The soluble terminal complement complex (SC5b-9) up-regulates osteoprotegerin expression and release by endothelial cells: implications in rheumatoid arthritis. Rheumatology (Oxford) 48:293–298. doi:10.1093/rheumatology/ken495
54. Dome B, Timar J, Dobos J, Meszaros L, Raso E, Paku S et al (2006) Identification and clinical significance of circulating endothelial progenitor cells in human non-small cell lung cancer. Cancer Res 66:7341–7347
55. Drukala J, Paczkowska E, Kucia M, Mlynska E, Krajewski A, Machalinski B et al (2006) Stem cells, including a population of very small embryonic-like stem cells, are mobilized into peripheral blood in patients after skin burn injury. Stem Cell Rev 8:184–194
56. Gao D, Nolan D, McDonnell K, Vahdat L, Benezra R, Altorki N et al (2009) Bone marrow-derived endothelial progenitor cells contribute to the angiogenic switch in tumor growth and metastatic progression. Biochim Biophys Acta 1796:33–40
57. Lee RH, Seo MJ, Reger RL, Spees JL, Pulin AA, Olson SD et al (2006) Multipotent stromal cells from human marrow home to and promote repair of pancreatic islets and renal glomeruli in diabetic NOD/scid mice. Proc Natl Acad Sci U S A 103:17438–17443
58. Ortiz LA, Dutreil M, Fattman C, Pandey AC, Torres G, Go K et al (2007) Interleukin 1 receptor antagonist mediates the antiinflammatory and antifibrotic effect of mesenchymal stem cells during lung injury. Proc Natl Acad Sci U S A 104:11002–11007
59. Ratajczak MZ, Liu R, Ratajczak J, Kucia M, Shin DM (2011) The role of pluripotent embryonic-like stem cells residing in adult tissues in regeneration and longevity. Differentiation 81:153–161
60. Wojakowski W, Kucia M, Liu R, Zuba-Surma E, Jadczyk T, Bachowski R et al (2011) Circulating very small embryonic-like stem cells in cardiovascular disease. J Cardiovasc Transl Res 4:138–144
61. Bruno S, Grange C, Deregibus MC, Calogero RA, Saviozzi S, Collino F et al (2009) Mesenchymal stem cell-derived microvesicles protect against acute tubular injury. J Am Soc Nephrol 20:1053–1067
62. Caplan AI, Dennis JE (2006) Mesenchymal stem cells as trophic mediators. J Cell Biochem 98:1076–1084
63. Ratajczak MZ, Zuba-Surma E, Kucia M, Reca R, Wojakowski W, Ratajczak J (2006) The pleiotropic effects of the SDF-1-CXCR4 axis in organogenesis, regeneration and tumorigenesis. Leukemia 20:1915–1924
64. Hoggatt J, Pelus LM (2010) Eicosanoid regulation of hematopoiesis and hematopoietic stem and progenitor trafficking. Leukemia 24:1993–2002
65. Pyne NJ, Pyne S (2010) Sphingosine 1-phosphate and cancer. Nat Rev Cancer 10:489–503

66. Vassilakopoulou M, Psyrri A, Argiris A (2015) Targeting angiogenesis in head and neck cancer. Oral Oncol 51:409–415
67. Samadi N, Bekele R, Capatos D, Venkatraman G, Sariahmetoglu M, Brindley DN (2011) Regulation of lysophosphatidate signaling by autotaxin and lipid phosphate phosphatases with respect to tumor progression, angiogenesis, metastasis and chemo-resistance. Biochimie 93:61–70
68. Rivera IG, Ordonez M, Presa N, Gangoiti P, Gomez-Larrauri A, Trueba M et al (2016) Ceramide 1-phosphate regulates cell migration and invasion of human pancreatic cancer cells. Biochem Pharmacol 102:107–119
69. Wijesinghe DS, Brentnall M, Mietla JA, Hoeferlin LA, Diegelmann RF, Boise LH et al (2014) Ceramide kinase is required for a normal eicosanoid response and the subsequent orderly migration of fibroblasts. J Lipid Res 55:1298–1309

Chapter 8
The Emerging Role of Sphingolipids in Cancer Stem Cell Biology

Alexander C. Lewis, Jason A. Powell, and Stuart M. Pitson

Abbreviations

ALDH1	Aldehyde dehydrogenase 1
AML	Acute myeloid leukaemia
BCSC	Breast cancer stem cell
C1P	Ceramide-1-phosphate
CERT	Ceramide transport protein
CSC	Cancer stem cell
EGFR	Epidermal growth factor receptor
ENL	Eleven-nineteen-leukaemia
ER	Endoplasmic reticulum
ERK	Extracellular signal-regulated kinase
FLK-1	Foetal liver kinase-1
GCS	Glucosylceramide synthase
HDAC	Histone deacetylase
HSC	Haematopoietic stem cell
HSPC	Haematopoietic stem and progenitor cells
MEF	Mouse embryonic fibroblast
MLL	Mixed lineage leukaemia
MSC	Mesenchymal stem cells
PDGF	Platelet-derived growth factor
ROS	Reactive oxygen species

A.C. Lewis • J.A. Powell • S.M. Pitson (✉)
Centre for Cancer Biology, University of South Australia and SA Pathology,
Frome Road, Adelaide, SA 5000, Australia
e-mail: stuart.pitson@unisa.edu.au

A. Pébay, R.C.B. Wong (eds.), *Lipidomics of Stem Cells*, Stem Cell Biology and
Regenerative Medicine, DOI 10.1007/978-3-319-49343-5_8

S1P	Sphingosine 1-phosphate
$S1P_{1-5}$	S1P receptors 1–5
SK	Sphingosine kinase
TMZ	Temozolomide

8.1 Introduction

Prior to the paradigm shifting cancer stem cell (CSC) work by Bonnet and Dick [1], it was unknown how reformation of tumours occurred following eradication of the tumour population to undetectable levels. In these studies, the sorting of acute myeloid leukaemia (AML) patient cells based on $CD34^+$ and $CD38^-$ expression led to the isolation of a small population of functionally and morphologically distinct leukaemia cells that alone could form human leukaemia in immune-compromised mice: the leukaemia CSC [1]. This initial finding accelerated the discovery of functionally similar cell populations in a number of solid tumours [2–4] and propagated the theory that many tumours are organised into a functional hierarchy whereby a small number of CSC alone can generate the bulk of the tumour population [5].

8.1.1 Criteria Defining Cancer Stem Cells

By definition, CSCs are classified by several criteria that distinguish them from the bulk of the tumour population. Isolated CSCs are capable of generating a xenograft in immune-compromised mice that recapitulates the heterogeneity observed within the primary human tumour based on comparative genetic analysis. From the propagating CSC population, daughter cells can acquire further mutations, forming subclonal populations that vary through the bulk population. However, each of these cells should retain the mutational profile of the initiating CSC population. Directed sequencing approaches have shown patient material and xenograft samples retain similar mutational landscapes [6] suggesting this model recapitulates primary disease with genetic drift akin to that observed in relasped patients [7].

As the head of a hierarchical organisation, CSCs must exhibit the ability to self-renew and undergo differentiation to maintain the bulk tumour population. CSCs produce daughter cells that lack the ability to form tumours when transplanted into immune-compromised mice. Pioneering work by Bonnet and Dick showed that when flow cytometry isolated $CD34^+$, $CD38^-$ leukemic stem cells were transplanted into immunocompromised mice they were able to form human AML [1]. However, purified $CD34^+$, $CD38^+$ AML cells were unable to engraft confirming that the formation of human leukaemia in immunocompromised mice arose purely from the $CD34^+$, $CD38^-$ population. Confirmation of the leukaemia propagating ability of the $CD34^+$, $CD38^-$ population was performed by serial transplantation of sorted human AML cells harvested from the primary recipient. Recently, Reinisch et al. utilised an elegant in vivo model where a humanised bone marrow microenvironment was generated by subcutaneous injection of human bone marrow-derived mesenchymal stromal cells [8]. Using this model, single cell engraftment could be achieved using CD34$^+$, CD38$^-$ cells isolated from AML patient samples reaffirming the CSC as the head of the hierarchy.

8.1.2 Markers for the Isolation of Cancer Stem Cells

Since the initial discovery by Bonnet and Dick identifying the CD34$^+$, CD38$^-$ leukemic stem cell [1], other markers such as CD123 have further enhanced the characterisation of the leukemic stem cell population [9]. These findings also propagated the discovery of similar cell populations in other cancers including, breast, prostate, glioblastoma, and lung cancer replicating the initial findings and proposing this cell population is a common player across many cancer types [5]. Cell surface markers such as CD44 and CD133, as well as aldehyde dehydrogenase 1 (ALDH1) have been identified across multiple cancer types as cancer stem cell-specific markers, as summarised in Table 8.1 [2, 3, 10–12]. Whilst there are a range of cell surface markers that can be employed to isolate CSCs from the majority of patient samples (Table 8.1), there are problems that require consideration when sorting for CSCs. Among those issues noted by Medema, sorting CSCs based on cell surface markers requires a set of optimal markers such as CD34$^+$, CD38$^-$, and CD123$^+$ in the case of AML [13]. However, identifying optimal sets of cell surface markers for other malignancies remains an issue with variables such as methylation patterns and certain mutations capable of affecting cell surface expression [13]. Indeed, using mouse models of lung cancer, Curtis et al. noted that sorting cells by Sca1 cell surface expression resulted in varying rates of secondary transplantation [14]. Variations in transplantation success were proposed to be dictated by driver mutation status such as K-Ras, p53, or epidermal growth factor receptor (EGFR). These findings question

Table 8.1: Summary of cancer stem cell markers. A list of markers commonly used to identify and isolate CSCs

Cancer type	Marker	Reference
AML	CD34$^+$, CD38$^-$, CD123$^+$, CD47$^+$, CLL-1$^+$, CD96$^+$, TIM-3	[1, 9, 110–113]
Breast	CD44$^+$, CD24^{low}, ALDH1$^+$, CD133$^+$	[2, 12, 114]
Cervical	CD49f, ALDH1$^+$	[115, 116]
Colorectal	CD24$^+$, CD44$^+$	[117, 118]
Glioblastoma	CD44$^+$, CD133$^+$, ALDH1$^+$	[3, 97, 119]
Liver	CD24$^+$, CD133$^+$, ALDH1$^+$	[120–122]
Lung	CD44$^+$, CD133$^+$, ALDH1$^+$	[123–125]
Pancreas	CD24$^+$, CD44$^+$, CD133$^+$	[11, 123, 126, 127]
Prostate	CD44$^+$, CD24$^-$, CD133$^+$, ALDH1$^+$	[126, 128, 129]

whether isolating CSCs by cell surface markers is the most suitable selection criteria, although since these observations were generated using mouse models, it remains important to assess whether similar a situation occurs with primary patient samples.

Low levels of reactive oxygen species (ROS) are associated with low metabolic activity, typical of quiescence, a common feature of cancer stem cells. In light of this, recent studies have suggested the use of redox-sensitive probes to sort leukemic stem cells based on their ROS levels [15]. Leukemic stem cells associated with low ROS exhibited a greater level of engraftment compared with leukemic stem cells with higher ROS [15]. These findings with AML warrant investigation in other malignancies. Kreso and Dick also eluded to the potential idea of using miRNA signatures to identify CSCs based on the findings from two separate studies [16, 17]. The proof-of-principle concept arose from findings by Lechman et al. with long-term haematopoietic stem cells (HSCs) isolated based on mir-126 levels alone exhibiting engraftment potential [17]. Recently, the same group further expanded this phenomenon to AML whereby leukemic stem cells exhibited high mir-126 levels in patient samples when compared to normal HSCs [18]. Collectively, this utilisation of metabolic state as well as miRNA signatures provides impetus to investigate other novel, non-surface markers for isolating CSCs.

8.1.3 Cancer Stem Cells May Not Occur in All Cancers

The concept of the cancer stem cell is still debated, with many questions remaining to be answered. Some groups favour a stochastic model whereby each cell has the potential to form a tumour [19]. Stochastic models provide an argument against the work by Bonnet and Dick by suggesting that the self-renewal ability is applicable to all cells as tumour cells have overcome the Hayflick limit resulting in limitless replicative potential [20]. Thus, whilst there is some acceptance that AML follows a classical hierarchical CSC model, some tumours appear to lack this hierarchy although it should be noted that establishing a hierarchical organisation in patient samples of epithelial origin is difficult with a current lack of CSC-specific markers for a number of solid tumour types.

Intriguingly, B-cell acute lymphoblastic leukaemia appears to not fit with the hierarchical model, with a high level of cancer stem cell frequency noted by several groups [21, 22]. In the absence of a hierarchy, tumour cell plasticity as a consequence of selective pressures and mutation gain may explain how the tumour population can be replenished [23]. Activation of stem cell genes, commonly through changes in the epigenetic landscape, can revert a differentiated cell to a stem-like state [24]. For example, murine multi-potent progenitors can be transformed with the Mixed Lineage Leukaemia (MLL) oncogene, a histone methyltransferase and positive global regulator of gene transcription [25]. The acquisition of MLL-ENL (Eleven-nineteen-leukaemia) fusion protein confers self-renewal properties to committed progenitors, allowing tumour formation [24]. Similar findings have been reported with other MLL fusion partners suggesting this is a common feature in

MLL-driven leukaemogenesis [24, 26–28]. Similar approaches using viral transduction of oncogenes, such as human telomerase, into human mammary epithelial cells have shown to induce spontaneous de-differentiation into $CD44^{+}$ $CD24^{lo}$ breast cancer stem cells [29]. Similar observations have been seen using colon cancer models with NF-κB signalling inducing de-differentiation through β-catenin activation of stem cell genes in intestinal epithelial cells [30].

In each of these cases, acquisition of certain oncogenes can de-differentiate normal cells into CSCs, supporting a stochastic model where every cell is capable of sustaining tumour growth. Yet if one were to analyse the overall tumour population, a hierarchical organisation may still be present, as the transformation event may occur in a single clone, positioning it at the head of the hierarchy. Furthermore, in cases where the resistant clone after selection exhibits a stem-like phenotype, the bulk of the tumour population may be phenotypically and functionally similar to a CSC, thus resembling a stochastic model. With this is mind, the amalgamation of these models proposed by Kreso and Dick deserves considerable thought with the idea that a model is not static but rather a dynamic process that may favour one model over another under certain conditions or different malignancies [5]. The complexity is such that this represents one point of contention at this stage. Whilst this chapter is unable to expand on the intricacies of the CSC debate, this area has been summarised well by Kreso and Dick [5].

8.2 Sphingolipids

Sphingolipids are a highly diverse class of lipids, defined by the presence of a sphingoid backbone, that serve biological roles both as structural components of cell membranes and as mediators of cell signalling [31]. Ceramide is a central sphingolipid that cellular levels can be altered in response to stimuli such as growth factors and chemotherapy (Fig. 8.1) [32]. As a potent inducer of cell cycle arrest and apoptosis, maintaining non-lethal levels of ceramide in the cell is required [33]. Three main mechanisms appear to control the maintenance of ceramide levels in cells: a degradation pathways involving sphingosine kinase (SK), glycosylation via glucosylceramide synthase (GCS), and conversion to sphingomyelin (Fig. 8.1) [32, 34]. In addition to ceramide, a number of the other sphingolipids function as second messengers, with the most well-studied example being sphingosine 1-phosphate (S1P) produced by SK [35]. S1P functions as a ligand for a family of five G protein-coupled receptors ($S1P_{1-5}$) to activate a number of signalling responses including survival, proliferation, migration, and differentiation [36–38]. Sphingomyelins and glucosylceramides are also indirectly involved in signal transduction by aggregating in the plasma membrane with protein receptors to form lipid rafts [39, 40]. Although the processes by which sphingolipids activate the various signalling pathways has been mapped out, much of their roles in both normal and cancer stem cell biology are only beginning to emerge.

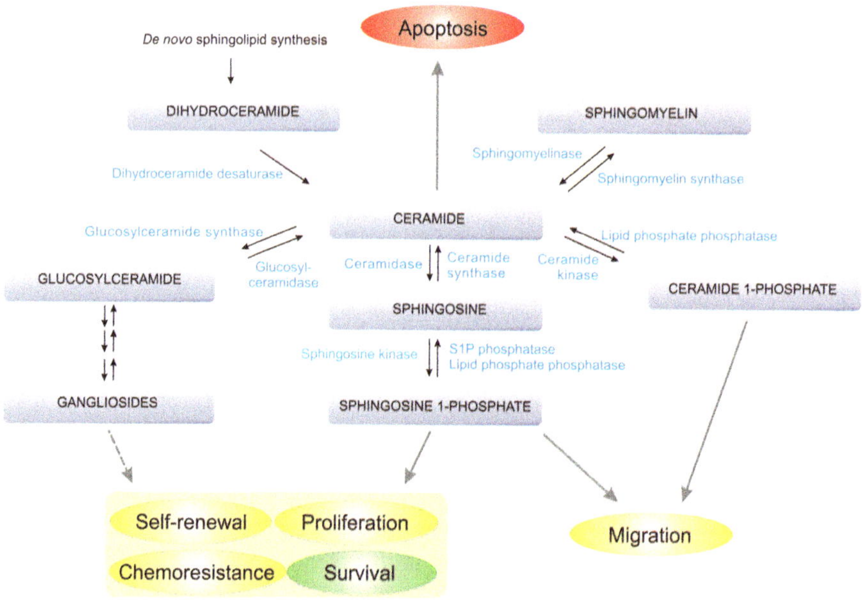

Fig. 8.1: Overview of sphingolipid metabolism and its roles in stem cell biology. Ceramide functions as an intermediate and can be metabolised in response to stimuli to induce specific cellular outcomes. Maintaining non-lethal levels of ceramide revolves around three main pathways involving conversion to other sphingolipids such as sphingomyelin, degradation into sphingosine 1-phosphate (S1P), and modifications such as glucosylation. Phosphorylation by ceramide kinase can promote a migratory role by ceramide 1-phosphate acting as a chemoattractant in a similar fashion to S1P. In addition to its chemoattractant role, S1P can act as a second messenger to promote stem cell biology such as self-renewal, proliferation, and drug resistance. Synthesis of complex glycosphingolipids such as gangliosides via glucosylceramide can also indirectly promote stem cell pathways. Targeting of multiple enzymes such as SK and GCS can promote the accumulation of ceramide as well as block stem cell-signalling pathway and consequently apoptosis

8.3 Maintenance of "Stemness" and Self-Renewal by Sphingolipids

Much of our understanding of the role of sphingolipids in stem cell biology has come from studies examining their contribution in normal homeostasis. For example, S1P along with platelet-derived growth factor (PDGF) have been shown to collectively maintain the primitive state of human embryonic stem cells and the expression of stem cell markers such as OCT-4 [41]. PDGF and S1P were found to maintain stemness through activation of extracellular signal-regulated kinase (ERK) signalling which in turn activates SK [41]. There is also some evidence to suggest a role for S1P in preservation of neural stem and progenitor cells [42]. S1P treatment of neural progenitor cells from rat embryos could also induce upregulation of ERK and telomerase activity suggesting a similar role for stem cell maintenance across different species [43]. In mouse embryonic stem cells, S1P can activate β-arrestin and c-Src through $S1P_1$ and $S1P_3$ to promote proliferation [44]. S1P-induced proliferation was dependent on the accumulation of $S1P_1$ and $S1P_3$ and foetal liver kinase-1 (FLK-1) in lipid rafts, enabling ERK activation [44]. ERK also appears to have a pro-proliferative role in mesenchymal stem cells (MSC) with inhibition of $S1P_2$ associated with proliferation and differentiation of MSCs [45]. Whilst not explicitly mentioned, inhibition of $S1P_2$ has been previously reported to enhance the migration of mouse embryonic fibroblasts (MEFs) towards PDGF [46]. Furthermore, knockdown of $S1P_1$ inhibited the migratory response in $S1P_2$ knockout MEFs implicating crosstalk between $S1P_1$ and $S1P_2$ [46]. Thus, it could be suggested that inhibiting $S1P_2$ in other stem cell types may exhibit similar signalling responses, by promoting $S1P_1$ signalling such as self-renewal and proliferation. Notably, many other stem cell types express S1P receptors although their function in the biology of these cells remains largely unknown at present [47]. Based on the findings elucidated thus far, it appears that $S1P_1$ and $S1P_3$ appear to promote self-renewal and proliferation pathways in normal stem cells.

Despite the evidence for the contribution of sphingolipids in normal stem cell biology, their contribution towards maintaining CSCs is only just emerging. CSCs exhibit traits similar to normal stem cells in that they show self-replicative potential essential for long-term maintenance of the tumour population [48]. Therefore, it is likely that many of the pathways extensively characterised in normal tissue maintenance can be hijacked in malignancy. Maintaining a stem cell-like state requires a number of signalling pathways such as those involving Wnt/β-Catenin, Notch, and Hedgehog [49]. Liu et al. first reported that in breast cancer cell lines, increases in GCS and ganglioside production upregulated β-catenin signalling in response to doxorubicin treatment [50]. Following on from these findings, this same group observed that doxorubicin treatment resulted in expansion of the $CD44^+$ $CD24^{lo}$ breast cancer stem cell (BCSC) population in vivo using cell line xenografts [51]. Further investigation into the mechanism behind this phenomenon revealed GCS increases ganglioside synthesis upon doxorubicin treatment, promoting β-catenin signalling and activating stem cell program genes such as CD44 and Oct-4 to ultimately promote BCSC expansion [52].

Similar signalling responses were also seen in ovary, cervical, and colon cancer cells suggesting this pathway may be highly expressed in CSCs irrespective of tumour type [53]. The observed increase in GCS expression in BCSCs suggests that resistant cells may revert to a more stem-like state by upregulating stem cell program genes through GCS and consequently creating a more stem cell-based tumour population. As the initiating step of ceramide glycosylation, GCS not only lowers cellular ceramide levels to enhance cell survival, but is also essential to the formation of gangliosides which have documented roles in both normal and cancer stem cells [54]. Gangliosides are thought to assemble in sites of signal transduction as a complex of glycosphingolipids and receptor tyrosine kinases and have been touted to be essential to signalling responses [55]. Changes in the lipid composition and structure of these “rafts” have been proposed as the mechanism as to how receptors activate signal transduction in response to ligand binding [40]. Mass spectrometric analysis of the changes in glycosphingolipid profiles revealed distinct preferences for gangliosides GD2 and GD3 [56] in BCSC. Sorted GD2$^+$ cells were highly enriched for CD44$^+$, CD24^{lo} BCSCs suggesting GD2 to be a BCSC marker [57]. Battula et al. also identified GD3 synthase, upstream of GD2, as essential to mammary tumour formation with shRNA knockdown completely abolishing tumours in vivo [57]. With the necessity of GCS in forming these gangliosides, this has the potential to make GCS a novel CSC-specific target.

Although ceramide glycosylation represents one mechanism by which cancer cells maintain ceramide levels at non-apoptotic levels, a potential role for other sphingolipid enzymes such as SK has also been reported in breast cancer. For instance, S1P can promote ligand-independent Notch signalling through $S1P_3$ to expand ALDH1$^+$ BCSCs [58]. Enforced expression of SK1 in ALDH1$^+$ BCSCs enhanced tumour development in vivo with blockade of $S1P_3$ signalling reducing tumour size [58]. In addition, both SK1/ALDH1$^+$ and $S1P_3^+$/ALDH1$^+$ CSCs could be isolated from breast cancer patient samples suggesting $S1P_3$ to be a BCSC marker [58]. The role of $S1P_3$ in breast cancer stem cells was further expanded upon recently with the carcinogen, benzyl butyl phthalate promoting breast cancer stem cell tumourigenesis through BCSC expansion by aryl hydrocarbon receptor-driven expression of $S1P_3$ [59].

Whilst the majority of above work has been performed in breast cancer, it is possible that sphingolipids play a similar role in other solid tumours. Analysis of glioblastoma patient samples found S1P to be secreted by glioma stem cells [60], and acts in both an auto- and paracrine manner to promote expression of stem cell markers such as CD133, implicating a role for S1P in maintaining "stemness" in glioma stem cells [60]. Interestingly, heterogeneity between isolated glioma stem cells from different patients was observed particularly in metabolic processing of sphingolipids [60], with increased S1P production associated with a greater level of glioma stem cell proliferation. Whilst the small number of patient samples may question the validity of these findings ($n = 2$), it prompts the question as to whether enhanced processing of ceramide or SK activity is associated with more aggressive disease by expansion of the cancer stem cell pool.

The activation of stem cell program genes by S1P suggests that the sphingolipids could play a role in cell plasticity. Whilst sphingolipids can maintain "stemness" of cancer stem cells, whether or not they can induce de-differentiation into a stem cell is unknown. As a point of interest, sphingosine kinase 2 (SK2) has demonstrated in acute lymphoblastic leukaemia to promote Myc expression through histone deacetylase (HDAC)1/2 inhibition [61]. This raises the possibility that SK2 may elicit large-scale epigenetic changes [62]. Although the targets regulated by SK2 through HDAC1/2 inhibition have been largely unexplored, activation of stem cell program genes allowing de-differentiation of mature cells to more stem-like cells is worth consideration. Of those characterised thus far, Myc is known to have a crucial role in both self-renewal and de-differentiation in normal stem cells. Myc target genes such as human telomerase reverse transcriptase (hTERT) are essential to the self-renewal capacity of both normal and cancer stem cells suggesting in may activate other stem cell genes [63]. CSC marker CD133 has been demonstrated to upregulate Myc in glioma stem cells suggesting it may have a role in CSC biology [64]. Recently, SK2 has also demonstrated to enhance telomerase stability in A549 lung cancer cells, albeit through a transcriptional independent mechanism [65]. Yet is conceivable that SK2 could transcriptionally upregulate hTERT through HDAC inhibition and Myc transcription. Whilst the initial findings into understanding stem cell maintenance by sphingolipids have focussed on breast cancer and glioblastoma, these findings provide an impetus to examine this system in other solid tumour stem cells.

8.4 Altered Sphingolipid Metabolism as a Mechanism of Drug Resistance

Being responsible for long-term maintenance of the tumour population, CSCs maintain a quiescent state by self-renewal and differentiate when necessary. Conventional chemotherapeutics such as DNA-damaging agents and microtubule poisons target rapidly dividing cells that compose the bulk of the tumour. The inability for these agents to effectively target the CSC population, however, provides a reservoir of drug resistance [66]. Many of the chemotherapeutics in use promote cell death, in part, by increasing pro-apoptotic ceramides [67]. Although not studied as extensively in CSCs specifically, increases in SK1 expression enhance chemotherapeutic resistance in numerous solid and haematological cancers [34, 68–70]. Interestingly, a study into glioma stem cells revealed S1P as a mechanism of resistance to the chemotherapeutic, temozolomide (TMZ), independent of the DNA repair protein MGMT [71]. Whilst enhanced processing of sphingosine as well as S1P transporter activity were implicated as mediators of S1P-mediated TMZ resistance, this avenue of investigating was not further explored [71]. Given exogenous S1P could reverse TMZ cytotoxicity, the involvement of the S1P receptors in TMZ resistance warrants consideration [71].

Ceramide glycosylation provides a rapid mechanism to escape cell death by blocking the accumulation of ceramide and the activation of pro-apoptotic pathways. Notably, GCS was shown to promote the expansion of BCSCs in response to doxorubicin as previously mentioned. Doxorubicin-resistant MCF7 cells were re-sensitised to doxorubicin by knockdown of GCS, which also reduced the ability of these cells to form colonies in soft agar [52]. Reductions in $CD44^{+}$, $CD24^{-}$ BCSCs were also observed in the doxorubicin-resistant cell population following GCS knockdown. These findings also translated to reductions in $CD44^{+}$, $CD24^{-}$ BCSC numbers in vivo when combining GCS knockdown with doxorubicin treatment [51, 52].

Retrospective meta-analysis of clinical trials revealed GCS as well as the ceramide transport protein (CERT) as potential markers of triple negative breast cancer patient response to the chemotherapeutic paclitaxel [72]. CERT is important in the maintenance of homeostatic levels of ceramide species via promoting the transport of ceramide from the endoplasmic reticulum (ER) to the golgi for conversion to sphingomyelin [73]. As a marker of paclitaxel sensitivity, upregulation of CERT likely contributes to the ability of malignant cells to bypass the apoptotic effects by preventing the accumulation of lethal levels of ceramide. Other potential targets such as sphingomyelin synthase the enzyme that converts ceramide to sphingomyelin have also demonstrated to govern resistance to ceramide-induced apoptosis when sorted CD55^{hi} BCSCs were treated with C2-ceramide [74].

Whilst much of the contribution of sphingolipids towards drug resistance in CSCs has been largely overlooked, it is reasonable to suggest that some of the pathways described above could be present in other malignancies given their role in both cell lines and primary patient samples [72].

8.5 Sphingolipids in Normal and CSC Migration: Following the Lipid Drops

Over the past decade, much of the work involving sphingolipids and their contribution towards cell migration has focussed on the S1P receptors. In particular, the importance of the S1P receptor 1 (S1P$_1$) has been demonstrated to facilitate lymphocyte egress following development within the lymphoid organs. Upregulation of S1P$_1$ allows lymphocytes to respond to the circulating S1P in the peripheral blood as a chemotactic factor with S1P$_1$ knockout lymphocytes accumulating within lymphoid organs due their inability to respond to the S1P gradient [75]. Similar findings have also been reported with a requirement of S1P$_1$ in the trafficking of haematopoietic stem and progenitor cells (HSPC) from the tissues and into the lymphatics as part of normal immune surveillance [76]. Complement activation can promote HSPC egress from the bone marrow by complement cascade-dependent release of S1P from erythrocytes into the peripheral blood [77]. Forced overexpression of S1P$_1$ in human CD34$^+$ haematopoietic progenitor cells promoted chemotaxis towards an S1P gradient as well as preventing in vivo homing to the bone marrow by preventing the expression of inhibitory receptors such as CD69 [75, 78]. Conditional knockout of SK1 in bone marrow cells, displayed a homing and engraftment defect implicating a role for S1P in bone marrow homing of circulating haematopoietic cells [79]. Enhancement of this defect was observed when HSPCs from *CXCR4* knockout mice exhibited short- and long-term defects in bone marrow engraftment in agreeance with its role as an adhesion molecule for retention of haematopoietic cells in the bone marrow [79].

The initial findings into the involvement of the S1P receptors in cell migration stemmed from the use of the S1P receptor modulator, FTY720. Normally, engagement of S1P$_1$ by S1P results in internalisation of the receptor followed by rapid recycling back to the cell surface. Engagement of S1P$_1$ by FTY720, however results in intracellular retention and degradation of the receptor, preventing cells from responding to the S1P gradient [80]. The use of FTY720 to study cell migration recapitulates the phenotypes using S1P$_1$ knockout haematopoietic cells with lymphocyte accumulation within the lymph nodes [75] as well as retention of HSPCs in the bone marrow [76]. Due to the role of S1P$_1$ in HSPC egress from the bone marrow, S1P$_1$ agonists in combination CXCR4 antagonists exhibit therapeutic potential for transplantation purposes [81].

The interplay between sphingolipids, ceramide-1-phosphate (C1P) and S1P controls the movement of HSPCs in a bidirectional manner. Ratajczak and Kim proposed a potential mechanism whereby myeloablative conditioning can induce the release of C1P, attracting circulating HPSCs to home to the bone marrow [82]. Homing of circulating HSPCs was dependent on the activation of the complement cascade and the membrane attack complex in a retrograde manner in contrast to the mechanism described above [83]. C1P has also been reported to be induced in response to radiation, promoting a pro-metastatic environment away from the site of radiation [84]. These findings support the notion that C1P may function as a chemoattractant, as well as its generation acting as a pro-survival pathway via reducing the accumulation of ceramide induced in response to radiation [85].

In the context of cancer biology, S1P as a chemotactic factor appears to have a more sinister role particularly within the tumour microenvironment and as a promoter of metastasis [86, 87]. Tumour cells respond to S1P as a chemotactic factor to migrate from the primary tumour site promoting secondary metastasis [88]. What remains unknown is how sphingolipids can influence CSCs in the context of migration and metastasis. Many of the studies involving sphingolipids as mediators of migration, homing, and metastasis in cancer have predominantly used cell lines. S1P has been shown to induce migration in multiple cancer types including ovarian, glioma, and thyroid [89–91]. As normal stem cells can respond to S1P as a chemotactic factor, it is conceivable to suggest CSCs may also respond in a similar fashion.

What is known thus far is that secreted S1P from both the stroma and tumour cells themselves appear to induce structural changes in the microenvironment as well as the production of enzymes such as matrix metalloproteases [92, 93]. CD133$^+$ glioblastoma stem cells isolated from U87 glioblastoma cells exhibited elevated levels of S1P receptors 1 and 2 with an enhanced migratory response to S1P [86]. Enhanced migration of CD133$^+$ glioblastoma stem cells was coupled with increases in membrane type 1 matrix metalloprotease (MT1-MMP) production to collectively promote migration and metastasis [86]. Knockdown of MT1-MMP was shown to block S1P-induced migration of glioma stem cells, which may have applications for secondary organ metastasis given the role of MT1-MMP in the breakdown of extracellular matrix [86, 94]. Interestingly, SK1 has shown to induce expression of CSC marker CD44 in colon cancer [95], a known target of MT1-MMP protease activity [96]. The expression of CD44 in other solid tumours (Table 8.1) including glioblastoma [97] proposes that SK1 may also have a role in promoting both CD44 and MT1-MMP in glioblastoma.

As the majority of disseminated tumour cells succumb to anoikis, there is a perception that the ability to successfully metastasise is exclusive to CSCs [98]. The success of metastasis also hinges on the ability of immune cells such as myeloid suppressor cells to provide a permissive microenvironment for secondary tumour growth [99]. S1P has shown to promote the invasion of myeloid-derived suppressor cells in secondary pre-metastatic sites, essential for the successful metastasise of tumour cells [87]. Given the roles of S1P in the modelling of the microenvironment [86] as well as migration [89–92], it is tempting to speculate that secreted S1P from secondary metastatic sites creates an environment permissible for CSC propagation. With the diverse roles sphingolipids regulate, it comes as no surprise that deletion of $S1P_1$ in CD11b+ myeloid cells, reduced secondary metastasis [87, 88] although further work is required to determine the stage of metastasis that is impacted on most by deleting of these genes. Clearly analysis of whether S1P is a significant chemotactic factor for CSCs requires further investigation.

8.6 Modulating Sphingolipid Metabolism to Target CSCs

The future implementation of CSC-specific therapies is predicted to greatly increase the percentage of patients that could achieve complete and deep molecular responses [100]. Thus given the number of roles sphingolipids appear to play in CSC biology, it is conceivable that the modulation of sphingolipid metabolism could play a crucial role in effectively targeting the CSC population. Ultimately, targeting sphingolipid metabolism focusses on increases in lethal levels of pro-apoptotic ceramides to induce cell death. Refractory responses towards CSCs in patients require development of rationale drug combinations to effectively increase ceramides and prevent their modification. As a proof of principle, targeting GCS prevents the accumulation of BCSCs in vivo when mice were treated with doxorubicin [51, 52], suggesting this approach may be effective in breast cancer patients. Tamoxifen whilst exclusively utilised in oestrogen receptor-positive breast cancer, displays off-target activity by inhibiting GCS [101]. With this in mind, Morad and Cabot proposed the addition of tamoxifen to the standard chemotherapeutics used in breast cancer such as doxorubicin and paclitaxel [101]. Based on the in vitro and mouse model findings, GCS inhibition has the potential to enhance patient response and 5-year survival by effectively targeting the BCSC population. As mentioned previously, GCS has been identified as a marker of paclitaxel sensitivity in breast cancer patients further advocating the use of tamoxifen in oestrogen receptor negative breast cancer.

As SK can cooperate to prevent ceramide accumulation, the use of SK inhibitors alongside chemotherapy could provide added stimuli to more effectively induce apoptosis in the CSC population. Indeed, several groups have shown synergistic killing when combining sphingosine kinase inhibitors with chemotherapeutics [34, 68, 102]. Currently, evidence for combining SK inhibitors with chemotherapy to effectively target CSCs remains largely unexplored. In glioblastoma, combining the SK/dihydroceramide desaturase inhibitor, SKI-II with TMZ, induced synergistic cell death in glioblastoma stem cells [71].

FTY720 has displayed anti-cancer activity against glioma stem cells by blocking S1P-induced cell cycle progression [71]. Combining FTY720 with TMZ also displayed enhanced survival in vivo using orthotopic intracranial glioblastoma xenograft models resulting in enhanced mouse survival [103]. S1P receptor modulators such as FTY720 may also inhibit CSC metastasis by blocking responses to the S1P gradient that promotes the migration of metastatic potential, particularly in glioblastoma [86]. Furthermore, FTY720 has also displayed anti-cancer activity independent of the S1P receptors by eradicating chronic myeloid leukaemia stem cells, through a mechanism reported to be mediated by the reactivation of protein phosphatase 2 by repressing the negative regulators of this enzyme [104–106], although FTY720 has numerous other cellular targets that may contribute to this effect [107].

The ability of S1P to invoke many of the signalling responses through the S1P receptors has attracted interest of drug development to target these receptors. Currently, several S1P receptor antagonists are in phase I/II clinical trials for a number of malignancies [108]. Indeed, targeting $S1P_3$ was shown to block the self-renewal pathways invoked by SK1 in BCSCs [58]. Similar findings were reported with knockdown of either SK1 or $S1P_3$ blocking the expansion of BCSCs upon benzyl butyl phthalate treatment [59].

S1P receptor antagonists also may be utilised to reduce the incidence of metastasis particularly in glioblastoma. Whilst the targeting of MT1-MMP in $CD133^+$ glioma stem cells reduced the migratory response to S1P, the elevated levels of $S1P_1$ and $S1P_2$ receptors suggest receptor antagonists could reduce the incidence of metastasis in glioblastoma [86].

8.7 Perspectives and Future Directions

Whilst extensively studied for their contribution towards biological signalling responses, there is an underappreciation for the potential roles of sphingolipids in CSC biology. The majority of studies investigating sphingolipids in the context of cancer biology have focussed predominately on in vitro studies using laboratory adapted cell lines that show little resemblance to CSCs. Although they provide an insight into basic biological processes, translating these potential findings into CSCs requires the use of patient biopsies. However, the handful of studies performed have exhibited similar trends in both cell lines and cancer stem cells such as the discovery of GCS as a mechanism of drug resistance by preventing fatal accumulation of ceramide species in response to chemotherapy. Do CSCs upregulate genes such as GCS as seen in breast cancer to prevent ceramide accumulation and does this represent an "oncogenic addiction"? Future investigation of potential roles of sphingolipids in cancer stem cells will likely require high throughput screenings such as those used by Hirata et al. to discover the role for $S1P_3$ in maintaining breast cancer stem cells [58]. Based on the findings thus far, GCS appears to present the most promising target in terms of stem cell-specific therapies. The implementation of GCS inhibitors, which include tamoxifen [109] alongside chemotherapy, would likely induce a deeper molecular response by preventing ceramide glycosylation and simultaneously preventing BCSC expansion [51, 52]. Whilst much of the work focussing on sphingolipid metabolism has focussed predominately on a handful of enzymes such as SK, ceramidase, and GCS in cancer, future research should encompass a broader examination of sphingolipid metabolism as a whole. Dysregulation of multiple enzymes can equally contribute to cell survival such as increased GCS and SK activity to collectively reduce ceramide accumulation in response to drug treatment. Thus, targeting of multiple enzymes in sphingolipid metabolism may prove to be the most beneficial approach. In the coming era of novel drug combinations, understanding how targeting certain aspects of sphingolipid metabolism and the changes in cell signalling associated with it will assist in targeting the CSC and inducing deeper and sustained molecular responses in cancer therapy.

Acknowledgements This work was supported by the Fay Fuller Foundation, Neurosurgical Research Foundation, Cancer Council SA Beat Cancer Project Grant (1086295), an Australian Postgraduate Award and Royal Adelaide Hospital Dawes Top-up scholarship to ACL, and a National Health and Medical Research Council of Australia Senior Research Fellowship to SMP.

References

1. Bonnet D, Dick JE (1997) Human acute myeloid leukemia is organized as a hierarchy that originates from a primitive hematopoietic cell. Nat Med 3(7):730–737
2. Al-Hajj M, Wicha MS, Benito-Hernandez A, Morrison SJ, Clarke MF (2003) Prospective identification of tumorigenic breast cancer cells. Proc Natl Acad Sci U S A 100(7):3983–3988
3. Singh SK, Clarke ID, Terasaki M, Bonn VE, Hawkins C, Squire J et al (2003) Identification of a cancer stem cell in human brain tumors. Cancer Res 63(18):5821–5828
4. Quintana E, Shackleton M, Sabel MS, Fullen DR, Johnson TM, Morrison SJ (2008) Efficient tumour formation by single human melanoma cells. Nature 456(7222):593–598
5. Kreso A, Dick JE (2014) Evolution of the cancer stem cell model. Cell Stem Cell 14(3):275–291
6. Lee EM, Yee D, Busfield SJ, McManus JF, Cummings N, Vairo G et al (2015) Efficacy of an Fc-modified anti-CD123 antibody (CSL362) combined with chemotherapy in xenograft models of acute myelogenous leukemia in immunodeficient mice. Haematologica 100(7):914–926
7. Shih LY, Huang CF, Wu JH, Lin TL, Dunn P, Wang PN et al (2002) Internal tandem duplication of FLT3 in relapsed acute myeloid leukemia: a comparative analysis of bone marrow samples from 108 adult patients at diagnosis and relapse. Blood 100(7):2387–2392
8. Reinisch A, Thomas D, Corces MR, Zhang X, Gratzinger D, Hong W-J, et al (2016) A humanized bone marrow ossicle xenotransplantation model enables improved engraftment of healthy and leukemic human hematopoietic cells. Nat Med
9. Jordan C, Upchurch D, Szilvassy S, Guzman M, Howard D, Pettigrew A et al (2000) The interleukin-3 receptor alpha chain is a unique marker for human acute myelogenous leukemia stem cells. Leukemia 14(10):1777–1784
10. Jiang F, Qiu Q, Khanna A, Todd NW, Deepak J, Xing L et al (2009) Aldehyde dehydrogenase 1 is a tumor stem cell-associated marker in lung cancer. Mol Cancer Res 7(3):330–338
11. Dalerba P, Dylla SJ, Park I-K, Liu R, Wang X, Cho RW et al (2007) Phenotypic characterization of human colorectal cancer stem cells. Proc Natl Acad Sci U S A 104(24):10158–10163
12. Ginestier C, Hur MH, Charafe-Jauffret E, Monville F, Dutcher J, Brown M et al (2007) ALDH1 is a marker of normal and malignant human mammary stem cells and a predictor of poor clinical outcome. Cell Stem Cell 1(5):555–567
13. Medema JP (2013) Cancer stem cells: The challenges ahead. Nat Cell Biol 15(4):338–344
14. Curtis SJ, Sinkevicius KW, Li D, Lau AN, Roach RR, Zamponi R et al (2010) Primary tumor genotype is an important determinant in identification of lung cancer propagating cells. Cell Stem Cell 7(1):127–133
15. Lagadinou Eleni D, Sach A, Callahan K, Rossi Randall M, Neering Sarah J, Minhajuddin M et al (2013) BCL-2 inhibition targets oxidative phosphorylation and selectively eradicates quiescent human leukemia stem cells. Cell Stem Cell 12(3):329–341

16. Gentner B, Visigalli I, Hiramatsu H, Lechman E, Ungari S, Giustacchini A et al (2010) Identification of hematopoietic stem cell-specific miRNAs enables gene therapy of globoid cell leukodystrophy. Sci Transl Med 2(58):58ra84
17. Lechman Eric R, Gentner B, van Galen P, Giustacchini A, Saini M, Boccalatte Francesco E et al (2012) Attenuation of miR-126 activity expands HSC in-vivo without exhaustion. Cell Stem Cell 11(6):799–811
18. Lechman ER, Gentner B, Ng SW, Schoof EM, van Galen P, Kennedy JA et al (2016) miR-126 regulates distinct self-renewal outcomes in normal and malignant hematopoietic stem cells. Cancer Cell 29(2):214–228
19. Shackleton M, Quintana E, Fearon ER, Morrison SJ (2009) Heterogeneity in cancer: cancer stem cells versus clonal evolution. Cell 138(5):822–829
20. Gupta PB, Chaffer CL, Weinberg RA (2009) Cancer stem cells: mirage or reality? Nat Med 15(9):1010–1012
21. Williams RT, den Besten W, Sherr CJ (2007) Cytokine-dependent imatinib resistance in mouse BCR-ABL+, Arf-null lymphoblastic leukemia. Genes Dev 21(18):2283–2287
22. Rehe K, Wilson K, Bomken S, Williamson D, Irving J, den Boer ML et al (2013) Acute B lymphoblastic leukaemia-propagating cells are present at high frequency in diverse lymphoblast populations. EMBO Mol Med 5(1):38–51
23. Chaffer CL, Brueckmann I, Scheel C, Kaestli AJ, Wiggins PA, Rodrigues LO et al (2011) Normal and neoplastic nonstem cells can spontaneously convert to a stem-like state. Proc Natl Acad Sci U S A 108(19):7950–7955
24. Cozzio A, Passegué E, Ayton PM, Karsunky H, Cleary ML, Weissman IL (2003) Similar MLL-associated leukemias arising from self-renewing stem cells and short-lived myeloid progenitors. Genes Dev 17(24):3029–3035
25. Krivtsov AV, Armstrong SA (2007) MLL translocations, histone modifications and leukaemia stem-cell development. Nat Rev Cancer 7(11):823–833
26. Lavau C, Du C, Thirman M, Zeleznik-Le N (2000) Chromatin-related properties of CBP fused to MLL generate a myelodysplastic-like syndrome that evolves into myeloid leukemia. EMBO J 19(17):4655–4664
27. So CW, Karsunky H, Passegué E, Cozzio A, Weissman IL, Cleary ML (2003) MLL-GAS7 transforms multipotent hematopoietic progenitors and induces mixed lineage leukemias in mice. Cancer Cell 3(2):161–171
28. Krivtsov AV, Twomey D, Feng Z, Stubbs MC, Wang Y, Faber J et al (2006) Transformation from committed progenitor to leukaemia stem cell initiated by MLL-AF9. Nature 442(7104):818–822
29. Zhao X, Malhotra GK, Lele SM, Lele MS, West WW, Eudy JD et al (2010) Telomerase-immortalized human mammary stem/progenitor cells with ability to self-renew and differentiate. Proc Natl Acad Sci U S A 107(32):14146–14151
30. Schwitalla S, Fingerle Alexander A, Cammareri P, Nebelsiek T, Göktuna Serkan I, Ziegler Paul K et al (2012) Intestinal tumorigenesis initiated by dedifferentiation and acquisition of stem-cell-like properties. Cell 152(1):25–38
31. van Meer G, Voelker DR, Feigenson GW (2008) Membrane lipids: where they are and how they behave. Nat Rev Mol Cell Biol 9(2):112–124
32. Veldman RJ, Klappe K, Hinrichs J, Hummel I, van der Schaaf G, Sietsma H et al (2002) Altered sphingolipid metabolism in multidrug-resistant ovarian cancer cells is due to uncoupling of glycolipid biosynthesis in the Golgi apparatus. FASEB J 16(9):1111–1113
33. Pitson SM (2011) Regulation of sphingosine kinase and sphingolipid signaling. Trends Biochem Sci 36(2):97–107
34. Bonhoure E, Pchejetski D, Aouali N, Morjani H, Levade T, Kohama T et al (2006) Overcoming MDR-associated chemoresistance in HL-60 acute myeloid leukemia cells by targeting sphingosine kinase-1. Leukemia 20(1):95–102
35. Olivera A, Spiegel S (1993) Sphingosine-1-phosphate as second messenger in cell proliferation induced by PDGF and FCS mitogens. Nature 365(6446):557–560

36. An S, Zheng Y, Bleu T (2000) Sphingosine 1-phosphate-induced cell proliferation, survival, and related signaling events mediated by G protein-coupled receptors Edg3 and Edg5. J Biol Chem 275(1):288–296
37. Nincheri P, Luciani P, Squecco R, Donati C, Bernacchioni C, Borgognoni L et al (2009) Sphingosine 1-phosphate induces differentiation of adipose tissue-derived mesenchymal stem cells towards smooth muscle cells. Cell Mol Life Sci 66(10):1741–1754
38. Kimura T, Watanabe T, Sato K, Kon J, Tomura H, Tamama K et al (2000) Sphingosine 1-phosphate stimulates proliferation and migration of human endothelial cells possibly through the lipid receptors, Edg-1 and Edg-3. Biochem J 348(Pt 1):71–76
39. Simons K, Gerl MJ (2010) Revitalizing membrane rafts: new tools and insights. Nat Rev Mol Cell Biol 11(10):688–699
40. Simons K, Toomre D (2000) Lipid rafts and signal transduction. Nat Rev Mol Cell Biol 1(1):31–39
41. Pebay A, Wong RC, Pitson SM, Wolvetang EJ, Peh GS, Filipczyk A et al (2005) Essential roles of sphingosine-1-phosphate and platelet-derived growth factor in the maintenance of human embryonic stem cells. Stem Cells 23(10):1541–1548
42. Pitson SM, Pebay A (2009) Regulation of stem cell pluripotency and neural differentiation by lysophospholipids. Neurosignals 17(4):242–254
43. Harada J, Foley M, Moskowitz MA, Waeber C (2004) Sphingosine-1-phosphate induces proliferation and morphological changes of neural progenitor cells. J Neurochem 88(4):1026–1039
44. Ryu JM, Baek YB, Shin MS, Park JH, Park SH, Lee JH et al (2014) Sphingosine-1-phosphate-induced Flk-1 transactivation stimulates mouse embryonic stem cell proliferation through S1P1/S1P3-dependent β-arrestin/c-Src pathways. Stem Cell Res 12(1):69–85
45. Price ST, Beckham TH, Cheng JC, Lu P, Liu X, Norris JS (2015) Sphingosine 1-phosphate. Receptor 2 regulates the migration, proliferation, and differentiation of mesenchymal stem cells. Int J Stem Cell Res Ther 2(2):014
46. Goparaju SK, Jolly PS, Watterson KR, Bektas M, Alvarez S, Sarkar S et al (2005) The S1P2 receptor negatively regulates platelet-derived growth factor-induced motility and proliferation. Mol Cell Biol 25(10):4237–4249
47. Pébay A, Bonder CS, Pitson SM (2007) Stem cell regulation by lysophospholipids. Prostaglandins Other Lipid Mediat 84(3–4):83–97
48. O'Brien CA, Kreso A, Jamieson CHM (2010) Cancer stem cells and self-renewal. Clin Cancer Res 16(12):3113–3120
49. Borah A, Raveendran S, Rochani A, Maekawa T, Kumar DS (2015) Targeting self-renewal pathways in cancer stem cells: clinical implications for cancer therapy. Oncogenesis 4:e177
50. Liu Y-Y, Gupta V, Patwardhan GA, Bhinge K, Zhao Y, Bao J et al (2010) Glucosylceramide synthase upregulates MDR1 expression in the regulation of cancer drug resistance through cSrc and β-catenin signaling. Mol Cancer 9(1):1–15
51. Bhinge KN, Gupta V, Hosain SB, Satyanarayanajois SD, Meyer SA, Blaylock B et al (2012) The opposite effects of doxorubicin on bone marrow stem cells versus breast cancer stem cells depend on glucosylceramide synthase. Int J Biochem Cell Biol 44(11):1770–1778
52. Gupta V, Bhinge KN, Hosain SB, Xiong K, Gu X, Shi R et al (2012) Ceramide glycosylation by glucosylceramide synthase selectively maintains the properties of breast cancer stem cells. J Biol Chem 287(44):37195–37205
53. Gouaze V, Yu JY, Bleicher RJ, Han TY, Liu YY, Wang H et al (2004) Overexpression of glucosylceramide synthase and P-glycoprotein in cancer cells selected for resistance to natural product chemotherapy. Mol Cancer Ther 3(5):633–639
54. Liu Y-Y, Li Y-T (2013) Ceramide glycosylation catalyzed by glucosylceramide synthase and cancer drug resistance. Adv Cancer Res 117:59–89
55. Sonnino S, Prinetti A (2013) Membrane domains and the "lipid raft" concept. Curr Med Chem 20(1):4–21
56. Liang YJ, Ding Y, Levery SB, Lobaton M, Handa K, Hakomori SI (2013) Differential expression profiles of glycosphingolipids in human breast cancer stem cells vs. cancer non-stem cells. Proc Natl Acad Sci U S A 110(13):4968–4973

57. Battula VL, Shi Y, Evans KW, Wang RY, Spaeth EL, Jacamo RO et al (2012) Ganglioside GD2 identifies breast cancer stem cells and promotes tumorigenesis. J Clin Invest 122(6):2066–2078
58. Hirata N, Yamada S, Shoda T, Kurihara M, Sekino Y, Kanda Y (2014) Sphingosine-1-phosphate promotes expansion of cancer stem cells via S1PR3 by a ligand-independent Notch activation. Nat Commun 5:4806
59. Wang Y-C, Tsai C-F, Chuang H-L, Chang Y-C, Chen H-S, Lee J-N, et al (2016) Benzyl butyl phthalate promotes breast cancer stem cell expansion via SPHK1/S1P/S1PR3 signaling. Oncotarget. doi:10.18632/oncotarget.9007
60. Marfia G, Campanella R, Navone SE, Di Vito C, Riccitelli E, Hadi LA et al (2014) Autocrine/paracrine sphingosine-1-phosphate fuels proliferative and stemness qualities of glioblastoma stem cells. Glia 62(12):1968–1981
61. Wallington-Beddoe CT, Powell JA, Tong D, Pitson SM, Bradstock KF, Bendall LJ (2014) Sphingosine kinase 2 promotes acute lymphoblastic leukemia by enhancing MYC expression. Cancer Res 74(10):2803–2815
62. Hait NC, Allegood J, Maceyka M, Strub GM, Harikumar KB, Singh SK et al (2009) Regulation of histone acetylation in the nucleus by sphingosine-1-phosphate. Science 325(5945):1254–1257
63. Murphy MJ, Wilson A, Trumpp A (2005) More than just proliferation: Myc function in stem cells. Trends Cell Biol 15(3):128–137
64. Wang J, Wang H, Li Z, Wu Q, Lathia JD, McLendon RE et al (2008) c-Myc is required for maintenance of glioma cancer stem cells. PLoS One 3(11):e3769
65. Panneer Selvam S, De Palma RM, Oaks JJ, Oleinik N, Peterson YK, Stahelin RV et al (2015) Binding of the sphingolipid S1P to hTERT stabilizes telomerase at the nuclear periphery by allosterically mimicking protein phosphorylation. Sci Signal 8(381):ra58-ra
66. Ishikawa F, Yoshida S, Saito Y, Hijikata A, Kitamura H, Tanaka S et al (2007) Chemotherapy-resistant human AML stem cells home to and engraft within the bone-marrow endosteal region. Nat Biotechnol 25(11):1315–1321
67. Ogretmen B, Hannun YA (2001) Updates on functions of ceramide in chemotherapy-induced cell death and in multidrug resistance. Drug Resist Updates 4(6):368–377
68. Pchejetski D, Golzio M, Bonhoure E, Calvet C, Doumerc N, Garcia V et al (2005) Sphingosine kinase-1 as a chemotherapy sensor in prostate adenocarcinoma cell and mouse models. Cancer Res 65(24):11667–11675
69. Bektas M, Jolly PS, Muller C, Eberle J, Spiegel S, Geilen CC (2005) Sphingosine kinase activity counteracts ceramide-mediated cell death in human melanoma cells: role of Bcl-2 expression. Oncogene 24(1):178–187
70. Song L, Xiong H, Li J, Liao W, Wang L, Wu J et al (2011) Sphingosine kinase-1 enhances resistance to apoptosis through activation of PI3K/Akt/NF-κB pathway in human non–small cell lung cancer. Clin Cancer Res 17(7):1839–1849
71. Riccitelli E, Giussani P, Di Vito C, Condomitti G, Tringali C, Caroli M et al (2013) Extracellular sphingosine-1-phosphate: a novel actor in human glioblastoma stem cell survival. PLoS One 8(6):e68229
72. Juul N, Szallasi Z, Eklund AC, Li Q, Burrell RA, Gerlinger M et al (2010) Assessment of an RNA interference screen-derived mitotic and ceramide pathway metagene as a predictor of response to neoadjuvant paclitaxel for primary triple-negative breast cancer: a retrospective analysis of five clinical trials. Lancet Oncol 11(4):358–365
73. Kumagai K, Yasuda S, Okemoto K, Nishijima M, Kobayashi S, Hanada K (2005) CERT mediates intermembrane transfer of various molecular species of ceramides. J Biol Chem 280(8):6488–6495
74. Xu JX, Morii E, Liu Y, Nakamichi N, Ikeda J, Kimura H et al (2007) High tolerance to apoptotic stimuli induced by serum depletion and ceramide in side-population cells: high expression of CD55 as a novel character for side-population. Exp Cell Res 313(9):1877–1885
75. Matloubian M, Lo CG, Cinamon G, Lesneski MJ, Xu Y, Brinkmann V et al (2004) Lymphocyte egress from thymus and peripheral lymphoid organs is dependent on S1P receptor 1. Nature 427(6972):355–360

76. Massberg S, Schaerli P, Knezevic-Maramica I, Köllnberger M, Tubo N, Moseman EA et al (2007) Immunosurveillance by hematopoietic progenitor cells trafficking through blood, lymph, and peripheral tissues. Cell 131(5):994–1008
77. Ratajczak MZ, Lee H, Wysoczynski M, Wan W, Marlicz W, Laughlin MJ et al (2010) Novel insight into stem cell mobilization-plasma sphingosine-1-phosphate is a major chemoattractant that directs the egress of hematopoietic stem progenitor cells from the bone marrow and its level in peripheral blood increases during mobilization due to activation of complement cascade/membrane attack complex. Leukemia 24(5):976–985
78. Ryser MF, Ugarte F, Lehmann R, Bornhauser M, Brenner S (2008) S1P(1) overexpression stimulates S1P-dependent chemotaxis of human CD34+ hematopoietic progenitor cells but strongly inhibits SDF-1/CXCR4-dependent migration and in vivo homing. Mol Immunol 46(1):166–171
79. Adamiak M, Borkowska S, Wysoczynski M, Suszynska M, Kucia M, Rokosh G et al (2015) Evidence for the involvement of sphingosine-1-phosphate in the homing and engraftment of hematopoietic stem cells to bone marrow. Oncotarget 6(22):18819–18828
80. Graler MH, Goetzl EJ (2004) The immunosuppressant FTY720 down-regulates sphingosine 1-phosphate G-protein-coupled receptors. FASEB J 18(3):551–553
81. Juarez JG, Harun N, Thien M, Welschinger R, Baraz R, Dela Pena A et al (2012) Sphingosine-1-phosphate facilitates trafficking of hematopoietic stem cells and their mobilization by CXCR4 antagonists in mice. Blood 119(3):707–716
82. Ratajczak MZ, Kim C (2011) Bioactive sphingolipids and complement cascade as new emerging regulators of stem cell mobilization and homing. J Stem Cell Res Ther 1(2):102
83. Kim CH, Wu W, Wysoczynski M, Abdel-Latif A, Sunkara M, Morris A et al (2012) Conditioning for hematopoietic transplantation activates the complement cascade and induces a proteolytic environment in bone marrow: a novel role for bioactive lipids and soluble C5b-C9 as homing factors. Leukemia 26(1):106–116
84. Schneider G, Bryndza E, Abdel-Latif A, Ratajczak J, Maj M, Tarnowski M et al (2013) Bioactive lipids S1P and C1P are prometastatic factors in human rhabdomyosarcoma, and their tissue levels increase in response to radio/chemotherapy. Mol Cancer Res 11(7):793–807
85. Bowen C, Spiegel S, Gelmann EP (1998) Radiation-induced apoptosis mediated by retinoblastoma protein. Cancer Res 58(15):3275–3281
86. Annabi B, Lachambre MP, Plouffe K, Sartelet H, Beliveau R (2009) Modulation of invasive properties of CD133+ glioblastoma stem cells: a role for MT1-MMP in bioactive lysophospholipid signaling. Mol Carcinog 48(10):910–919
87. Deng J, Liu Y, Lee H, Herrmann A, Zhang W, Zhang C et al (2012) S1PR1-STAT3 signaling is crucial for myeloid cell colonization at future metastatic sites. Cancer Cell 21(5):642–654
88. Ponnusamy S, Selvam SP, Mehrotra S, Kawamori T, Snider AJ, Obeid LM et al (2012) Communication between host organism and cancer cells is transduced by systemic sphingosine kinase 1/sphingosine 1-phosphate signalling to regulate tumour metastasis. EMBO Mol Med 4(8):761–775
89. Bergelin N, Blom T, Heikkila J, Lof C, Alam C, Balthasar S et al (2009) Sphingosine kinase as an oncogene: autocrine sphingosine 1-phosphate modulates ML-1 thyroid carcinoma cell migration by a mechanism dependent on protein kinase C-alpha and ERK1/2. Endocrinology 150(5):2055–2063
90. Wang D, Zhao Z, Caperell-Grant A, Yang G, Mok SC, Liu J et al (2008) S1P differentially regulates migration of human ovarian cancer and human ovarian surface epithelial cells. Mol Cancer Ther 7(7):1993–2002
91. Van Brocklyn JR, Young N, Roof R (2003) Sphingosine-1-phosphate stimulates motility and invasiveness of human glioblastoma multiforme cells. Cancer Lett 199(1):53–60
92. Albinet V, Bats ML, Huwiler A, Rochaix P, Chevreau C, Segui B et al (2014) Dual role of sphingosine kinase-1 in promoting the differentiation of dermal fibroblasts and the dissemination of melanoma cells. Oncogene 33(26):3364–3373

93. Ko P, Kim D, You E, Jung J, Oh S, Kim J et al (2016) Extracellular matrix rigidity-dependent sphingosine-1-phosphate secretion regulates metastatic cancer cell invasion and adhesion. Sci Rep 6:21564
94. Seiki M (2003) Membrane-type 1 matrix metalloproteinase: a key enzyme for tumor invasion. Cancer Lett 194(1):1–11
95. Kawahara S, Otsuji Y, Nakamura M, Murakami M, Murate T, Matsunaga T et al (2013) Sphingosine kinase 1 plays a role in the upregulation of CD44 expression through extracellular signal-regulated kinase signaling in human colon cancer cells. Anticancer Drugs 24(5):473–483
96. Kajita M, Itoh Y, Chiba T, Mori H, Okada A, Kinoh H et al (2001) Membrane-Type 1 matrix metalloproteinase cleaves Cd44 and promotes cell migration. J Cell Biol 153(5):893–904
97. Anido J, Sáez-Borderías A, Gonzàlez-Juncà A, Rodón L, Folch G, Carmona MA et al (2010) TGF-β receptor inhibitors target the CD44high/Id1high glioma-initiating cell population in human glioblastoma. Cancer Cell 18(6):655–668
98. Oskarsson T, Batlle E, Massagué J (2014) Metastatic stem cells: sources, niches, and vital pathways. Cell Stem Cell 14(3):306–321
99. Khaled YS, Ammori BJ, Elkord E (2013) Myeloid-derived suppressor cells in cancer: recent progress and prospects. Immunol Cell Biol 91(8):493–502
100. Reya T, Morrison SJ, Clarke MF, Weissman IL (2001) Stem cells, cancer, and cancer stem cells. Nature 414(6859):105–111
101. Morad SA, Cabot MC (2015) Tamoxifen regulation of sphingolipid metabolism—therapeutic implications. Biochim Biophys Acta 1851(9):1134–1145
102. Noack J, Choi J, Richter K, Kopp-Schneider A, Régnier-Vigouroux A (2014) A sphingosine kinase inhibitor combined with temozolomide induces glioblastoma cell death through accumulation of dihydrosphingosine and dihydroceramide, endoplasmic reticulum stress and autophagy. Cell Death Dis 5(9):e1425
103. Estrada-Bernal A, Palanichamy K, Ray Chaudhury A, Van Brocklyn JR (2012) Induction of brain tumor stem cell apoptosis by FTY720: a potential therapeutic agent for glioblastoma. Neuro Oncol 14(4):405–415
104. Neviani P, Harb JG, Oaks JJ, Santhanam R, Walker CJ, Ellis JJ et al (2013) PP2A-activating drugs selectively eradicate TKI-resistant chronic myeloid leukemic stem cells. J Clin Invest 123(10):4144–4157
105. Liu Q, Zhao X, Frissora F, Ma Y, Santhanam R, Jarjoura D et al (2008) FTY720 demonstrates promising preclinical activity for chronic lymphocytic leukemia and lymphoblastic leukemia/lymphoma. Blood 111(1):275–284
106. Matsuoka Y, Nagahara Y, Ikekita M, Shinomiya T (2003) A novel immunosuppressive agent FTY720 induced Akt dephosphorylation in leukemia cells. Br J Pharmacol 138(7):1303–1312
107. Pitman MR, Woodcock JM, Lopez AF, Pitson SM (2012) Molecular targets of FTY720 (fingolimod). Curr Mol Med 12(10):1207–1219
108. Kunkel GT, Maceyka M, Milstien S, Spiegel S (2013) Targeting the sphingosine-1-phosphate axis in cancer, inflammation and beyond. Nat Rev Drug Discov 12(9):688–702
109. Lavie Y, Cao H-T, Volner A, Lucci A, Han T-Y, Geffen V et al (1997) Agents that reverse multidrug resistance, tamoxifen, verapamil, and cyclosporin A, block glycosphingolipid metabolism by inhibiting ceramide glycosylation in human cancer cells. J Biol Chem 272(3):1682–1687
110. Majeti R, Chao MP, Alizadeh AA, Pang WW, Jaiswal S, Gibbs KD Jr et al CD47 is an adverse prognostic factor and therapeutic antibody target on human acute myeloid leukemia stem cells. Cell 138(2):286–299
111. Hosen N, Park CY, Tatsumi N, Oji Y, Sugiyama H, Gramatzki M et al (2007) CD96 is a leukemic stem cell-specific marker in human acute myeloid leukemia. Proc Natl Acad Sci U S A 104(26):11008–11013
112. Kikushige Y, Shima T, Takayanagi S, Urata S, Miyamoto T, Iwasaki H et al (2010) TIM-3 is a promising target to selectively kill acute myeloid leukemia stem cells. Cell Stem Cell 7(6):708–717

113. Goardon N, Marchi E, Atzberger A, Quek L, Schuh A, Soneji S et al (2011) Coexistence of LMPP-like and GMP-like leukemia stem cells in acute myeloid leukemia. Cancer Cell 19(1):138–152
114. Wright MH, Calcagno AM, Salcido CD, Carlson MD, Ambudkar SV, Varticovski L (2008) Brca1 breast tumors contain distinct CD44+/CD24- and CD133+cells with cancer stem cell characteristics. Breast Cancer Res 10(1):1–16
115. Liu SY, Zheng PS (2013) High aldehyde dehydrogenase activity identifies cancer stem cells in human cervical cancer. Oncotarget 4(12):2462–2475
116. López J, Poitevin A, Mendoza-Martínez V, Pérez-Plasencia C, García-Carrancá A (2012) Cancer-initiating cells derived from established cervical cell lines exhibit stem-cell markers and increased radioresistance. BMC Cancer 12(1):1–14
117. Du L, Wang H, He L, Zhang J, Ni B, Wang X et al (2008) CD44 is of functional importance for colorectal cancer stem cells. Clin Cancer Res 14(21):6751–6760
118. Ke J, Wu X, Wu X, He X, Lian L, Zou Y et al (2012) A subpopulation of CD24(+) cells in colon cancer cell lines possess stem cell characteristics. Neoplasma 59(3):282–288
119. Rasper M, Schäfer A, Piontek G, Teufel J, Brockhoff G, Ringel F et al (2010) Aldehyde dehydrogenase 1 positive glioblastoma cells show brain tumor stem cell capacity. Neuro Oncol 12(10):1024–1033
120. Lee TK, Castilho A, Cheung VC, Tang KH, Ma S, Ng IO (2011) CD24(+) liver tumor-initiating cells drive self-renewal and tumor initiation through STAT3-mediated NANOG regulation. Cell Stem Cell 9(1):50–63
121. Ma S, Tang KH, Chan YP, Lee TK, Kwan PS, Castilho A et al (2010) miR-130b Promotes CD133(+) liver tumor-initiating cell growth and self-renewal via tumor protein 53-induced nuclear protein 1. Cell Stem Cell 7(6):694–707
122. Ma S, Chan KW, Lee TK-W, Tang KH, Wo JY-H, Zheng B-J et al (2008) Aldehyde dehydrogenase discriminates the CD133 liver cancer stem cell populations. Mol Cancer Res 6(7):1146–1153
123. Leung EL-H, Fiscus RR, Tung JW, Tin VP-C, Cheng LC, Sihoe AD-L et al (2010) Non-small cell lung cancer cells expressing CD44 are enriched for stem cell-like properties. PLoS One 5(11):e14062
124. Eramo A, Lotti F, Sette G, Pilozzi E, Biffoni M, Di Virgilio A et al (2007) Identification and expansion of the tumorigenic lung cancer stem cell population. Cell Death Differ 15(3):504–514
125. Sullivan JP, Spinola M, Dodge M, Raso MG, Behrens C, Gao B et al (2010) Aldehyde dehydrogenase activity selects for lung adenocarcinoma stem cells dependent on notch signaling. Cancer Res 70(23):9937–9948
126. Hurt EM, Kawasaki BT, Klarmann GJ, Thomas SB, Farrar WL (2008) CD44(+)CD24(−) prostate cells are early cancer progenitor/stem cells that provide a model for patients with poor prognosis. Br J Cancer 98(4):756–765
127. Li C, Heidt DG, Dalerba P, Burant CF, Zhang L, Adsay V et al (2007) Identification of pancreatic cancer stem cells. Cancer Res 67(3):1030–1037
128. Miki J, Furusato B, Li H, Gu Y, Takahashi H, Egawa S et al (2007) Identification of putative stem cell markers, CD133 and CXCR4, in hTERT–immortalized primary nonmalignant and malignant tumor-derived human prostate epithelial cell lines and in prostate cancer specimens. Cancer Res 67(7):3153–3161
129. van den Hoogen C, van der Horst G, Cheung H, Buijs JT, Lippitt JM, Guzmán-Ramírez N et al (2010) High aldehyde dehydrogenase activity identifies tumor-initiating and metastasis-initiating cells in human prostate cancer. Cancer Res 70(12):5163–5173

Chapter 9
Lysophosphatidic Acid Signalling Enhances Glioma Stem Cell Properties

Wayne Ng

Abbreviations

ATX	Autotaxin
BBB	Blood–brain barrier
BTSC	Brain tumour stem cell
CSC	Cancer stem cell
EGF	Epidermal growth factor
EGFR	Epidermal growth factor receptor
GBM	Glioblastoma multiforme
GFAP	Glial fibrillary acidic protein
GSC	Glioma stem cell
LPA	Lysophosphatidic acid
LPAR	LPA receptor
LPC	Lysophosphatidylcholine
LPL	Lysophospholipid
MAPK	Mitogen-activated protein kinase
MMP	Matrix metalloprotease
mTOR	Mammalian target of rapamycin
NSC	Neural stem cell
PI3K	Phosphoinositide 3-kinase
PTEN	Phosphatase tensin homologue

W. Ng (✉)
University of Melbourne, Parkville, VIC, Australia

Department of Surgery, Centre for Medical Research, Royal Melbourne Hospital, Grattan Street, Parkville, VIC 3050, Australia

Melbourne Brain Centre at Royal Melbourne Hospital, Parkville, VIC, Australia
e-mail: wayneng@y7mail.com

A. Pébay, R.C.B. Wong (eds.), *Lipidomics of Stem Cells*, Stem Cell Biology and Regenerative Medicine, DOI 10.1007/978-3-319-49343-5_9

RMS	Rostral migratory stream
S1P	Sphingosine 1-phosphate
SVZ	Subventricular zone
TMZ	Temozolomide

9.1 Introduction

Tissue-specific stem cells are a rare subpopulation of cells that are defined as having the ability to self-renew and differentiate into all cell types in that tissue (Fig. 9.1) [4–7]. Initial reports regarding stem-like cancer cells in haematopoietic malignancies (leukaemia) described a rare subpopulation of cells possessing the characteristics of stem cells [8–10]. The CSC theory attempts to explain intratumoural heterogeneity based on the existence of stem-like cells within solid tumours [11, 12]. Subsequent discoveries led to the description of stem-like cells within solid tumours such as breast cancer, paediatric neuro-malignancies (such as medulloblastoma), and GBM [11–17]. The discovery that NSCs exist within the RMS and SVZ of the lateral ventricle has aroused a major shift in neuroscience research [13, 18–31]. With respect to GBM, GFAP-positive NSCs residing within the SVZ have been suggested to more readily undergo neoplastic transformation [32–36]. There have been clinical studies

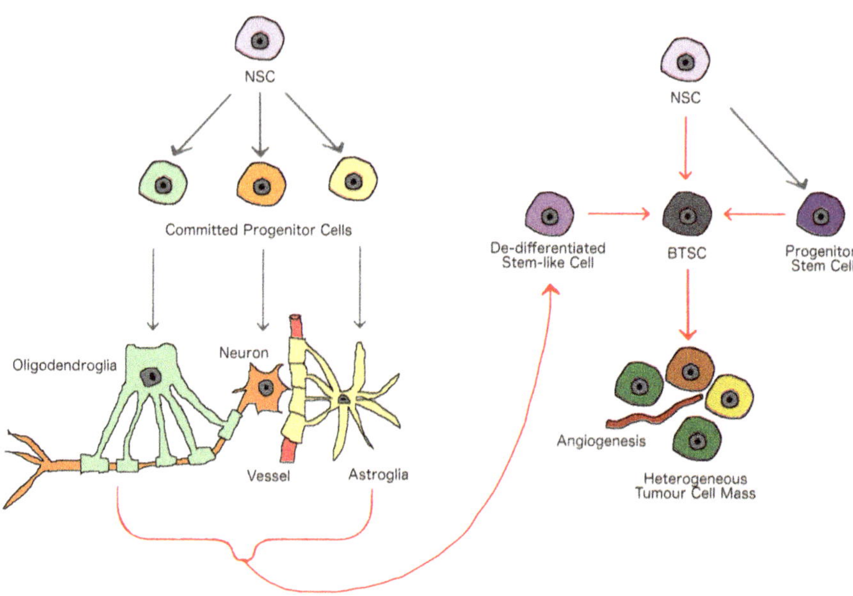

Fig. 9.1. Glioma stem cell theory. Normal neurogenesis displayed on the *left*. On the *right*, de-differentiation and/or dysregulation of NSCs/progenitor cells are shown as competing origins of glioma stem cells [1–3]. *Black lines*, normal progression of activity; *red lines*, mutational events leading to abnormal progression. BTSC, brain tumour stem cell; NSC, neural stem cell

suggesting a close relationship between GBM tumours and the SVZ, but some tumours are still found along the radial glial 'tract' (from SVZ to pial/cortical surface), suggesting that such stem cells may also migrate from the SVZ to the site of the tumour [4–7, 20, 37–40]. These concepts may be tied together by the radial glia theory which suggests that the radial glia cells (rather than the SVZ cells) are the source of adult NSCs [8–10, 18, 34, 39, 41–43]. A competing concept (de-differentiation) that attempts to explain intratumoural heterogeneity describes cells as existing in a continuum from stem cell to intermediate/progenitor cell to mature/differentiated cell with appropriate triggers altering their state within this continuum (Fig. 9.1) [11, 12, 40, 44–49]. Therefore, the origins of the GSCs are not yet fully elucidated [11–24, 50–54].

Regardless, the discovery of CSCs suggests cancer treatment failure occurs because certain cancers contain a rare subpopulation of stem-like cells which are able to fuel the tumour's immortality and recurrence [13, 18–31, 55, 56]. There is good evidence that higher levels of CSCs correlate with more aggressive tumours and poorer outcomes [11, 32–36, 57–59]. Therefore, dysregulation of pathways that can influence stem cell migration, motility, differentiation, growth, and proliferation are likely to be involved in gliomagenesis. LPA signalling is involved in normal CNS development through modulating cell processes such as cell migration, adhesion, apoptosis, and proliferation [60].

9.2 Lysophosphatidic Acid Signalling in the CNS

LPA is a species of lipid involved in LPL signalling, and there are a number of pathways that produce LPA from LPC; the ATX pathway appears to be the key enzymatic pathway [60–64]. LPA normally signals in an autocrinic/paracrinic fashion [61, 62, 65]. LPA negatively regulates its own production (feedback inhibition of ATX), such that under physiological conditions only small amounts of LPA (<1 μM) are present in tissues [66–68].

LPA signalling is mediated through 6 recognised G protein-coupled LPA receptors (collectively LPAR; individually LPA_{1-6}) that are collected into two families: Endothelial differentiation gene (Edg) and non-Edg (purinergic) [65, 69]. LPA_1 (Edg2), LPA_2 (Edg4), and LPA_3 (Edg7) are members of the Edg family and are the best characterised to date with LPA_1 being the dominant LPA receptor in the CNS [61, 70, 71]. Whilst LPA_2 is expressed in embryonic brain, there is little expression of LPA_2 in the adult CNS [72]. LPA_3 is expressed in brain and enhances cell motility [64, 73, 74].

The non-Edg or purinergic family of LPARs are genetically distinct from the Edg family [70]. An important functional difference lies in this family's preference for alkyl side chained LPA species compared to the acyl variants preferred by the Edg family [69, 70]. Current members of this family include LPA_4 (P2Y9), LPA_5 (GPR92), and LPA_6 (P2Y5). LPA_4 probably plays an inhibitory role in cell motility/migration and so it has been suggested that it might suppress LPA_1-mediated signalling [70, 75]. Therefore, the dominant LPARs involved in mediating cell migration appear to be

LPA_1 and LPA_4, with cell migration being an important function in some cancer cells [60, 75]. As yet there is no reported role for LPA_5 (GPR92) in tumourigenesis; however, LPA_6 (P2Y5) may play an indirect role in gliomagenesis via putative effects on vascular development [64, 69, 70]. LPARs signal by activating a myriad number of G-proteins that are coupled to a range of second messenger systems.

9.3 The Role of Lysophosphatidic Acid Signalling in Glioblastoma Multiforme

Since the discovery that ATX has promotile effects on melanoma cells, the LPA pathway has been investigated for its role in tumour invasion and metastasis [76–79]. Further, LPA signalling has been shown to have complex interactions with other cancer signalling pathways (including PI3K/MAPK/Rho/YAP/FAK) in ovarian, cervical, pancreatic and colorectal cancers, and osteosarcoma [64, 65, 80–92].

Studies suggest that LPA's effects on cell shape and motility may play a role in tumour cell invasiveness, and in the case of glioma, may play a role in tumour-associated epilepsy by modulating synaptic transmission through astrocytic cell shape and therefore synaptic cleft shape changes [93]. LPA's effects at a cellular level (modifying cell adhesion and cell shape) may also be important in GBM where the BBB has been shown to be less privileged [94–98].

LPAR overexpression has been described in sex hormone-linked cancers such as prostate and ovarian cancers [99]. This overexpression has been shown to occur concurrently with overproduction of LPA in ovarian cancer mouse models and has been postulated to create an autocrine loop promoting proliferation and suppressing apoptosis [100]. LPA_1 has been linked to increased stem cell apoptosis, astrocytic differentiation of stem cells, and inducing astrocyte proliferation via induction of GFAP [101–104]. LPA also has actions leading to differentiation of stem-like cells into neuronal and oligodendroglial lineages; however, there are reports that LPA can inhibit neuronal differentiation of embryonic-derived stem cells [102, 105–109]. The action of LPA in astrocytes appears to be mitogenic, but LPA_1 has also been linked to motility of glial tumour cells in vitro [93, 110]. LPA_1 has been found to be the predominant receptor subtype in GBM cells (SNB-78, SNB-75, SF-268, SF539, and SF-295) [110]. When ATX and LPA_1 are overexpressed (particularly ATX) in GBM, autocrine stimulation of GBM clearly contributes to increased cell motility [110]. The motile response of GBM cells can be completely abolished with the $LPA_{1/3}$ receptor antagonist, Ki16425 [110].

Knockdown studies of *atx* in *atx*-null murine embryos have a profound effect on vessel maturation in these embryos resulting in uniform lethality at day 9.5–10.5 [60, 111, 112]. This effect was probably exacerbated by the absence of ATX's other metabolite, sphingosine 1-phosphate (S1P) [60]. CNS-specific effects of murine *atx* knockdown included massive neural tube defects (due to the contribution of ATX and LPA to cell motility), supporting a role for ATX and LPA in neurogenesis. *Atx* levels were highest in the brain, and it was found to be secreted by secretory epithelial cells such as those in the choroid plexus [60]. In the adult brain, ATX levels are highest in white matter regions (associated with oligodendrocyte precursor cells), choroid plexus, and leptomeninges [93, 113]. This may relate to the putative SVZ from which neural stem/progenitor cells arise [114–118]. There is also evidence of a functional LPAR in the SVZ regulating cortical neurogenesis and apoptosis [93]. This might not be true of normal adult brains as there are some reports that LPARs are absent in the subventricular and ependymal zones [72]. Regardless, ATX and/or LPA blockade represents a novel opportunity for treating the grossly invasive GBM.

9.4 LPA Can Influence EGFR/PI3K Signalling

It is now well recognised that the development of GBM, like other cancers, depends on a series of alterations in key (tumour suppressor or onco-) genes. Gain of chromosome 7 and loss of chromosome 10 appear to be common events in molecular analyses of GBM [119, 120]. The EGFR pathway is a major growth factor signalling pathway that promotes malignancy by enhancing cell proliferation and survival via the Ras/MAPK and PI3K downstream signalling pathways, respectively [64, 65, 80–83]. EGFR signalling may be affected by these chromosomal abnormalities. EGFR is commonly amplified in GBM, and some mutations in EGFR result in a constitutively active receptor (designated EGFRvIII) [121–126]. EGFR, p53, and PTEN are amongst the most common mutations in GBM, and these genetic events have been suggested to be important in GBM maintenance and recurrence [120, 125–127]. EGFR has been reported as being overexpressed in up to 50% of GBMs and its signalling is upregulated in 40–60% of GBMs [128–130]. Further, anti-EGFR monotherapies have shown efficacy in the subgroup of GBM patients whose tumour cells depend on EGFR signalling [131–137]. However, results from phase II trials assessing EGFR targeting in GBM patients with erlotinib failed to show any significant improvement in OS [129, 136, 138]. These phase II trials demonstrated inconsistent inhibition of EGFR phosphorylation and no modulation of downstream Akt/MAPK signalling [125, 139, 140]. Whilst inadequate dosing or lack of BBB penetration can't be ruled out, the results when taken in conjunction with the results of the bevacizumab and cilengitide trials suggest that using targeted inhibitors in mechanistic isolation may be prone to failure [141–143].

Within the EGFR pathway, up to 70% of GBM patients have lost their PTEN tumour suppressor gene (via loss of chromosome 10q). PTEN normally downregulates PI3K signalling [144–148]. PTEN may also have transcriptional functions related to apoptosis in cells exposed to oxidative damage [125, 149]. PI3K mutations in GBM (PIK3CA, PIK3CB, PIK3CD, PIK3R1 mutations) lead to its constitutive activation in signalling [125, 127, 150–152]. It is likely that the EGFR, PTEN, and PI3K mutations interact in ways that limit success with monotherapies which target individual pathway components (Fig. 9.2) and partly explains why EGFR inhibition alone has not yielded more positive clinical trial results [129, 138].

Akt is a major intracellular signalling hub that is also a downstream signalling component of EGFR signalling, thus Akt has the potential to be a powerful target

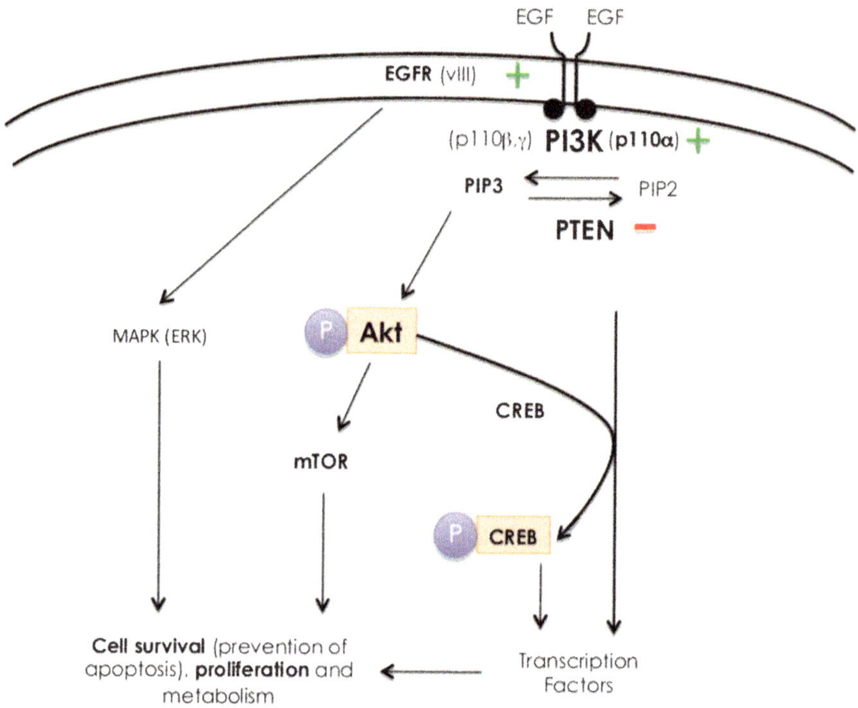

Fig. 9.2. EGFR signalling. CREB, cAMP response element-binding protein; EGF, epidermal growth factor; EGFR, EGF receptor; MAPK, mitogen activated protein kinase; mTOR, mammalian target of rapamycin; PI3K, phosphoinositide 3-kinase; PIP2, phosphatidylinositol 4,5-bisphosphate; PIP3, phosphatidylinositol 3,4,5 trisphosphate; PTEN, phosphatase tensin homologue

in treating malignancies such as GBM [153–162]. For example, two new Akt inhibitors (KP-372-1 and KP-372-2) have shown effective in vitro retardation of glioma growth and invasion [163]. Akt is downstream from the often times mutated EGFR/PTEN/PI3K components [120, 140, 163]. The EGFR/PI3K pathway is thought to act as a major conduit that phosphorylates Akt and has been reported to influence both glioma cell migration and survival (resistance to treatment) [120, 127, 140, 149]. In addition to the EGFR and Akt inhibitors, PI3K/mTOR inhibitors have also been developed for the treatment of GBM [164, 165]. The anti-tumour efficacy of PI3K inhibitors (e.g. BKM120) and dual PI3K/mTOR inhibitors (e.g. BEZ235, XL765) in GBM patients are currently being assessed in ongoing phase I/II clinical trials (clinicaltrials.gov NCT00704080, NCT01240460, NCT01339052, NCT01349660, NCT01870726, NCT01576666).

Much of the preclinical evidence supporting a role for EGFR/PI3K modulation in GBM has been conducted in traditional glioma cell lines that have a poor translational record. Although the translatability of the GSC model is yet to be validated, there are now some reports of testing within patient-derived tumour models [119, 154, 165–168]. In these studies, small molecule inhibitors appear only to be cytostatic, being mostly effective by sensitising cells to apoptosis-inducing treatments such as irradiation and chemotherapy [119, 120, 125, 169]. PI3K inhibitors appear to sensitise GBM cells to chemotherapy, regardless of PTEN mutational status [143, 170]. The PI3K inhibitors are probably best tested in combination with cytotoxic drugs such as TMZ or in combination with EGFR or mTOR inhibitors. Alternatively, interaction of the EGFR pathways with lipid signalling pathways (such as LPA) may be important.

LPA-mediated transactivation of EGFR upregulates cell proliferation via MAPK dependent mitogenic signalling and enhances cell survival via PI3K signalling (Fig. 9.3) [171, 172]. LPA can transactivate EGFR signalling independently of EGF and may also be involved in cleaving and activating growth factors like EGF [80–82, 173, 174]. LPA_1 has been putatively described as activating the MAPK and PI3K/Akt pathways, with LPA being reported to directly activate Akt and promote cell migration [60, 175]. This interaction appears to occur via the G_{ai}-protein coupled to the EGFRvIII receptor [173]. Intracellular cross-communication (transactivation) integrates the myriad cellular signals [80]. It has been reported that the p110b/g subunits of PI3K may be involved in the Gi-protein transactivation of PI3K, a process also influenced by LPA [176].

In contrast to PI3K transactivation, G protein-dependent MAPK transactivation appears to be dependent on simultaneous agonistic activation of EGFR [81, 82, 90, 177].

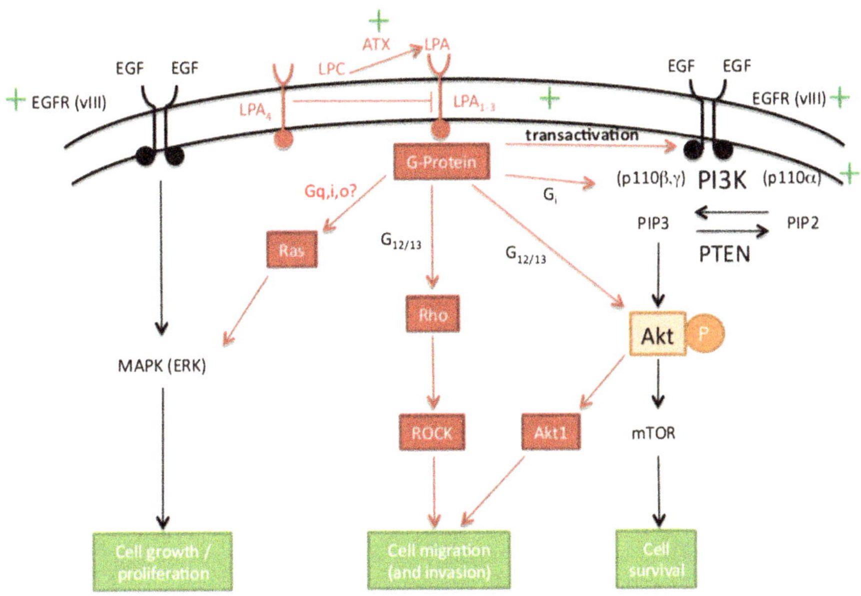

Fig. 9.3. Interactions in LPA and EGFR signalling. ATX, autotaxin; EGF, epidermal growth factor; EGFR, EGF receptor; LPA, lysophosphatidic acid; LPC, lysophosphatidylcholine; MAPK, mitogen activated protein kinase; mTOR, mammalian target of rapamycin; PI3K, phosphoinositide 3-kinase; PIP2, phosphatidylinositol 4,5-bisphosphate; PIP3, phosphatidylinositol 3,4,5 trisphosphate; PTEN, phosphatase tensin homologue; Rho, Rho GTPase; ROCK, Rho-associated kinase. *Red boxes*: key proteins and enzymes; *green boxes*: cellular processes; *red arrows/blockheads*: key pathway interactions; *green plus signs*: overexpressed proteins/enzymes

Transactivation of EGFR signalling by LPA may also be linked to PI3K activation [90]. However, transactivation of MAPK by LPA appears to be more tightly regulated than PI3K and in fact may also require concurrent MMP activity which is also upregulated in gliomas [174, 178]. MMPs are involved in cleaving membrane bound precursors of the active ligand (EGF) for EGFR [174]. Inhibition of MMPs or EGFR have both been shown to reduce phosphoactivation of EGFR [174]. Also, invadopodia have been implicated in invasion and metastasis and are enriched with MMPs [179–183]. MMPs have been extensively studied and implicated in invasion and metastasis by providing a favourable milieu for cells to invade and are also involved in cleaving and activating growth factors such as EGF [174, 178, 183]. LPA_4 has been shown to promote invadopodia formation in HT1080 fibrosarcoma cells [184]. ATX overexpression has also been shown to enhance the invasion of U87 and U251 GBM cells (autocrine effect) through in vitro oligodendrocyte monolayers whilst simultaneously reducing the adhesiveness of oligodendrocytes (paracrine effect) [185]. The in vitro effects of ATX to increase GBM cell motility and reduce oligodendrocyte (white matter cells) adhesiveness are consistent with the in vivo propensity of GBM to invade along white matter tracts [185]. Depletion or inactivation of ATX ameliorates the invasion of GBM cells through oligodendrocyte monolayers in vitro [185].

It has also been confirmed that ATX-induced motility in melanoma cells is mediated via the p110g subunit of PI3K and that PI3K inhibitors inhibit this response in a dose-dependent manner [176]. Also, simultaneous knockout of Akt1 and Akt2 abolishes LPA-induced motility in mouse embryonic fibroblasts [175]. Only re-expression of Akt1 following the double knockout restored the cell's motile response to LPA [175]. Both a pan-PI3K inhibitor (LY294002) and an LPAR antagonist (Ki16425) also completely abolished the motile response [175]. This suggests that PI3K activates Akt1 to mediate cell migration [175].

Recent reports also suggest that LPA signalling enables the murine glioma cell line GL-261 to be more radioresistant [186]. Inhibition of LPA signalling resulted in impaired survival in response to irradiation (3 gray); and knockdown of LPA_1 and LPA_3 with siRNA resulted in reduced phosphorylation of Akt, correlating with reduced tumour cell survival [186].

LPA also appears to mediate cell migration via Rho signalling [61, 187, 188]. Rho-dependent cytoskeletal rearrangement has been associated with producing discohesive (reduced adhesion) cells by causing cell rounding [188]. However, there have been some contradictory reports with regard to Rho-related LPA signalling. For example, LPA-induced glioma cell migration can be ameliorated by blocking Rho activation [92]. In contrast, stimulation of Rho activity has also been shown to cause immobilisation of glioma cells [189]. The Edg and non-Edg families of LPA receptors are evolutionarily distinct and therefore signalling differences between these families of LPAR may explain these inconsistencies. Activation of LPA_5 was recently reported to inhibit B16 melanoma cell migration, further supporting the notion that non-Edg family LPARs (LPA_{4-6}) may mitigate cancer cell properties whereas Edg family LPARs (LPA_{1-3}) are more likely to enhance tumourigenic effects [190].

ATX overexpression can be differentially higher at a glioma tumour's invasive edges [185]. Currently, there are no reports to address which tumour cell subpopulations overexpress ATX/LPAR. In glioma, there are suggestions that microglia are recruited to the invasive edge by various chemokines (including LPA) and subsequently facilitate invasion by producing ATX, LPA, and EGF [191]. As yet, there are no in vivo reports investigating the efficacy of LPA inhibition in GBM. There are however, reports that overexpression of ATX correlates with increased invasiveness of breast cancer cells compared to normal breast cells [192]. Nude *(nu/nu)* mouse metastatic breast cancer models have shown that in vivo treatment with a selective LPA_{1-3} receptor antagonist (Ki16425) can safely retard tumour growth [193, 194].

9.5 Conclusion and Future Directions

GBM is a grossly heterogeneous solid malignancy, whose hallmark is its aggressive and invasive biology. Glioma stem cell theory has significantly altered our approach to discovering efficacious treatments for GBM by postulating that a subpopulation of cells exist that are both difficult to kill and also give rise to the heterogeneity. Research that seeks to address the biology of glioma stem cells and modulation of pathways that can induce their death is still relatively new. EGFR signalling has been well documented to play an important part in gliomagenesis. However, there are also reports that having a simplistic single pathway approach to GBM treatment is unlikely to lead to a significant breakthrough to improving patient outcomes. LPA signalling is a complex signalling pathway that has been shown to have a putative role in gliomagenesis and has also been shown to interact with EGFR signalling. Currently, there is a dearth of published reports of LPA signalling in glioma stem cells. This represents an exciting opportunity moving forward.

References

1. Medical Research Council Brain Tumor Working Party (2001) Randomized trial of procarbazine, lomustine, and vincristine in the adjuvant treatment of high-grade astrocytoma: a Medical Research Council trial. J Clin Oncol 19(2):509–518
2. Singh SK, Clarke ID, Hide T, Dirks PB (2004) Cancer stem cells in nervous system tumors. Oncogene 23(43):7267–7273. doi:10.1038/sj.onc.1207946
3. Quinn JA, Phase II (2002) Trial of carmustine plus O6-benzylguanine for patients with nitrosourea-resistant recurrent or progressive malignant glioma. J Clin Oncol 20(9):2277–2283. doi:10.1200/JCO.2002.09.084
4. Colen CB, Allcut E (2012) Quality of life and outcomes in glioblastoma management. Neurosurg Clin N Am 23(3):507–513. doi:10.1016/j.nec.2012.04.010
5. Potten CS, Loeffler M (1990) Stem cells: attributes, cycles, spirals, pitfalls and uncertainties. Lessons for and from the crypt. Development 110(4):1001–1020
6. Heimans JJ, Taphoorn MJB (2002) Impact of brain tumour treatment on quality of life. J Neurol 249(8):955–960. doi:10.1007/s00415-002-0839-5

7. Reynolds BA, Weiss S (1992) Generation of neurons and astrocytes from isolated cells of the adult mammalian central nervous system. Science 255(5052):1707–1710
8. Quesenberry P, Levitt L (1979) Hematopoietic stem cells. N Engl J Med 301(14):755–761. doi:10.1056/NEJM197910043011404
9. Lewis JP, Trobaugh FE (1964) Hæmatopoietic stem cells. Nature 204(4958):589–590. doi:10.1038/204589a0
10. Zhang J, Niu C, Ye L et al (2003) Identification of the haematopoietic stem cell niche and control of the niche size. Nature 425(6960):836–841. doi:10.1038/nature02041
11. Singh SK, Clarke ID, Terasaki M et al (2003) Identification of a cancer stem cell in human brain tumors. Cancer Res 63(18):5821–5828
12. Hemmati HD, Nakano I, Lazareff JA et al (2003) Cancerous stem cells can arise from pediatric brain tumors. Proc Natl Acad Sci U S A 100(25):15178–15183. doi:10.1073/pnas.2036535100
13. Australian Institute of Health and Welfare (2013) Cancer in Australia: actual incidence data from 1991 to 2009 and mortality data from 1991 to 2010 with projections to 2012. Asia Pac J Clin Oncol 9(3):199–213. doi:10.1111/ajco.12127
14. Dobes M, Shadbolt B, Khurana VG et al (2011) A multicenter study of primary brain tumor incidence in Australia (2000–2008). Neuro Oncol 13(7):783–790. doi:10.1093/neuonc/nor052
15. Galli R, Binda E, Orfanelli U et al (2004) Isolation and characterization of tumorigenic, stem-like neural precursors from human glioblastoma. Cancer Res 64(19):7011–7021. doi:10.1158/0008-5472.CAN-04-1364
16. Al-Hajj M, Wicha MS, Benito-Hernandez A, Morrison SJ, Clarke MF (2003) Prospective identification of tumorigenic breast cancer cells. Proc Natl Acad Sci U S A 100(7):3983–3988. doi:10.1073/pnas.0530291100
17. Visvader JE, Lindeman GJ (2008) Cancer stem cells in solid tumours: accumulating evidence and unresolved questions. Nat Rev Cancer 8(10):755–768. doi:10.1038/nrc2499
18. Alvarez-Buylla A, Seri B, Doetsch F (2002) Identification of neural stem cells in the adult vertebrate brain. Brain Res Bull 57(6):751–758
19. Kohler BA, Ward E, McCarthy BJ et al (2011) Annual report to the nation on the status of cancer, 1975–2007, featuring tumors of the brain and other nervous system. JNCI J Natl Cancer Inst 103(9):714–736. doi:10.1093/jnci/djr077
20. Uchida N, Buck DW, He D et al (2000) Direct isolation of human central nervous system stem cells. Proc Natl Acad Sci U S A 97(26):14720–14725. doi:10.1073/pnas.97.26.14720
21. Buckner JC, Brown PD, O'Neill BP, Meyer FB, Wetmore CJ, Uhm JH (2007) Central nervous system tumors. Mayo Clin Proc 82(10):1271–1286. doi:10.4065/82.10.1271
22. Seri B, Alvarez-Buylla A (2002) Neural stem cells and the regulation of neurogenesis in the adult hippocampus. Clin Neurosci Res 2(1):11–16. doi:10.1016/S1566-2772(02)00004-X
23. Nogueira AB, Sogayar MC, Colquhoun A et al (2014) Existence of a potential neurogenic system in the adult human brain. J Transl Med 12:75. doi:10.1186/1479-5876-12-75
24. Gage FH (2002) Neurogenesis in the adult brain. J Neurosci 22(3):612–613
25. Ostrom QT, Barnholtz-Sloan JS (2011) Current state of our knowledge on brain tumor epidemiology. Curr Neurol Neurosci Rep 11(3):329–335. doi:10.1007/s11910-011-0189-8
26. Pencea V, Bingaman KD, Freedman LJ, Luskin MB (2001) Neurogenesis in the subventricular zone and rostral migratory stream of the neonatal and adult primate forebrain. Exp Neurol 172(1):1–16. doi:10.1006/exnr.2001.7768
27. Alvarez-Buylla A, García-Verdugo JM (2002) Neurogenesis in adult subventricular zone. J Neurosci 22(3):629–634
28. Cummings DM, Snyder JS, Brewer M, Cameron HA, Belluscio L (2014) Adult neurogenesis is necessary to refine and maintain circuit specificity. J Neurosci 34(41):13801–13810. doi:10.1523/JNEUROSCI.2463-14.2014
29. Sonego M, Oberoi M, Stoddart J et al (2015) Drebrin regulates neuroblast migration in the postnatal mammalian brain. PLoS One 10(5):e0126478. doi:10.1371/journal.pone.0126478
30. Heng YHE, Zhou B, Harris L, et al (2014) NFIX regulates proliferation and migration within the murine SVZ neurogenic niche. Cereb Cortex. doi:10.1093/cercor/bhu253

31. Curtis MA, Kam M, Nannmark U et al (2007) Human neuroblasts migrate to the olfactory bulb via a lateral ventricular extension. Science 315(5816):1243–1249. doi:10.1126/science.1136281
32. Kreth FW, Berlis A, Spiropoulou V, Faist M (1999) The role of tumor resection in the treatment of glioblastoma multiforme in adults. Cancer
33. Dirks PB (2010) Brain tumor stem cells: the cancer stem cell hypothesis writ large. Mol Oncol 4(5):420–430. doi:10.1016/j.molonc.2010.08.001
34. Lacroix M, Abi-Said D, Fourney DR et al (2001) A multivariate analysis of 416 patients with glioblastoma multiforme: prognosis, extent of resection, and survival. J Neurosurg 95(2):190–198
35. Wittko IM, Schanzer A, Kuzmichev A et al (2009) VEGFR-1 regulates adult olfactory bulb neurogenesis and migration of neural progenitors in the rostral migratory stream in vivo. J Neurosci 29(27):8704–8714. doi:10.1523/JNEUROSCI.5527-08.2009
36. Keles GE, Anderson B, Berger MS (1999) The effect of extent of resection on time to tumor progression and survival in patients with glioblastoma multiforme of the cerebral hemisphere. Surg Neurol 52(4):371–379
37. Chaichana K, McGirt M, Frazier J et al (2008) Relationship of glioblastoma multiforme to the lateral ventricles predicts survival following tumor resection. J Neurooncol 89(2):219–224
38. Barami K, Sloan AE, Rojiani A, Schell MJ, Staller A, Brem S (2009) Relationship of gliomas to the ventricular walls. J Clin Neurosci 16(2):195–201. doi:10.1016/j.jocn.2008.03.006
39. Mcgirt MJ, Chaichana KL, Gathinji M et al (2009) Independent association of extent of resection with survival in patients with malignant brain astrocytoma. J Neurosurg 110(1):156–162. doi:10.3171/2008.4.17536
40. Stummer W, Reulen H-J, Meinel T et al (2008) Extent of resection and survival in glioblastoma multiforme: identification of and adjustment for bias. Neurosurgery 62(3):564–576. doi:10.1227/01.neu.0000317304.31579.17
41. Sanai N, Berger MS (2008) Glioma extent of resection and its impact on patient outcome. Neurosurgery 62(4):753–764. doi:10.1227/01.neu.0000318159.21731.cf
42. Sanai N, Polley M-Y, McDermott MW, Parsa AT, Berger MS (2011) An extent of resection threshold for newly diagnosed glioblastomas. J Neurosurg 115(1):3–8. doi:10.3171/2010.9.JNS101437
43. Merkle FT, Tramontin AD, Garcia-Verdugo JM, Alvarez-Buylla A (2004) Radial glia give rise to adult neural stem cells in the subventricular zone. Proc Natl Acad Sci U S A 101(50):17528–17532. doi:10.1073/pnas.0407893101
44. Lassman AB, Holland EC (2005) Glioblastoma multiforme—past, present, and future. US Oncol Rev 1(1):109–111
45. Singh SR (2013) Cancer stem cells: recent developments and future prospects. Cancer Lett 338(1):1–2. doi:10.1016/j.canlet.2013.03.036
46. Stummer W, Pichlmeier U, Meinel T, Wiestler OD, Zanella F, Reulen HJ (2006) Fluorescence-guided surgery with 5-aminolevulinic acid for resection of malignant glioma: a randomised controlled multicentre phase III trial. Lancet Oncol 7(5):392–401
47. Meacham CE, Morrison SJ (2013) Tumour heterogeneity and cancer cell plasticity. Nature 501(7467):328–337. doi:10.1038/nature12624
48. Fulda S (2013) Regulation of apoptosis pathways in cancer stem cells. Cancer Lett 338(1):168–173. doi:10.1016/j.canlet.2012.03.014
49. Friedmann-Morvinski D, Bushong EA, Ke E et al (2012) Dedifferentiation of neurons and astrocytes by oncogenes can induce gliomas in mice. Science 338(6110):1080–1084. doi:10.1126/science.1226929
50. Groszer M, Erickson R, Scripture-Adams DD et al (2001) Negative regulation of neural stem/progenitor cell proliferation by the Pten tumor suppressor gene in vivo. Science 294(5549):2186–2189. doi:10.1126/science.1065518
51. Zheng H, Ying H, Yan H et al (2008) p53 and Pten control neural and glioma stem/progenitor cell renewal and differentiation. Nature 455(7216):1129–1133. doi:10.1038/nature07443

52. Kendall SE, Najbauer J, Johnston HF et al (2008) Neural stem cell targeting of glioma is dependent on phosphoinositide 3-Kinase signaling. Stem Cells 26(6):1575–1586. doi:10.1634/stemcells.2007-0887
53. Alcantara Llaguno S, Chen J, Kwon C-H et al (2009) Malignant astrocytomas originate from neural stem/progenitor cells in a somatic tumor suppressor mouse model. Cancer Cell 15(1):45–56. doi:10.1016/j.ccr.2008.12.006
54. Chen R, Nishimura MC, Bumbaca SM et al (2010) A hierarchy of self-renewing tumor-initiating cell types in glioblastoma. Cancer Cell 17(4):362–375. doi:10.1016/j.ccr.2009.12.049
55. McCullough AK, Dodson ML, Lloyd RS (1999) Initiation of base excision repair: glycosylase mechanisms and structures. Annu Rev Biochem 68:255–285. doi:10.1146/annurev.biochem.68.1.255
56. Magee JA, Piskounova E, Morrison SJ (2012) Cancer stem cells: impact, heterogeneity, and uncertainty. Cancer Cell 21(3):283–296. doi:10.1016/j.ccr.2012.03.003
57. Walker MD, Strike TA, Sheline GE (1979) An analysis of dose-effect relationship in the radiotherapy of malignant gliomas. Int J Radiat Oncol Biol Phys 5(10):1725–1731
58. Zhang M, Song T, Yang L et al (2008) Nestin and CD133: valuable stem cell-specific markers for determining clinical outcome of glioma patients. J Exp Clin Cancer Res 27:85. doi:10.1186/1756-9966-27-85
59. Zeppernick F, Ahmadi R, Campos B et al (2008) Stem cell marker CD133 affects clinical outcome in glioma patients. Clin Cancer Res 14(1):123–129. doi:10.1158/1078-0432.CCR-07-0932
60. Choi JW, Lee C-W, Chun J (2008) Biological roles of lysophospholipid receptors revealed by genetic null mice: an update. Biochim Biophys Acta 1781(9):531–539. doi:10.1016/j.bbalip.2008.03.004
61. Moolenaar WH (2002) Lysophospholipids in the limelight: autotaxin takes center stage. J Cell Biol 158(2):197–199. doi:10.1083/jcb.200206094
62. Umezu-Goto M, Kishi Y, Taira A et al (2002) Autotaxin has lysophospholipase D activity leading to tumor cell growth and motility by lysophosphatidic acid production. J Cell Biol 158(2):227–233. doi:10.1083/jcb.200204026
63. Tokumura A, Majima E, Kariya Y et al (2002) Identification of human plasma lysophospholipase D, a lysophosphatidic acid-producing enzyme, as autotaxin, a multifunctional phosphodiesterase. J Biol Chem 277(42):39436–39442. doi:10.1074/jbc.M205623200
64. Meyer zu Heringdorf D, Jakobs KH (2007) Lysophospholipid receptors: signalling, pharmacology and regulation by lysophospholipid metabolism. Biochim Biophys Acta 1768(4):923–940. doi:10.1016/j.bbamem.2006.09.026
65. Ishii I, Fukushima N, Ye X, Chun J (2004) Lysophospholipid receptors: signaling and biology. Annu Rev Biochem 73:321–354. doi:10.1146/annurev.biochem.73.011303.073731
66. van Meeteren LA, Ruurs P, Christodoulou E et al (2005) Inhibition of autotaxin by lysophosphatidic acid and sphingosine 1-phosphate. J Biol Chem 280(22):21155–21161. doi:10.1074/jbc.M413183200
67. Baker DL, Morrison P, Miller B et al (2002) Plasma lysophosphatidic acid concentration and ovarian cancer. JAMA 287(23):3081–3082
68. Das AK, Hajra AK (1989) Quantification, characterization and fatty acid composition of lysophosphatidic acid in different rat tissues. Lipids 24(4):329–333
69. Chun J, Hla T, Lynch KR, Spiegel S, Moolenaar WH (2010) International union of basic and clinical pharmacology. LXXVIII Lysophospholipid Receptor Nomenclature Pharmacological Reviews 62(4):579–587. doi:10.1124/pr.110.003111
70. Yanagida K, Ishii S (2011) Non-Edg family LPA receptors: the cutting edge of LPA research. J Biochem 150(3):223–232. doi:10.1093/jb/mvr087
71. Anliker B, Lysophospholipid G (2004) Protein-coupled receptors. J Biol Chem 279(20):20555–20558. doi:10.1074/jbc.R400013200
72. Goldshmit Y, Munro K, Leong SY, Pébay A, Turnley AM (2010) LPA receptor expression in the central nervous system in health and following injury. Cell Tissue Res 341(1):23–32. doi:10.1007/s00441-010-0977-5

73. Chan LC, Peters W, Xu Y, Chun J, Farese RV, Cases S (2007) LPA3 receptor mediates chemotaxis of immature murine dendritic cells to unsaturated lysophosphatidic acid (LPA). J Leukoc Biol 82(5):1193–1200. doi:10.1189/jlb.0407221
74. Lin C-I, Chen C-N, Lin P-W, Chang K-J, Hsieh F-J, Lee H (2007) Lysophosphatidic acid regulates inflammation-related genes in human endothelial cells through LPA1 and LPA3. Biochem Biophys Res Commun 363(4):1001–1008. doi:10.1016/j.bbrc.2007.09.081
75. Lee Z, Cheng C-T, Zhang H et al (2008) Role of LPA4/p2y9/GPR23 in negative regulation of cell motility. Mol Biol Cell 19(12):5435–5445. doi:10.1091/mbc.E08-03-0316
76. Pope WB, Young JR, Ellingson BM (2011) Advances in MRI assessment of gliomas and response to anti-VEGF therapy. Curr Neurol Neurosci Rep 11(3):336–344. doi:10.1007/s11910-011-0179-x
77. Stracke ML, Krutzsch HC, Unsworth EJ et al (1992) Identification, purification, and partial sequence analysis of autotaxin, a novel motility-stimulating protein. J Biol Chem 267(4):2524–2529
78. Narayana A, Gruber D, Kunnakkat S et al (2012) A clinical trial of bevacizumab, temozolomide, and radiation for newly diagnosed glioblastoma. J Neurosurg 116(2):341–345. doi:10.3171/2011.9.JNS11656
79. Spasic M, Chow F, Tu C, Nagasawa DT, Yang I (2012) Molecular characteristics and pathways of Avastin for the treatment of glioblastoma multiforme. Neurosurg Clin N Am 23(3):417–427. doi:10.1016/j.nec.2012.05.002
80. Prenzel N, Zwick E, Daub H et al (1999) EGF receptor transactivation by G-protein-coupled receptors requires metalloproteinase cleavage of proHB-EGF. Nature 402(6764):884–888. doi:10.1038/47260
81. Daub H, Weiss FU, Wallasch C, Ullrich A (1996) Role of transactivation of the EGF receptor in signalling by G-protein-coupled receptors. Nature 379(6565):557–560. doi:10.1038/379557a0
82. Daub H, Wallasch C, Lankenau A, Herrlich A, Ullrich A (1997) Signal characteristics of G protein-transactivated EGF receptor. EMBO J 16(23):7032–7044. doi:10.1093/emboj/16.23.7032
83. Keller JN, Steiner MR, Holtsberg FW, Mattson MP, Steiner SM (1997) Lysophosphatidic acid-induced proliferation-related signals in astrocytes. J Neurochem 69(3):1073–1084
84. Ng W, Pébay A, Drummond K, Burgess A, Kaye AH, Morokoff A (2014) Complexities of lysophospholipid signalling in glioblastoma. J Clin Neurosci 21(6):893–898. doi:10.1016/j.jocn.2014.02.013
85. Liao Y, Mu G, Zhang L, Zhou W, Zhang J, Yu H (2013) Lysophosphatidic acid stimulates activation of focal adhesion kinase and paxillin and promotes cell motility, via LPA1–3, in human pancreatic cancer. Dig Dis Sci 58(12):3524–3533. doi:10.1007/s10620-013-2878-4
86. Korkina O, Dong Z, Marullo A, Warshaw G, Symons M, Ruggieri R (2013) The MLK-related kinase (MRK) is a novel RhoC effector that mediates lysophosphatidic acid (LPA)-stimulated tumor cell invasion. J Biol Chem 288(8):5364–5373. doi:10.1074/jbc.M112.414060
87. Leve F, Marcondes TGC, Bastos LGR, Rabello SV, Tanaka MN, Morgado-Díaz JA (2011) Lysophosphatidic acid induces a migratory phenotype through a crosstalk between RhoA–Rock and Src–FAK signalling in colon cancer cells. Eur J Pharmacol 671(1–3):7–17. doi:10.1016/j.ejphar.2011.09.006
88. Xu X, Yang G, Zhang H, Prestwich GD (2009) Evaluating dual activity LPA receptor pan-antagonist/autotaxin inhibitors as anti-cancer agents in vivo using engineered human tumors. Prostaglandins Other Lipid Mediat 89(3–4):140–146. doi:10.1016/j.prostaglandins.2009.07.006
89. Sengupta S, Kim KS, Berk MP et al (2006) Lysophosphatidic acid downregulates tissue inhibitor of metalloproteinases, which are negatively involved in lysophosphatidic acid-induced cell invasion. Oncogene 26(20):2894–2901. doi:10.1038/sj.onc.1210093
90. Shah BH, Neithardt A, Chu DB, Shah FB, Catt KJ (2006) Role of EGF receptor transactivation in phosphoinositide 3-kinase-dependent activation of MAP kinase by GPCRs. J Cell Physiol 206(1):47–57. doi:10.1002/jcp.20423

91. Bian D, Mahanivong C, Yu J et al (2005) The G12/13-RhoA signaling pathway contributes to efficient lysophosphatidic acid-stimulated cell migration. Oncogene 25(15):2234–2244. doi:10.1038/sj.onc.1209261
92. Manning TJ, Parker JC, Sontheimer H (2000) Role of lysophosphatidic acid and rho in glioma cell motility. Cell Motil Cytoskeleton 45(3):185–199. doi:10.1002/(SICI)1097-0169(200003)45:3<185::AID-CM2>3.0.CO;2-G
93. Tabuchi S, Kume K, Aihara M, Shimizu T (2000) Expression of lysophosphatidic acid receptor in rat astrocytes: mitogenic effect and expression of neurotrophic genes. Neurochem Res 25(5):573–582
94. Alves TR, Lima FRS, Kahn SA et al (2011) Glioblastoma cells: a heterogeneous and fatal tumor interacting with the parenchyma. Life Sci 89(15–16):532–539. doi:10.1016/j.lfs.2011.04.022
95. Charles NA, Holland EC, Gilbertson R, Glass R, Kettenmann H (2011) The brain tumor microenvironment. Glia 59(8):1169–1180. doi:10.1002/glia.21136
96. Schneider SW, Ludwig T, Tatenhorst L et al (2004) Glioblastoma cells release factors that disrupt blood-brain barrier features. Acta Neuropathol 107(3):272–276. doi:10.1007/s00401-003-0810-2
97. Rascher G, Fischmann A, Kröger S, Duffner F, Grote E-H, Wolburg H (2002) Extracellular matrix and the blood-brain barrier in glioblastoma multiforme: spatial segregation of tenascin and agrin. Acta Neuropathol 104(1):85–91. doi:10.1007/s00401-002-0524-x
98. Coomberl BL, Stewart PA, Hayakawa K, Farrell CL, Del Maestros RF (1987) Quantitative morphology of human glioblastoma multiforme microvessels: structural basis of blood-brain barrier defect. J Neurooncol 5(4):299–307. doi:10.1007/BF00148386
99. Huang M-C, Lee H-Y, Yeh C-C, Kong Y, Zaloudek CJ, Goetzl EJ (2004) Induction of protein growth factor systems in the ovaries of transgenic mice overexpressing human type 2 lysophosphatidic acid G protein-coupled receptor (LPA2). Oncogene 23(1):122–129. doi:10.1038/sj.onc.1206986
100. E S, Lai Y-J, Tsukahara R et al (2009) Lysophosphatidic acid 2 receptor-mediated supramolecular complex formation regulates its antiapoptotic effect. J Biol Chem 284(21):14558–14571. doi:10.1074/jbc.M900185200
101. Shano S, Moriyama R, Chun J, Fukushima N (2008) Lysophosphatidic acid stimulates astrocyte proliferation through LPA1. Neurochem Int 52(1–2):216–220. doi:10.1016/j.neuint.2007.07.004
102. Matas-Rico E, García-Diaz B, Llebrez-Zayas P et al (2008) Deletion of lysophosphatidic acid receptor LPA1 reduces neurogenesis in the mouse dentate gyrus. Mol Cell Neurosci 39(3):342–355. doi:10.1016/j.mcn.2008.07.014
103. Zeng W, Zhang Z (2007) Lysophosphatidic acid induces astrocyte proliferation in hippocampus slices partially through activating extracellular signal-regulated kinases. Res Chem Intermed 33(6):567–577. doi:10.1163/156856707782565859
104. Sorensen SD, Nicole O, Peavy RD et al (2003) Common signaling pathways link activation of murine PAR-1, LPA, and S1P receptors to proliferation of astrocytes. Mol Pharmacol 64(5):1199–1209. doi:10.1124/mol.64.5.1199
105. Weiner JA, Hecht JH, Chun J (1998) Lysophosphatidic acid receptor gene vzg-1/lpA1/edg-2 is expressed by mature oligodendrocytes during myelination in the postnatal murine brain. J Comp Neurol 398(4):587–598
106. Spohr TCS, de Sampaio E, Spohr TC, Choi JW et al (2008) Lysophosphatidic acid receptor-dependent secondary effects via astrocytes promote neuronal differentiation. J Biol Chem 283(12):7470–7479. doi:10.1074/jbc.M707758200
107. Cui H-L, Qiao J-T (2007) Effect of lysophosphatidic acid on differentiation of embryonic neural stem cells into neuroglial cells in rats in vitro. Sheng Li Xue Bao 59(6):759–764
108. Fukushima N, Shano S, Moriyama R, Chun J (2007) Lysophosphatidic acid stimulates neuronal differentiation of cortical neuroblasts through the LPA1-G(i/o) pathway. Neurochem Int 50(2):302–307. doi:10.1016/j.neuint.2006.09.008

109. Dottori M, Leung J, Turnley AM, Pébay A (2008) Lysophosphatidic acid inhibits neuronal differentiation of neural stem/progenitor cells derived from human embryonic stem cells. Stem Cells 26(5):1146–1154. doi:10.1634/stemcells.2007-1118
110. Kishi Y, Okudaira S, Tanaka M et al (2006) Autotaxin is overexpressed in glioblastoma multiforme and contributes to cell motility of glioblastoma by converting lysophosphatidylcholine to lysophosphatidic acid. J Biol Chem 281(25):17492–17500. doi:10.1074/jbc.M601803200
111. van Meeteren LA, Ruurs P, Stortelers C et al (2006) Autotaxin, a secreted lysophospholipase D, is essential for blood vessel formation during development. Mol Cell Biol 26(13):5015–5022. doi:10.1128/MCB.02419-05
112. Tanaka M, Okudaira S, Kishi Y et al (2006) Autotaxin stabilizes blood vessels and is required for embryonic vasculature by producing lysophosphatidic acid. J Biol Chem 281(35):25822–25830. doi:10.1074/jbc.M605142200
113. Savaskan NE, Rocha L, Kotter MR et al (2006) Autotaxin (NPP-2) in the brain: cell type-specific expression and regulation during development and after neurotrauma. Cell Mol Life Sci 64(2):230–243. doi:10.1007/s00018-006-6412-0
114. Fuentealba LC, Rompani SB, Parraguez JI et al (2015) Embryonic origin of postnatal neural stem cells. Cell 161(7):1644–1655. doi:10.1016/j.cell.2015.05.041
115. Kriegstein A, Alvarez-Buylla A (2009) The glial nature of embryonic and adult neural stem cells. Annu Rev Neurosci 32(1):149–184. doi:10.1146/annurev.neuro.051508.135600
116. Doetsch F (2003) The glial identity of neural stem cells. Nat Neurosci 6(11):1127–1134. doi:10.1038/nn1144
117. Doetsch F, Caillé I, Lim DA, García-Verdugo JM, Alvarez-Buylla A (1999) Subventricular zone astrocytes are neural stem cells in the adult mammalian brain. Cell 97(6):703–716. doi:10.1016/S0092-8674(00)80783-7
118. Taupin P, Gage FH (2002) Adult neurogenesis and neural stem cells of the central nervous system in mammals. J Neurosci Res 69(6):745–749. doi:10.1002/jnr.10378
119. Sottoriva A, Spiteri I, Piccirillo SGM et al (2013) Intratumor heterogeneity in human glioblastoma reflects cancer evolutionary dynamics. Proc Natl Acad Sci U S A 110(10):4009–4014. doi:10.1073/pnas.1219747110
120. Ozawa T, Riester M, Cheng Y-K et al (2014) Most human non-GCIMP glioblastoma subtypes evolve from a common proneural-like precursor glioma. Cancer Cell 26(2):288–300. doi:10.1016/j.ccr.2014.06.005
121. Aldape KD, Ballman K, Furth A et al (2004) Immunohistochemical detection of EGFRvIII in high malignancy grade astrocytomas and evaluation of prognostic significance. J Neuropathol Exp Neurol 63(7):700–707
122. Fernandes H, Cohen S, Bishayee S (2001) Glycosylation-induced conformational modification positively regulates receptor-receptor association: a study with an aberrant epidermal growth factor receptor (EGFRvIII/DeltaEGFR) expressed in cancer cells. J Biol Chem 276(7):5375–5383. doi:10.1074/jbc.M005599200
123. Gan HK, Kaye AH, Luwor RB (2009) The EGFRvIII variant in glioblastoma multiforme. J Clin Neurosci 16(6):748–754. doi:10.1016/j.jocn.2008.12.005
124. Omidfar K, Shirvani Z (2012) Single domain antibodies: a new concept for epidermal growth factor receptor and EGFRvIII targeting. DNA Cell Biol 31(6):1015–1026. doi:10.1089/dna.2011.1529
125. Verhaak RGW, Hoadley KA, Purdom E et al (2010) Integrated genomic analysis identifies clinically relevant subtypes of glioblastoma characterized by abnormalities in PDGFRA, IDH1, EGFR, and NF1. Cancer Cell 2012(1):98–110. doi:10.1016/j.ccr.2009.12.020
126. Cancer Genome Atlas Research Network (2008) Comprehensive genomic characterization defines human glioblastoma genes and core pathways. Nature 455(7216):1061–1068. doi:10.1038/nature07385
127. Lino MM, Merlo A (2010) PI3Kinase signaling in glioblastoma. J Neurooncol 103(3):417–427. doi:10.1007/s11060-010-0442-z

128. Pelloski CE, Ballman KV, Furth AF et al (2007) Epidermal growth factor receptor variant III status defines clinically distinct subtypes of glioblastoma. J Clin Oncol 25(16):2288–2294. doi:10.1200/JCO.2006.08.0705
129. Prados MD, Chang SM, Butowski N et al (2008) Phase II study of erlotinib plus temozolomide during and after radiation therapy in patients with newly diagnosed glioblastoma multiforme or gliosarcoma. J Clin Oncol 27(4):579–584. doi:10.1200/JCO.2008.18.9639
130. Szerlip NJ, Pedraza A, Chakravarty D et al (2012) Intratumoral heterogeneity of receptor tyrosine kinases EGFR and PDGFRA amplification in glioblastoma defines subpopulations with distinct growth factor response. Proc Natl Acad Sci U S A 109(8):3041–3046. doi:10.1073/pnas.1114033109
131. Stupp R, Hegi ME, Mason WP et al (2009) Effects of radiotherapy with concomitant and adjuvant temozolomide versus radiotherapy alone on survival in glioblastoma in a randomised phase III study: 5-year analysis of the EORTC-NCIC trial. Lancet Oncol 10(5):459–466. doi:10.1016/S1470-2045(09)70025-7
132. Preusser M, Gelpi E, Rottenfusser A et al (2008) Epithelial growth factor receptor inhibitors for treatment of recurrent or progressive high grade glioma: an exploratory study. J Neurooncol 89(2):211–218. doi:10.1007/s11060-008-9608-3
133. Stupp R, Mason WP, Van den Bent MJ et al (2005) Radiotherapy plus concomitant and adjuvant temozolomide for glioblastoma. N Engl J Med 352(10):987–996. doi:10.1056/NEJMoa043330
134. Mellinghoff IK, Wang MY, Vivanco I et al (2005) Molecular determinants of the response of glioblastomas to EGFR kinase inhibitors. N Engl J Med 353(19):2012–2024. doi:10.1056/NEJMoa051918
135. Hegi ME, Diserens A-C, Gorlia T et al (2005) MGMT gene silencing and benefit from temozolomide in glioblastoma. N Engl J Med 352(10):997–1003. doi:10.1056/NEJMoa043331
136. Camara-Quintana JQ, Nitta RT, Li G (2012) Pathology: commonly monitored glioblastoma markers: EFGR, EGFRvIII, PTEN, and MGMT. Neurosurg Clin N Am 23(2):237–246. doi:10.1016/j.nec.2012.01.011
137. Hegi ME, Liu L, Herman JG et al (2008) Correlation of O6-methylguanine methyltransferase (MGMT) promoter methylation with clinical outcomes in glioblastoma and clinical strategies to modulate MGMT activity. J Clin Oncol 26(25):4189–4199. doi:10.1200/JCO.2007.11.5964
138. Brown PD, Krishnan S, Sarkaria JN et al (2008) Phase I/II trial of erlotinib and temozolomide with radiation therapy in the treatment of newly diagnosed glioblastoma multiforme: north central cancer treatment group study N0177. J Clin Oncol 26(34):5603–5609. doi:10.1200/JCO.2008.18.0612
139. Raizer JJ, Abrey LE, Lassman AB et al (2010) A phase II trial of erlotinib in patients with recurrent malignant gliomas and nonprogressive glioblastoma multiforme postradiation therapy. Neuro Oncol 12(1):95–103. doi:10.1093/neuonc/nop015
140. Huse JT, Phillips HS, Brennan CW (2011) Molecular subclassification of diffuse gliomas: seeing order in the chaos. Glia 59(8):1190–1199. doi:10.1002/glia.21165
141. Nabors LB, Fink KL, Mikkelsen T et al (2015) Two cilengitide regimens in combination with standard treatment for patients with newly diagnosed glioblastoma and unmethylated MGMT gene promoter: results of the open-label, controlled, randomized phase II CORE study. Neuro Oncol 17(5):708–717. doi:10.1093/neuonc/nou356
142. Stupp R, Hegi ME, Gorlia T et al (2014) Cilengitide combined with standard treatment for patients with newly diagnosed glioblastoma with methylated MGMT promoter (CENTRIC EORTC 26071-22072 study): a multicentre, randomised, open-label, phase 3 trial. Lancet Oncol 15(10):1100–1108. doi:10.1016/S1470-2045(14)70379-1
143. Gilbert MR, Dignam JJ, Armstrong TS et al (2014) A randomized trial of bevacizumab for newly diagnosed glioblastoma. N Engl J Med 370(8):699–708. doi:10.1056/NEJMoa1308573
144. Wu X, Senechal K, Neshat MS, Whang YE, Sawyers CL (1998) The PTEN/MMAC1 tumor suppressor phosphatase functions as a negative regulator of the phosphoinositide 3-kinase/Akt pathway. Proc Natl Acad Sci U S A 95(26):15587–15591. doi:10.1073/pnas.95.26.15587

145. Myers MP, Pass I, Batty IH et al (1998) The lipid phosphatase activity of PTEN is critical for its tumor supressor function. Proc Natl Acad Sci U S A 95(23):13513–13518. doi:10.1073/pnas.95.23.13513
146. Cantley LC, Neel BG (1999) New insights into tumor suppression: PTEN suppresses tumor formation by restraining the phosphoinositide 3-kinase/AKT pathway. Proc Natl Acad Sci U S A 96(8):4240–4245. doi:10.1073/pnas.96.8.4240
147. Maehama T, Dixon JE (1999) PTEN: a tumour suppressor that functions as a phospholipid phosphatase. Trends Cell Biol 9(4):125–128
148. Maehama T, Dixon JE (1998) The tumor suppressor, PTEN/MMAC1, dephosphorylates the lipid second messenger, phosphatidylinositol 3,4,5-trisphosphate. J Biol Chem 273(22):13375–13378. doi:10.1074/jbc.273.22.13375
149. Ming M, He Y-Y (2012) PTEN in DNA damage repair. Cancer Lett 319(2):125–129. doi:10.1016/j.canlet.2012.01.003
150. Wen PY, Lee EQ, Reardon DA, Ligon KL, Alfred Yung WK (2012) Current clinical development of PI3K pathway inhibitors in glioblastoma. Neuro Oncol 14(7):819–829. doi:10.1093/neuonc/nos117
151. Bambury RM, Teo MY, Power DG et al (2013) The association of pre-treatment neutrophil to lymphocyte ratio with overall survival in patients with glioblastoma multiforme. Cell Cycle 114(10):1221–1224. doi:10.1007/s11060-013-1164-9
152. Berenjeno IM, Guillermet-Guibert J, Pearce W, Gray A, Fleming S, Vanhaesebroeck B (2012) Both p110α and p110β isoforms of PI3K can modulate the impact of loss-of-function of the PTEN tumour suppressor. Biochem J 442(1):151–159. doi:10.1042/BJ20111741
153. Manning BD, Cantley LC (2007) AKT/PKB signaling: navigating downstream. Cell 129(7):1261–1274. doi:10.1016/j.cell.2007.06.009
154. Wei Y, Jiang Y, Zou F et al (2013) Activation of PI3K/Akt pathway by CD133-p85 interaction promotes tumorigenic capacity of glioma stem cells. Proc Natl Acad Sci U S A 110(17):6829–6834. doi:10.1073/pnas.1217002110
155. Datta SR, Dudek H, Tao X et al (1997) Akt phosphorylation of BAD couples survival signals to the cell-intrinsic death machinery. Cell 91(2):231–241. doi:10.1016/S0092-8674(00)80405-5
156. Sekulić A, Hudson CC, Homme JL et al (2000) A direct linkage between the phosphoinositide 3-kinase-AKT signaling pathway and the mammalian target of rapamycin in mitogen-stimulated and transformed cells. Cancer Res 60(13):3504–3513
157. Zhong H, Chiles K, Feldser D et al (2000) Modulation of hypoxia-inducible factor 1alpha expression by the epidermal growth factor/phosphatidylinositol 3-kinase/PTEN/AKT/FRAP pathway in human prostate cancer cells: implications for tumor angiogenesis and therapeutics. Cancer Res 60(6):1541–1545
158. Potter CJ, Pedraza LG, Xu T (2002) Akt regulates growth by directly phosphorylating Tsc2. Nat Cell Biol 4(9):658–665. doi:10.1038/ncb840
159. Kane LP, Shapiro VS, Stokoe D, Weiss A (1999) Induction of NF-κB by the Akt/PKB kinase. Curr Biol 9(11):601–S601. doi:10.1016/S0960-9822(99)80265-6
160. Davies MA, Koul D, Dhesi H et al (1999) Regulation of Akt/PKB activity, cellular growth, and apoptosis in prostate carcinoma cells by MMAC/PTEN. Cancer Res 59(11):2551–2556
161. Lawlor MA, Alessi DR (2001) PKB/Akt: a key mediator of cell proliferation, survival and insulin responses? J Cell Sci 114(Pt 16):2903–2910
162. Scheid MP, Woodgett JR (2001) PKB/AKT: functional insights from genetic models. Nat Rev Mol Cell Biol 2(10):760–768. doi:10.1038/35096067
163. Koul D, Shen R, Bergh S et al (2006) Inhibition of Akt survival pathway by a small-molecule inhibitor in human glioblastoma. Mol Cancer Ther 5(3):637–644. doi:10.1158/1535-7163.MCT-05-0453
164. LoPiccolo J, Blumenthal GM, Bernstein WB, Dennis PA (2008) Targeting the PI3K/Akt/mTOR pathway: effective combinations and clinical considerations. Drug Resist Updat 11(1–2):32–50. doi:10.1016/j.drup.2007.11.003

165. Gil del Alcazar CR, Hardebeck MC, Mukherjee B et al (2013) Inhibition of DNA double-strand break repair by the dual PI3K/mTOR inhibitor NVP-BEZ235 as a strategy for radiosensitization of glioblastoma. Clin Cancer Res 20(5):1235–1248. doi:10.1158/1078-0432.CCR-13-1607
166. Ping Y-F, Yao X-H, Jiang J-Y et al (2011) The chemokine CXCL12 and its receptor CXCR4 promote glioma stem cell-mediated VEGF production and tumour angiogenesis via PI3K/AKT signalling. J Pathol 224(3):344–354. doi:10.1002/path.2908
167. Gallia GL, Tyler BM, Hann CL et al (2009) Inhibition of Akt inhibits growth of glioblastoma and glioblastoma stem-like cells. Mol Cancer Ther 8(2):386–393. doi:10.1158/1535-7163.MCT-08-0680
168. Sunayama J, Matsuda K-I, Sato A et al (2010) Crosstalk between the PI3K/mTOR and MEK/ERK pathways involved in the maintenance of self-renewal and tumorigenicity of glioblastoma stem-like cells. Stem Cells 28(11):1930–1939. doi:10.1002/stem.521
169. Chen JS, Zhou LJ, Entin-Meer M et al (2008) Characterization of structurally distinct, isoform-selective phosphoinositide 3′-kinase inhibitors in combination with radiation in the treatment of glioblastoma. Mol Cancer Ther 7(4):841–850. doi:10.1158/1535-7163.MCT-07-0393
170. Opel D, Westhoff MA, Bender A, Braun V, Debatin KM, Fulda S (2008) Phosphatidylinositol 3-kinase inhibition broadly sensitizes glioblastoma cells to death receptor- and drug-induced apoptosis. Cancer Res 68(15):6271–6280. doi:10.1158/0008-5472.CAN-07-6769
171. Yart A, Roche S, Wetzker R et al (2002) A function for phosphoinositide 3-kinase beta lipid products in coupling beta gamma to Ras activation in response to lysophosphatidic acid. J Biol Chem 277(24):21167–21178. doi:10.1074/jbc.M110411200
172. Lopez-Ilasaca M (1997) Linkage of G protein-coupled receptors to the MAPK signaling pathway through PI 3-kinase gamma. Science 275(5298):394–397. doi:10.1126/science.275.5298.394
173. Hernández M, Barrero MJ, Crespo MS, Nieto ML (2000) Lysophosphatidic acid inhibits Ca2+ signaling in response to epidermal growth factor receptor stimulation in human astrocytoma cells by a mechanism involving phospholipase C(gamma) and a G(alphai) protein. J Neurochem 75(4):1575–1582
174. Gschwind A, Prenzel N, Ullrich A (2002) Lysophosphatidic acid-induced squamous cell carcinoma cell proliferation and motility involves epidermal growth factor receptor signal transactivation. Cancer Res 62(21):6329–6336
175. Kim EK, Yun SJ, Do KH et al (2008) Lysophosphatidic acid induces cell migration through the selective activation of Akt1. Exp Mol Med 40(4):445. doi:10.3858/emm.2008.40.4.445
176. Lee HY, Bae GU, Jung ID et al (2002) Autotaxin promotes motility via G protein-coupled phosphoinositide 3-kinase gamma in human melanoma cells. FEBS Lett 515(1–3):137–140
177. Gschwind A, Zwick E, Prenzel N, Leserer M, Ullrich A (2001) Cell communication networks: epidermal growth factor receptor transactivation as the paradigm for interreceptor signal transmission. Oncogene 20(13):1594–1600. doi:10.1038/sj.onc.1204192
178. Pullen NA, Anand M, Cooper PS, Fillmore HL (2011) Matrix metalloproteinase-1 expression enhances tumorigenicity as well as tumor-related angiogenesis and is inversely associated with TIMP-4 expression in a model of glioblastoma. J Neurooncol 106(3):461–471. doi:10.1007/s11060-011-0691-5
179. Buccione R, Caldieri G, Ayala I (2009) Invadopodia: specialized tumor cell structures for the focal degradation of the extracellular matrix. Cancer Metastasis Rev 28(1–2):137–149. doi:10.1007/s10555-008-9176-1
180. Poincloux R, Lizarraga F, Chavrier P (2009) Matrix invasion by tumour cells: a focus on MT1-MMP trafficking to invadopodia. J Cell Sci 122(17):3015–3024. doi:10.1242/jcs.034561
181. Clark ES, Whigham AS, Yarbrough WG, Weaver AM (2007) Cortactin is an essential regulator of matrix metalloproteinase secretion and extracellular matrix degradation in invadopodia. Cancer Res 67(9):4227–4235. doi:10.1158/0008-5472.CAN-06-3928

182. Artym VV, Zhang Y, Seillier-Moiseiwitsch F, Yamada KM, Mueller SC (2006) Dynamic interactions of cortactin and membrane Type 1 Matrix metalloproteinase at invadopodia: defining the stages of invadopodia formation and function. Cancer Res 66(6):3034–3043. doi:10.1158/0008-5472.CAN-05-2177
183. Kelly T, Yan Y, Osborne RL et al (1998) Proteolysis of extracellular matrix by invadopodia facilitates human breast cancer cell invasion and is mediated by matrix metalloproteinases. Clin Exp Metastasis 16(6):501–512. doi:10.1023/A:1006538200886
184. Harper K, Arsenault D, Boulay-Jean S, Lauzier A, Lucien F, Dubois CM (2010) Autotaxin promotes cancer invasion via the lysophosphatidic acid receptor 4: participation of the cyclic AMP/EPAC/Rac1 signaling pathway in invadopodia formation. Cancer Res 70(11):4634–4643. doi:10.1158/0008-5472.CAN-09-3813
185. Hoelzinger DB, Nakada M, Demuth T, Rosensteel T, Reavie LB, Berens ME (2007) Autotaxin: a secreted autocrine/paracrine factor that promotes glioma invasion. J Neurooncol 86(3):297–309. doi:10.1007/s11060-007-9480-6
186. Schleicher SM, Thotala DK, Linkous AG et al (2011) Autotaxin and LPA receptors represent potential molecular targets for the radiosensitization of murine glioma through effects on tumor vasculature. PLoS One 6(7):e22182. doi:10.1371/journal.pone.0022182.g007
187. Clair T, Lee HY, Liotta LA, Stracke ML (1997) Autotaxin is an exoenzyme possessing 5'-nucleotide phosphodiesterase/ATP pyrophosphatase and ATPase activities. J Biol Chem 272(2):996–1001
188. Seasholtz TM, Radeff-Huang J, Sagi SA et al (2004) Rho-mediated cytoskeletal rearrangement in response to LPA is functionally antagonized by Rac1 and PIP2. J Neurochem 91(2):501–512. doi:10.1111/j.1471-4159.2004.02749.x
189. Khalil BD, El-Sibai M (2012) Rho GTPases in primary brain tumor malignancy and invasion. J Neurooncol 108(3):333–339. doi:10.1007/s11060-012-0866-8
190. Jongsma M, Matas-Rico E, Rzadkowski A, Jalink K, Moolenaar WH (2011) LPA is a chemorepellent for B16 melanoma cells: action through the cAMP-elevating LPA5 receptor. PLoS One 6(12):e29260. doi:10.1371/journal.pone.0029260
191. Hoelzinger DB, Demuth T, Berens ME (2007) Autocrine factors that sustain glioma invasion and paracrine biology in the brain microenvironment. JNCI J Natl Cancer Inst 99(21):1583–1593. doi:10.1093/jnci/djm187
192. Yang SY, Lee J, Park CG et al (2002) Expression of autotaxin (NPP-2) is closely linked to invasiveness of breast cancer cells. Clin Exp Metastasis 19(7):603–608
193. Boucharaba A, Serre C-M, Guglielmi J, Bordet J-C, Clézardin P, Peyruchaud O (2006) The type 1 lysophosphatidic acid receptor is a target for therapy in bone metastases. Proc Natl Acad Sci U S A 103(25):9643–9648. doi:10.1073/pnas.0600979103
194. Prestwich GD, Gajewiak J, Zhang H, Xu X, Yang G, Serban M (2008) Phosphatase-resistant analogues of lysophosphatidic acid: agonists promote healing, antagonists and autotaxin inhibitors treat cancer. Biochim Biophys Acta 1781(9):588–594. doi:10.1016/j.bbalip.2008.03.008

Chapter 10
New Developments in Free Fatty Acids and Lysophospholipids: Decoding the Role of Phospholipases in Exocytosis

Vinod K. Narayana, David Kvaskoff, and Frederic A. Meunier

Abbreviations

AA	Arachidonic acid
Ca^{2+}	Calcium
DHA	Docosahexaenoic acid
ER	Endoplasmic reticulum
ESI	Electrospray ionisation
FFAs	Free fatty acids
GC/MS	Gas chromatography/mass spectrometry
GPCRs	G-protein coupled receptors
LC/MS	Liquid chromatography mass spectrometry
LPA	Lysophosphatidic acid
LPAAT	Lysophosphatidic acid acyl transferase
LPC	Lysophosphatidylcholine
LPL	Lysophospholipids
MS	Mass spectrometry
PA	Phosphatidic acid
PC	Phosphatidylcholine
PH	Pleckstrin homology domain

V.K. Narayana • F.A. Meunier (✉)
Clem Jones Centre for Ageing Dementia Research, Queensland Brain Institute,
The University of Queensland, Brisbane, QLD, Australia
e-mail: f.meunier@uq.edu.au

D. Kvaskoff (✉)
Clem Jones Centre for Ageing Dementia Research, Queensland Brain Institute,
The University of Queensland, Brisbane, QLD, Australia

Heidelberg University Biochemistry Centre (BZH),
Im Neuenheimer Feld 328, Heidelberg 69120, Germany
e-mail: david.kvaskoff@bzh.uni-heidelberg.de

A. Pébay, R.C.B. Wong (eds.), *Lipidomics of Stem Cells*, Stem Cell Biology and Regenerative Medicine, DOI 10.1007/978-3-319-49343-5_10

PIPs	Phosphatidylinositol phosphates
PKC	Protein kinase C
PL	Phospholipase
PtdIns(4,5)P_2	Phosphatidylinositol (4,5) bisphosphate
SM	Sphingomyelin
SNARE	Soluble N-ethylmaleimide-sensitive factor activating protein receptor

10.1 Introduction

The staggering diversity of lipids, currently exceeding 40,000 species, underpins their critical importance in all aspects of cellular life. The specific distribution of lipids in cells is related to their distinct properties in membranes, their polarity, fluidity and curvature, which characterise different organelles such as the ER, Golgi, and mitochondria (Fig. 10.1).

Phospholipids are the building blocks of cell membranes, and contain different polar head groups defining their function, location and properties in the cell. For example, phosphatidylcholine (PC), sphingomyelin (SM) and glycolipids mostly constitute the outer leaflet of the membrane bilayer in eukaryotes, while the negatively

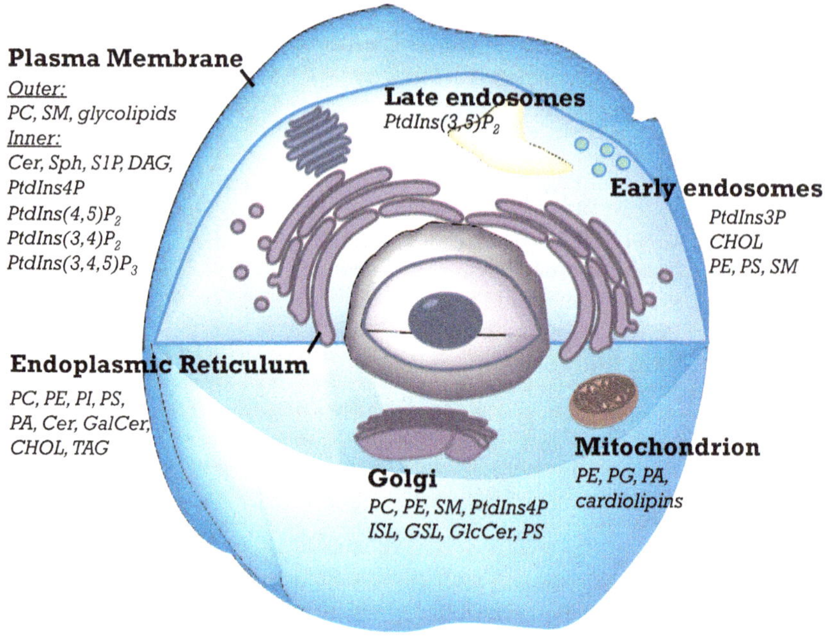

Fig. 10.1 The functional diversity and localisation of lipids within different membrane compartments outlines the specific role of molecular species at the cellular level. The figure shows the site of synthesis of the major phospholipids and other lipids that are involved in signalling and organelle recognition pathways. Adapted from [1–5]

charged phosphatidylinositol phosphates (PIPs) localise on the inner leaflet to recruit proteins kinases to the plasma membrane, and lipid microdomains enriched with sphingolipids and cholesterol are important to stabilise protein complexes, underpinning their dynamic role [6]. Very little is known about the significance of the diversity in the lipid acyl chains, arising from the degree of unsaturation and chain length. Nevertheless, some proteins have evolved to recognise specific lipid features to modulate their activity, e.g. PH domains, cholesterol consensus motifs in GPCRs [7], or sphingolipid-binding motifs [8].

No longer regarded as bystanders, lipids modulate the function of proteins through lipid–protein interactions. Accordingly, changes in membrane lipid composition can affect membrane-associated proteins, membrane fusion, vesicle transport, neurotransmitter uptake, and have been implicated in the pathophysiology of many neurodegenerative disorders [9, 10]. Although lipids are major constituents of the brain, their roles in the pathogenesis of several neurodegenerative disorders are not fully understood. Lipid mediators such as prostaglandins, leukotrienes, lysophosphatidic acid (LPA), and sphingosine1-phosphate play essential roles in immune regulation, brain function, cell proliferation and development [11]. Therefore, understanding the role of lipids in neuronal communication has important implications for human health, drug development and early diagnosis and treatment of disease.

Membrane lipid composition and lipid-based signalling are fundamental to neural function. Chemical synaptic transmission between neurons relies on the fusion of neurotransmitter-containing secretory vesicles with the plasma membrane upon influx of calcium (Ca^{2+}) ions, following an action potential. This process forms the basis of neuronal communication and the formation of memory. Secretory vesicles are storage compartments for neurotransmitters such as neuropeptides and hormones. The exocytic mechanism which lead these vesicles to release their content into the extracellular space includes a series of trafficking, tethering, docking, priming and fusion events, underpinned by complex protein–protein interactions (e.g. N-ethylmaleimide-sensitive factor attachment proteins receptors (SNARE) proteins). The role played by protein–lipid interactions in this process is not yet fully understood. The lipid composition of secretory vesicles and the plasma membrane are both important for neurotransmitter release, as they control the spatial coordination of proteins and the fusion reaction. Recent reports point to the dysregulation of lipid metabolism as triggering factors in a number of neurological disorders such as Alzheimer's [10] and Parkinson's diseases [9]. These insights are advancing our understanding of the action of lipids in cellular processes. Several categories of lipids have been implicated in exocytosis, including polyunsaturated fatty acids [12], phosphatidic acids [13], cholesterol [14] and phospholipids [15]—particularly sphingolipids [16] and low-abundance signalling lipids such as phosphoinositides [17–24] .

To understand the lipidome, and how lipids affect diverse cellular processes, it is important to characterise and quantify them, both collectively and individually. We therefore turned to a rapid and sensitive monitoring of the lipid composition in of complex tissue extracts such as cells and brain tissue using mass spectrometry (MS) as the method of choice for lipidomics. Exocytosis is defined as the fusion of an intracellular trafficking vesicle with the target plasma membrane and is a fundamental cellular process involved in many physiological functions including neurotransmission and hormone release. Regulated exocytosis occurs when an appropriate secondary messenger, such as an increase in intracellular Ca^{2+}, triggers their fusion with the plasma membrane [25]. Secretory vesicles formed at the trans-Golgi network are transported to the plasma membrane through various stages. Among the docking and priming stages, a number of protein–protein and protein–lipid interactions are involved [20, 26]. The two main groups of protein families involved are SNARE proteins and Sec1/Munc18 proteins [27]. These groups of proteins are evolutionarily conserved and are involved in all known membrane fusion events of eukaryotic cells [28]. As the primary constituents of the plasma membrane, lipids are intrinsically linked to the modifications required for membrane fusion during exocytosis. The plasma membrane is made up of a combination of glyco-sphingolipids, glycerophospholipids, cholesterol and protein receptors organised in glycolipoprotein domains termed as lipid rafts [29–31]. These specialised domains serve as organising centres for the assembly of signalling molecules capable of regulating several essential functions such as neurotransmission [30]. The key role of lipid rafts in regulated exocytosis is evidenced by their association with the SNARE proteins [32]. Several categories of phospholipids present in lipid rafts have been implicated in exocytosis, including phosphoinositides [18, 33], or polyunsaturated fatty acids [34] that are clipped from phospholipids by different types of phospholipases A. Moreover, the addition of exogenous lipids also affects membrane fusion reactions [35, 36]. Hence these observations clearly point to the involvement of phospholipids and phospholipases in controlling the exocytic mechanism. It appears that cells have developed complex mechanisms where proteins regulate phospholipids and phospholipases, which in turn regulate proteins, to provide a precisely controlled exocytosis process, where and when it is needed [26]. Moreover, a precise understanding of the regulation of the lipid environment during exocytosis is critical because changes in the dynamic balance of lipids and the cascade of downstream events could affect fundamental processes such as learning and memory [37], as well as pathophysiological conditions such as in Parkinson's [9] and Alzheimer's diseases [10]. Therefore, we aimed to unravel the changes in the membrane lipid landscape associated with neuroexocytosis.

10.2 Role of Phospholipases and Their Lipid Products in Exocytosis

Phospholipases (PL) are hydrolysing enzymes that hydrolyse various components of membrane phospholipid molecules. There are four major classes of PL, A, B, C and D, among them PLA_2, PLC and PLD are known to be involved in membrane exocytosis [34, 38–40]. PLA_1 and PLA_2 cleave the fatty acyl chain at the *sn*-1 and *sn-2* position of glycerophospholipids the membrane phospholipid leading to the release of free fatty acids (FFAs) [41–43]. PLD catalyses the cleavage of the terminal phosphodiester bond of phosphatidylcholine to release phosphatidic acid [44]. PLC hydrolyse PtdIns(4,5)P_2 on the glycerol side of the phosphodiester bond for the formation of diacylglycerides (DAG) and inositol triphosphate (IP_3) [45]. Here we will review the literature on what is known about the role of phospholipases and their products in exocytosis.

10.3 Phosphatidic Acids (PA)

Phosphatidic acids are a class of glycerophospholipids, which have a small and negatively charged head group. They are the precursors of many other lipids but are only present in small amounts in mammalian cells and were shown as key metabolites in lipid biosynthesis [46]. PA can be synthesised through three alternative pathways: first by breaking the phosphatidylcholine through a PLD enzyme, second by the phosphorylation of diacylglycerol through a diacylglycerol kinase enzyme, and third by the acylation of LPA through LPA acyl transferase (LPAAT) [47]. Among them, PLD is the only phospholipase to be involved in the late stages of exocytosis [48]. PLD knockdown studies have shown the inhibition of secretory vesicle fusion to the membrane [44], which suggests that phosphatidic acid synthesis is closely related to membrane fusion. A possible explanation comes from the accumulation of phosphatidic acid at the fusion site with the SNARE complex that has been associated to it [49]. When the associated SNARE protein region is mutated (i.e. the polybasic juxtamembrane region of syntaxin-1), the binding of phosphatidic acid is prevented [49]. With the help of PLA_2 enzyme catalysis, phosphatidic acids can be the precursors of polyunsaturated fatty acids such as arachidonic acid, another potentiator of exocytosis that also interacts with the SNARE proteins [41]. In addition, phosphatidic acid can be hydrolysed to be a precursor for diacylglycerol [34], a secondary messenger involved in regulating Munc-13 proteins with high affinity [50], suggesting a complex interaction between proteins and lipids during exocytosis.

10.4 Diacylglycerides (DAG)

DAG contain two fatty acyl chains covalently bonded to a glycerol molecule through ester linkages. DAG are important intermediates in the synthesis and degradation of triglycerides, glycerophospholipids and glycerol-glycolipids [50]. Unsubstituted diacylglycerol is an essential secondary messenger in mammalian cells. Diacylglycerol can be synthesised by hydrolysis of phosphatidic acid [34] and cleavage of PtdIns(4,5) P_2 by PI-PLC enzyme [45]. This can regulate several target proteins, one of which belongs to the calcium-dependent protein kinase C (PKC) family [50, 51] and this PKC family protein is known to phosphorylate Munc-18 upon stimulation [52]. Diacylglycerol also regulates Munc-13 proteins with high affinity at the C1 domain [50]. Mutation to this C1 domain decreases vesicle priming by inhibiting neuronal exocytosis [53]. Rolling blackout protein, which is enriched at the neuromuscular junction, can also regulate diacylglycerol [54].

10.5 Free Fatty Acids: Polyunsaturated Fatty Acids

Fatty acids are the primary building blocks of more structurally complex lipids such as triglycerides, phospholipids and cholesterol esters. Polyunsaturated fatty acids are considered to be beneficial to human diets due to their role in human health and disease. The brain contains large amount of fatty acids, 50% of which are polyunsaturated fatty acids [55]. Polyunsaturated fatty acids such as arachidonic (AA) and docosahexaenoic (DHA) acids are presumed to be present in equal amounts in the brain and also linked in neuronal exocytosis [55]. The interaction and incorporation of fatty acids with the cell membrane leads to membrane fluidity, membrane-bound enzyme activity, ion channel permeability, membrane fusion and neurotransmitter release [56]. The release of fatty acids in the cell membrane is catalysed by PLA_2 [34]. Examples of such mechanisms are seen in secretory cells where they release fatty acids from the *sn*-2 position of glycerophospholipids [26]. PLA_2 can occur in multiple ways, either cytosolic or secretory or Ca^{2+}-dependent [38, 57, 58]. Among them, Ca^{2+}-dependent stimulus is the major external signal for the PLA_2 enzyme to catalyse arachidonic release. Although polyunsaturated fatty acids are linked to neurotransmitter exocytosis [12], the molecular mechanisms underlying this process have been under intense scrutiny, and this also involves fatty acid-mediated protein-lipid interactions. Interestingly, early studies revealed that stimulation of exocytosis was accompanied by a parallel increase of arachidonic acid in chromaffin cells, also showing that AA was not found to be directly linked to exocytosis, but through an alternative pathway [59]. Later, AA was linked to Munc-18a and syntaxin 1a binding, which prevents the formation of a stable SNARE complex and subsequent membrane fusion [17].

Another example is docosahexaenoic acid (DHA), which is most abundant in the cortex [60, 61]. The main source of the DHA for neural cells is through diet and biosynthesis from essential linolenic acid. It can also be obtained from the cleavage of membrane phospholipids by phospholipases. DHA is the precursor for neuroprotectin D1 [62], which activates neurotrophins [63, 64], suggesting DHA's role in modulating cell signalling and survival [61, 65]. Since the discovery of the significance of polyunsaturated fatty acids in exocytosis, there has been great interest into the mechanisms underlying their involvement, including fatty acid-protein interactions.

10.6 Lysophospholipids in Exocytosis

Lysophospholipids (LPLs) are bioactive lipids with detergent properties, composed of a single fatty acid bound to glycerol and a polar phosphatidyl head group. LPLs are generated along with FFAs following cleavage of glycerophospholipids by various phospholipases activity of phospholipid substrates. There is good evidence that the activity of PLA_2 is required for the generation of LPLs during exocytosis, by hydrolysing phospholipids at the *sn*-2 position [41, 66, 67]. Phosphatidylcholine (PC) is a major substrate for the PLA_2 enzyme, which cleaves PC by releasing FFAs and lysophosphatidylcholine (LPC) [22, 67, 68]. These LPCs are inverted cone shaped lipids capable of generating positive curvature in the membrane bilayer, which either facilitate or inhibit exocytosis [42, 43]. However, LPCs are known to remain confined to the leaflet of the membrane bilayer in which they are produced and distributed asymmetrically with relevance to membrane dynamics, whereas the free FFA generated can equilibrate between the two sides of the membrane bilayer [22]. Examples of such mechanisms are seen in the snake presynaptic PLA_2 neurotoxins (SPANs), which hydrolyse the *sn-2* ester bond of PC to generate AA and LPC and lead to progressive paralysis at the neuromuscular junction by stimulating exocytosis and blocking endocytosis [22]. Importantly, the combined addition of LPC with FFA such as oleic acid shows similar effect as SPANs at the nerve terminals and mimics its paralytic effect [67, 68], suggesting the necessity of the LPLs in the outer leaflet for the fusion pore formation.

In recent years, PLD_1 has emerged as a major player in several cellular processes including the production of phosphatidic acids (PA) through hydrolysis of PC during membrane trafficking and cell signalling [44]. PA are central bioactive lipids that have been shown to promote negative curvature in plasma membranes [41, 44], and can be further metabolised into LPA by phospholipases and LPAAT activity [69]. PLA_2 and PLA_1 produce either 1-acyl-2-LPL or 2-acyl-1-LPL that are linked to the glycerol backbone either in the *sn*-1 or *sn*-2 position of phospholipid, respectively. Importantly, it has been shown that phosphatidic acid-specific PLA_1 uses PA as a preferred substrate to generate LPA [70, 71]. Moreover, LPL receptors are able to discriminate between 1-acyl and 2-acyl LPL species [72], suggesting that these pathways are likely to strongly impact on the landscape of phospholipids and LPLs, thereby significantly altering the fusogenicity of secretory vesicles.

10.7 Emergence of Lipidomics Impact

Understanding the role of the highly heterogeneous array of lipid species that are involved in multiple and sometimes overlapping biological functions is a huge and technically challenging task [73, 74]. The emerging field of lipidomics, based on advances in mass spectrometry, is starting to provide some answers to this problem through detection and characterisation of all lipid classes and species in cells, organism tissue and even subcellular fractions. Mass spectrometry (MS) has acquired a well-deserved importance in biology, particularly since the development of 'soft' ionisation techniques such as electrospray (ESI), an invention duly rewarded with the Nobel Prize in Chemistry in 2002 attributed to John Bennett Fenn and Koichi Tanaka [75]. Their work enabled the characterisation of intact biomolecules particularly proteins and peptides, as well as lipids. The advantages of electrospray combined with tandem MS can be summarised by its high specificity and sensitivity (down to fmol). Molecules are ionised in the gas phase and selected according to their mass-to-charge (*m/z*) ratio. A typical mass spectrum shows the relative abundance of detected ions as a function of their *m/z* ratio. As a result of the mass overlap of many lipids, accurate mass alone is not sufficient to identify species, and collision-induced fragmentation is used to characterise their structure. This provides a very specific *mass signature*.

A tandem mass spectrometer consists of three main components, ion source, mass analyser and detector. The ion source converts molecules to ions, which can be manipulated by alternating electric fields along a quadrupole, and stabilised by resonance (depending on *m/z*). A mass spectrum typically gives the abundance of ions across the mass range. Additional structural information is obtained using an intermediate collision cell filled with an inert collision gas, such as nitrogen or argon, to break down the molecule into smaller characteristic mass fragments. This feature enables a series of scanning modes for lipid profiling experiments (Fig. 10.2).

Thanks to the development and advances in mass spectrometry, the field has started to move from an inferably biased view of particular lipid molecules to a

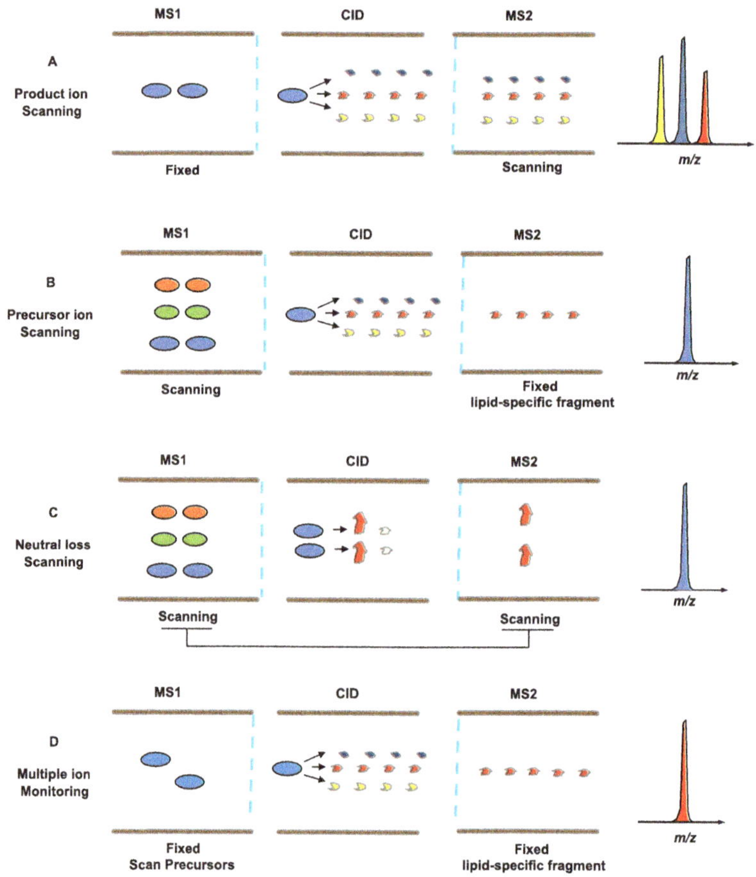

Fig. 10.2 Schematic representation of tandem mass spectrometry experiments. Adapted from [76]. Product scan (**a**) can help determine the fatty acyl fragments of phospholipids in the negative ion mode. On the other hand, a precursor scan (**b**) or neutral loss scan (**c**) can be used to profile a large number of phospholipid precursors, which contain any particular fatty acid fragment, or a specific head group [77]. A targeted approach using specific mass pairs can be used to identify several lipids of interest concomitantly (**d**)

comprehensive and deeper profiling of the lipidome, which will ultimately lead to a better understanding of the phenotypes and molecular mechanisms of disease [10, 78, 79].

10.8 State of Affair for the Detection of Free Fatty Acids

Profiling and quantification of carboxylic acid-containing lipid intermediates such as fatty acids and their metabolites (e.g. eicosanoids) is of major significance to understand a number of diseases involving phospholipases. Therefore, identification and characterisation of these compounds has both important physiological and clinical implications. Traditionally, FFAs are measured as their methyl esters (FAME) by gas chromatography/mass spectrometry (GC/MS) using electron impact ionisation [81–85] because FFAs are too polar and GC/MS is better suited for volatile compounds. Advantages of this technique are the high resolution of gas chromatography and the large number of species analysed concomitantly [84–86]. However, electron impact ionisation leads to substantial fragmentation, where the molecular ion is mostly absent and identification is based on matching the mass spectrum fingerprint to a database. Recently, liquid chromatography mass spectrometry (LC/MS) and the advent of soft ionisation (electrospray), has allowed the analysis of many lipid classes including FFAs by identifying the intact molecular ion or its adduct [87, 88]. Nevertheless, the LC/MS analysis of FFAs in their native form is deceiving due to their high polarity and their tendency to lose water or decarboxylate, and limited ionisation of the carboxylic group leading to poor sensitivity [89–92]. Moreover, analysing samples separately and comparing signal intensities of different conditions could result in inter-assay variability from differences in injection amounts, analyte stability and instrument sensitivity [93]. Therefore, several recent studies have concentrated on chemical derivatisation of carboxylic group of fatty acids to improve the LC/MS detection, specificity and sensitivity [80, 93–96]. The advantage of the derivatisation approach is that internal standards have the same chromatographic properties as the analytes but can still be differentiated from the analyte of interest on the basis of the isotopic mass difference. However, these methodologies are not amenable to multiplexing and were limited to the comparison of two separate conditions. As a result, we developed a multiplex approach aiming at providing both absolute and relative measurements of more than two samples simultaneously in complex matrices with internal standards in one analytical run [97].

10.9 State of Affair for the Detection of Lysophospholipids

LPLs are composed of a glycerol backbone connected to a polar phosphatidyl head group and a single fatty acid, differing in either its chain length and/or degree of unsaturation [72]. The phospholipase enzymes such as PLA_1 and PLA_2 produce either at the *sn*-1 or *sn*-2 position of glycerophospholipids, respectively (Fig. 10.3).

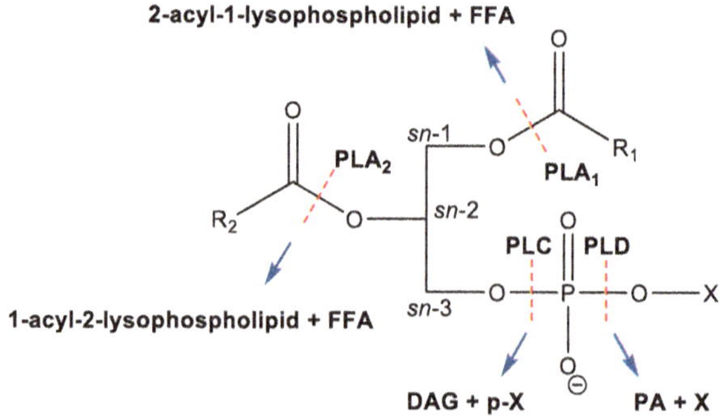

Fig. 10.3 Specificity of phospholipases in the hydrolysis of glycerophospholipids. PLA$_1$ and PLA$_2$ release free fatty acids (FFAs) by hydrolysing the *sn*-1 and *sn*-2 fatty acyl ester bonds leading to 2-acyl- and 1-acyl-lysophospholipids, respectively, while PLC cleaves the glycerophosphoester bond to form diacylglycerols (DAG) and the phosphorylated head group (p-X), and PLD hydrolyses off the head group (X) to release phosphatidic acids (PA)

These bioactive lipids can either facilitate or inhibit exocytosis [42, 43] according to their biophysical properties defining membrane curvature (head group, acyl chain composition and position).

Conventional methods used to measure phospholipase activity in biological samples include bioassays using radiolabelled substrates [98], indirect measurement of LPL by analysing hydrolysed fatty acids by GC/MS after thin layer chromatography (TLC) purification [99], ESI-MS through syringe infusion [73] and two-dimensional TLC [100]. LC/MS methods have been developed for more targeted sensitive and reproducible procedures to quantify LPLs, although they do not provide information about the regioisomers of LPLs [101, 102]. This is mainly due to the high diversity of molecular species in each LPL class and co-elution of the 1- and 2-acyl isomers on reverse phase (C18) columns, even by 2D chromatography [103, 104]. However, it was discovered that their separation could be achieved by hydrophilic interaction liquid chromatography (HILIC) [105, 106]. This type of chromatography is particularly well suited for the analysis of polar lipids such as LPLs. LPLs are labile and prone to intramolecular acyl conversion between *sn*-1 and *sn*-2 positions within minutes [72], which means care is necessary when handling them (snap freezing in liquid N_2 and acidified extraction). A novel LC/MS method was developed to measure LPL species and to determine their fatty acyl chain composition and *sn*-position on the glycerol backbone with high accuracy, adapted from a method previously described [105]. This procedure also utilises the recent procedure developed by Baker and colleagues to efficiently recover and preserve the LPL content [107]. We adapted this unbiased method to carry out a comprehensive profiling of different LPLs and FFAs during neuroexocytosis [97, and unpublished results]. The role of PA-PLA$_1$ and 2-acyl-1-LPLs has been discussed in terms of regulating vesicle formation and trafficking, although the exact mechanisms of inducing membrane curvature remain unclear [34]. This underpins the importance to accurately measure the levels of LPL and FFA to understand which and how these species might recruit effector proteins to the membrane and modify membrane properties to induce vesicle fusion [108].

10.10 Conclusion

Exocytosis is a multidimensional process involved in the release of neurotransmitters but also a myriad of other intra- and intercellular communication processes such as exosome release. It involves a complex series of protein–protein and protein–lipid interactions. Our understanding of the exocytotic mechanisms has been hampered by the lack of specific lipid changes occurring during this process. Recent findings suggest that LPA appears to play a major role in the fusogenicity of secretory vesicles. MS lipid profiling is likely to play a critical role in unravelling the changes occurring in the lipidome during stimulation of neuroexocytosis. Furthermore, MS lipid profiling is increasingly seen as a powerful tool to gain a deeper understanding of physiologic and pathogenic mechanisms affecting neuronal and more generally cellular functions. With neurodegenerative diseases on the rise, research in this field has tremendous physiological and clinical implications.

Acknowledgements The authors would like to thank Rachel Gormal and Rowan Tweedale for helping in the preparation of the chapter. This work was supported by a National Health and Medical Research Council (NHMRC) project grant (APP1058769). F.A.M. is a NHMRC Senior Research Fellow (APP1060075).

References

1. van Meer G, Voelker DR, Feigenson GW (2008) Membrane lipids: where they are and how they behave. Nat Rev Mol Cell Biol 9(2):112–124
2. Kobayashi T, Yamaji-Hasegawa A, Kiyokawa E (2001) Lipid domains in the endocytic pathway. Semin Cell Dev Biol 12(2):173–182
3. Haucke V, Di Paolo G (2007) Lipids and lipid modifications in the regulation of membrane traffic. Curr Opin Cell Biol 19(4):426–435
4. Iaea DB, Maxfield FR (2015) Cholesterol trafficking and distribution. Essays Biochem 57:43–55
5. Xu P et al (2013) Phosphatidylserine flipping enhances membrane curvature and negative charge required for vesicular transport. J Cell Biol 202(6):875–886
6. Shevchenko A, Simons K (2010) Lipidomics: coming to grips with lipid diversity. Nat Rev Mol Cell Biol 11(8):593–598
7. Paila YD, Tiwari S, Chattopadhyay A (2009) Are specific nonannular cholesterol binding sites present in G-protein coupled receptors? Biochim Biophys Acta 1788(2):295–302
8. Contreras FX et al (2012) Molecular recognition of a single sphingolipid species by a protein's transmembrane domain. Nature 481(7382):525–529
9. Jankovic J, Stacy M (2007) Medical management of levodopa-associated motor complications in patients with Parkinson's disease. CNS Drugs 21(8):677–692
10. Chan RB et al (2012) Comparative lipidomic analysis of mouse and human brain with Alzheimer disease. J Biol Chem 287(4):2678–2688
11. Shimizu T (2009) Lipid mediators in health and disease: enzymes and receptors as therapeutic targets for the regulation of immunity and inflammation. Annu Rev Pharmacol Toxicol 49:123–150
12. Lesa GM et al (2003) Long chain polyunsaturated fatty acids are required for efficient neurotransmission in *C. elegans*. J Cell Sci 116(Pt 24):4965–4975

13. Raghu P et al (2009) Rhabdomere biogenesis in Drosophila photoreceptors is acutely sensitive to phosphatidic acid levels. J Cell Biol 185(1):129–145
14. Cutler RG et al (2004) Involvement of oxidative stress-induced abnormalities in ceramide and cholesterol metabolism in brain aging and Alzheimer's disease. Proc Natl Acad Sci U S A 101(7):2070–2075
15. Wen PJ, Osborne SL, Meunier FA (2011) Dynamic control of neuroexocytosis by phosphoinositides in health and disease. Prog Lipid Res 50(1):52–61
16. Rohrbough J et al (2004) Ceramidase regulates synaptic vesicle exocytosis and trafficking. J Neurosci 24(36):7789–7803
17. Rickman C, Davletov B (2005) Arachidonic acid allows SNARE complex formation in the presence of Munc 18. Chem Biol 12(5):545–553
18. Osborne SL, Meunier FA, Schiavo G (2001) Phosphoinositides as key regulators of synaptic function. Neuron 32(1):9–12
19. Meunier FA et al (2005) Phosphatidylinositol 3-kinase C2alpha is essential for ATP-dependent priming of neurosecretory granule exocytosis. Mol Biol Cell 16(10):4841–4851
20. Osborne SL, Wen PJ, Meunier FA (2006) Phosphoinositide regulation of neuroexocytosis: adding to the complexity. J Neurochem 98(2):336–342
21. Osborne SL et al (2008) PIKfyve negatively regulates exocytosis in neurosecretory cells. J Biol Chem 283(5):2804–2813
22. Osborne SL, Meunier FA (2008) Lipids and secretory vesicle exocytosis. In: Wang Z-W (ed) Molecular mechanisms of neurotransmitter release. Springer, New York, pp 239–261
23. Wen PJ et al (2008) Ca2+–regulated pool of phosphatidylinositol-3-phosphate produced by phosphatidylinositol 3-kinase C2alpha on neurosecretory vesicles. Mol Biol Cell 19(12):5593–5603
24. Wen PJ, Osborne SL, Meunier FA (2012) Phosphoinositides in neuroexocytosis and neuronal diseases. Curr Top Microbiol Immunol 362:87–98
25. Jahn R, Fasshauer D (2012) Molecular machines governing exocytosis of synaptic vesicles. Nature 490(7419):201–207
26. Rituper B, Davletov B, Zorec R (2010) Lipid–protein interactions in exocytotic release of hormones and neurotransmitters. Clin Lipidol 5(5):747–761
27. Gerber SH et al (2008) Conformational switch of syntaxin-1 controls synaptic vesicle fusion. Science 321(5895):1507–1510
28. Hong W (2005) SNAREs and traffic. Biochim Biophys Acta 1744(3):493–517
29. Lucero HA, Robbins PW (2004) Lipid rafts-protein association and the regulation of protein activity. Arch Biochem Biophys 426(2):208–224
30. Salaun C, James DJ, Chamberlain LH (2004) Lipid rafts and the regulation of exocytosis. Traffic 5(4):255–264
31. Levental I, Grzybek M, Simons K (2010) Greasing their way: lipid modifications determine protein association with membrane rafts. Biochemistry 49(30):6305–6316
32. Lang T et al (2001) SNAREs are concentrated in cholesterol-dependent clusters that define docking and fusion sites for exocytosis. EMBO J 20(9):2202–2213
33. Wenk MR, De Camilli P (2004) Protein-lipid interactions and phosphoinositide metabolism in membrane traffic: insights from vesicle recycling in nerve terminals. Proc Natl Acad Sci U S A 101(22):8262–8269
34. Darios F, Connell E, Davletov B (2007) Phospholipases and fatty acid signalling in exocytosis. J Physiol 585(Pt 3):699–704
35. Verhage M (2005) Fatty acids add grease to exocytosis. Chem Biol 12(5):511–512
36. Amatore C et al (2006) Regulation of exocytosis in chromaffin cells by trans-insertion of lysophosphatidylcholine and arachidonic acid into the outer leaflet of the cell membrane. Chembiochem 7(12):1998–2003
37. Lin RC, Scheller RH (2000) Mechanisms of synaptic vesicle exocytosis. Annu Rev Cell Dev Biol 16:19–49
38. Brown WJ, Chambers K, Doody A (2003) Phospholipase A2 (PLA2) enzymes in membrane trafficking: mediators of membrane shape and function. Traffic 4(4):214–221

39. Vitale N et al (2001) Phospholipase D1: a key factor for the exocytotic machinery in neuroendocrine cells. EMBO J 20(10):2424–2434
40. Rossetto O et al (2006) Presynaptic enzymatic neurotoxins. J Neurochem 97(6):1534–1545
41. Latham CF et al (2007) Arachidonic acid potentiates exocytosis and allows neuronal SNARE complex to interact with Munc 18a. J Neurochem 100(6):1543–1554
42. Megighian A et al (2007) A lysolecithin/fatty acid mixture promotes and then blocks neurotransmitter release at the *Drosophila melanogaster* larval neuromuscular junction. Neurosci Lett 416(1):6–11
43. Kozlov MM, McMahon HT, Chernomordik LV (2010) Protein-driven membrane stresses in fusion and fission. Trends Biochem Sci 35(12):699–706
44. Zeniou-Meyer M et al (2007) Phospholipase D1 production of phosphatidic acid at the plasma membrane promotes exocytosis of large dense-core granules at a late stage. J Biol Chem 282(30):21746–21757
45. Poccia D, Larijani B (2009) Phosphatidylinositol metabolism and membrane fusion. Biochem J 418(2):233–246
46. Chasserot-Golaz S et al (2010) Lipid dynamics in exocytosis. Cell Mol Neurobiol 30(8):1335–1342
47. Freyberg Z, Siddhanta A, Shields D (2003) "Slip, sliding away": phospholipase D and the Golgi apparatus. Trends Cell Biol 13(10):540–546
48. Humeau Y et al (2001) A role for phospholipase D1 in neurotransmitter release. Proc Natl Acad Sci U S A 98(26):15300–15305
49. Lam AD et al (2008) SNARE-catalyzed fusion events are regulated by Syntaxin1A-lipid interactions. Mol Biol Cell 19(2):485–497
50. Brose N, Rosenmund C (2002) Move over protein kinase C, you've got company: alternative cellular effectors of diacylglycerol and phorbol esters. J Cell Sci 115(Pt 23):4399–4411
51. Wakelam MJ (1998) Diacylglycerol—when is it an intracellular messenger? Biochim Biophys Acta 1436(1–2):117–126
52. Burgoyne RD et al (2009) The functions of Munc18-1 in regulated exocytosis. Ann N Y Acad Sci 1152:76–86
53. Rhee JS et al (2002) Beta phorbol ester- and diacylglycerol-induced augmentation of transmitter release is mediated by Munc13s and not by PKCs. Cell 108(1):121–133
54. Huang FD et al (2004) Rolling blackout, a newly identified PIP2-DAG pathway lipase required for Drosophila phototransduction. Nat Neurosci 7(10):1070–1078
55. Denis I et al (2013) Omega-3 fatty acids and brain resistance to ageing and stress: body of evidence and possible mechanisms. Ageing Res Rev 12(2):579–594
56. Horrocks LA, Farooqui AA (2004) Docosahexaenoic acid in the diet: its importance in maintenance and restoration of neural membrane function. Prostaglandins Leukot Essent Fatty Acids 70(4):361–372
57. Creutz CE (1981) cis-Unsaturated fatty acids induce the fusion of chromaffin granules aggregated by synexin. J Cell Biol 91(1):247–256
58. Farooqui AA et al (1997) Phospholipase A2 and its role in brain tissue. J Neurochem 69(3):889–901
59. Sontag JM et al (1991) A pertussis-toxin-sensitive protein controls exocytosis in chromaffin cells at a step distal to the generation of second messengers. Biochem J 274(Pt 2):339–347
60. Bazan NG (2006) Cell survival matters: docosahexaenoic acid signaling, neuroprotection and photoreceptors. Trends Neurosci 29(5):263–271
61. Scott BL, Bazan NG (1989) Membrane docosahexaenoate is supplied to the developing brain and retina by the liver. Proc Natl Acad Sci U S A 86(8):2903–2907
62. Calandria JM et al (2009) Selective survival rescue in 15-lipoxygenase-1-deficient retinal pigment epithelial cells by the novel docosahexaenoic acid-derived mediator, neuroprotectin D1. J Biol Chem 284(26):17877–17882
63. Marcheselli VL et al (2010) Neuroprotectin D1/protectin D1 stereoselective and specific binding with human retinal pigment epithelial cells and neutrophils. Prostaglandins Leukot Essent Fatty Acids 82(1):27–34

64. Mukherjee PK et al (2004) Neuroprotectin D1: a docosahexaenoic acid-derived docosatriene protects human retinal pigment epithelial cells from oxidative stress. Proc Natl Acad Sci U S A 101(22):8491–8496
65. Bazan NG (2003) Synaptic lipid signaling: significance of polyunsaturated fatty acids and platelet-activating factor. J Lipid Res 44(12):2221–2233
66. Wei S et al (2003) Group IIA secretory phospholipase A2 stimulates exocytosis and neurotransmitter release in pheochromocytoma-12 cells and cultured rat hippocampal neurons. Neuroscience 121(4):891–898
67. Rigoni M et al (2005) Equivalent effects of snake PLA2 neurotoxins and lysophospholipid-fatty acid mixtures. Science 310(5754):1678–1680
68. Caccin P et al (2006) Reversible skeletal neuromuscular paralysis induced by different lysophospholipids. FEBS Lett 580(27):6317–6321
69. Gallop JL, Butler PJ, McMahon HT (2005) Endophilin and CtBP/BARS are not acyl transferases in endocytosis or Golgi fission. Nature 438(7068):675–678
70. Higgs HN, Glomset JA (1996) Purification and properties of a phosphatidic acid-preferring phospholipase A1 from bovine testis. Examination of the molecular basis of its activation. J Biol Chem 271(18):10874–10883
71. Baba T et al (2014) Phosphatidic acid (PA)-preferring phospholipase A1 regulates mitochondrial dynamics. J Biol Chem 289(16):11497–11511
72. Okudaira M et al (2014) Separation and quantification of 2-acyl-1-lysophospholipids and 1-acyl-2-lysophospholipids in biological samples by LC-MS/MS. J Lipid Res 55(10):2178–2192
73. Han X, Gross RW (1996) Structural determination of lysophospholipid regioisomers by electrospray ionization tandem mass spectrometry. J Am Chem Soc 118:451–457
74. Fahy E et al (2005) A comprehensive classification system for lipids. J Lipid Res 46(5):839–861
75. Fenn JB et al (1989) Electrospray ionization for mass spectrometry of large biomolecules. Science 246(4926):64–71
76. Domon B, Aebersold R (2006) Mass spectrometry and protein analysis. Science 312(5771):212–217
77. Ozbalci C, Sachsenheimer T, Brugger B (2013) Quantitative analysis of cellular lipids by nano-electrospray ionization mass spectrometry. Methods Mol Biol 1033:3–20
78. Wenk MR (2010) Lipidomics: new tools and applications. Cell 143(6):888–895
79. Narayanaswamy P et al (2014) Lipidomic "deep profiling": an enhanced workflow to reveal new molecular species of signaling lipids. Anal Chem 86(6):3043–3047
80. Leng J et al (2013) A highly sensitive isotope-coded derivatization method and its application for the mass spectrometric analysis of analytes containing the carboxyl group. Anal Chim Acta 758:114–121
81. Li J et al (2009) Rapid transmethylation and stable isotope labeling for comparative analysis of fatty acids by mass spectrometry. Anal Chem 81(12):5080–5087
82. Griffiths WJ (2003) Tandem mass spectrometry in the study of fatty acids, bile acids, and steroids. Mass Spectrom Rev 22(2):81–152
83. Moldovan Z, Jover E, Bayona JM (2002) Systematic characterisation of long-chain aliphatic esters of wool wax by gas chromatography-electron impact ionisation mass spectrometry. J Chromatogr A 952(1–2):193–204
84. Christie WW (1998) Gas chromatography-mass spectrometry methods for structural analysis of fatty acids. Lipids 33(4):343–353
85. Eder K (1995) Gas chromatographic analysis of fatty acid methyl esters. J Chromatogr B Biomed Appl 671(1–2):113–131
86. Dodds ED et al (2005) Gas chromatographic quantification of fatty acid methyl esters: flame ionization detection vs. electron impact mass spectrometry. Lipids 40(4):419–428
87. Rezanka T (2000) Analysis of very long chain polyunsaturated fatty acids using high-performance liquid chromatography—atmospheric pressure chemical ionization mass spectrometry. Biochem Syst Ecol 28(9):847–856

88. Lee SH, Pettinella C, Blair IA (2006) LC/ESI/MS analysis of saturated and unsaturated fatty acids in rat intestinal epithelial cells. Curr Drug Metab 7(8):929–937
89. Hellmuth C et al (2012) Nonesterified fatty acid determination for functional lipidomics: comprehensive ultrahigh performance liquid chromatography-tandem mass spectrometry quantitation, qualification, and parameter prediction. Anal Chem 84(3):1483–1490
90. Johnson DW (2005) Contemporary clinical usage of LC/MS: analysis of biologically important carboxylic acids. Clin Biochem 38(4):351–361
91. Kanawati B, Schmitt-Kopplin P (2010) Exploring rearrangements along the fragmentation of glutaric acid negative ion: a combined experimental and theoretical study. Rapid Commun Mass Spectrom 24(8):1198–1206
92. Johnson DW, Trinh MU (2003) Analysis of isomeric long-chain hydroxy fatty acids by tandem mass spectrometry: application to the diagnosis of long-chain 3-hydroxyacyl CoA dehydrogenase deficiency. Rapid Commun Mass Spectrom 17(2):171–175
93. Koulman A et al (2009) Comparative high-speed profiling of carboxylic acid metabolite levels by differential isotope-coded MALDI mass spectrometry. Anal Chem 81(18):7544–7551
94. Yang WC, Adamec J, Regnier FE (2007) Enhancement of the LC/MS analysis of fatty acids through derivatization and stable isotope coding. Anal Chem 79(14):5150–5157
95. Lamos SM et al (2007) Relative quantification of carboxylic acid metabolites by liquid chromatography-mass spectrometry using isotopic variants of cholamine. Anal Chem 79(14):5143–5149
96. Li X, Franke AA (2011) Improved LC-MS method for the determination of fatty acids in red blood cells by LC-orbitrap MS. Anal Chem 83(8):3192–3198
97. Narayana VK et al (2015) Profiling of free fatty acids using stable isotope tagging uncovers a role for saturated fatty acids in neuroexocytosis. Chem Biol 22:1552–1561
98. Morikawa R et al (2007) Phospholipase A(1) assays using a radiolabeled substrate and mass spectrometry. Methods Enzymol 434:1–13
99. Tokumura A et al (1986) Involvement of lysophospholipase D in the production of lysophosphatidic acid in rat plasma. Biochim Biophys Acta 875(1):31–38
100. Yokoyama K, Shimizu F, Setaka M (2000) Simultaneous separation of lysophospholipids from the total lipid fraction of crude biological samples using two-dimensional thin-layer chromatography. J Lipid Res 41(1):142–147
101. Crack PJ et al (2014) Anti-lysophosphatidic acid antibodies improve traumatic brain injury outcomes. J Neuroinflammation 11:37
102. Salous AK et al (2013) Mechanism of rapid elimination of lysophosphatidic acid and related lipids from the circulation of mice. J Lipid Res 54(10):2775–2784
103. Lisa M, Cifkova E, Holcapek M (2011) Lipidomic profiling of biological tissues using offline two-dimensional high-performance liquid chromatography-mass spectrometry. J Chromatogr A 1218(31):5146–5156
104. Onorato JM et al (2014) Challenges in accurate quantitation of lysophosphatidic acids in human biofluids. J Lipid Res 55(8):1784–1796
105. Koistinen KM et al (2015) Quantitative lysophospholipidomics in human plasma and skin by LC-MS/MS. Anal Bioanal Chem 407(17):5091–5099
106. Cifkova E et al (2016) Hydrophilic interaction liquid chromatography-mass spectrometry of (lyso)phosphatidic acids, (lyso)phosphatidylserines and other lipid classes. J Chromatogr A
107. Baker DL et al (2001) Direct quantitative analysis of lysophosphatidic acid molecular species by stable isotope dilution electrospray ionization liquid chromatography-mass spectrometry. Anal Biochem 292(2):287–295
108. Bechler ME, de Figueiredo P, Brown WJ (2012) A PLA1-2 punch regulates the Golgi complex. Trends Cell Biol 22(2):116–124

Index

A
Acute myeloid leukaemia (AML), 152
Adult neurogenesis, 118, 121
  cannabinoids, 127–129
  overview, 126–130
Adult stem cells, 2, 15
Akt, 175, 177
Alzheimer's disease (AD), 17, 94, 102–104, 129, 130
aNSCs, 99, 100
ApoE, 98
Apolipoprotein (Apo), 97–98
Arachidonic acid (AA), 196, 197
2-Arachidonoylglycerol (2-AG), 118, 120, 123, 128, 130
Artery occlusion rat model, 130
Atypical PKC (aPKC), 15, 16
Autism, 104
Autocrine motility factor, 42
Autotaxin (ATX), 67, 75, 173, 174, 179
  B-like domains, 45
  cancer stem cells, 48–49
  cancer treatment, 44
  catalytic activity, 44
  embryonic stem cells, 46
  enzymatic activity, 45, 48, 49, 52, 53
  isoforms, 42
  knockout mice, 44
  neural stem cells, 50–51
  neurodevelopment, 43, 49–53
  neurogenesis, 50, 51
  NPPs, 42
  oligodendrogenesis, 51–53
  overview, 41
  pathophysiological process, 44
  protein products, 41
  role, 42, 43
  stem cell biology, 43, 45–49
  tissue-specific stem cells, 46–48
  tumor cells, 42, 43

B
Bioactive lipids, 2, 3, 143, 144
Biological effect, 139
B-like domains, 44, 45, 51
Blood–brain barrier (BBB), 98
Bone marrow (BM), 47, 48, 141, 142, 153, 160
Brain, 94–98, 101
Brain development, 121–125
Brain disease
  AD, 102–104
  autism, 104–105
Brain-derived neurotrophic factor (BDNF), 120
Breast cancer stem cell (BCSC), 157–159, 163
Bromodeoxyuridine (BrdU), 92

C
Cancer cells, 17, 69, 140, 144–145
Cancer stem cell (CSC), 48, 49, 172
  criteria, 152–153
  introduction, 152–155
  markers, 153–154
  oncogenes, 155
  sphingolipids, 156–163
Cannabidiol (CBD), 118, 128, 130
Cannabinergic drugs, 129
Cannabinoid receptor type 1 (CB_1), 118, 120, 121, 123–125, 128–130

A. Pébay, R.C.B. Wong (eds.), *Lipidomics of Stem Cells*, Stem Cell Biology and Regenerative Medicine, DOI 10.1007/978-3-319-49343-5

Cannabinoid receptor type 2 (CB_2), 118, 120, 121, 123, 129
Cannabinoids, 121–125
  adult neurogenesis, 127–129
  CB_1 receptor, 118, 120, 121, 124, 125, 128–130
  CB_2 receptor, 120, 121, 123, 129
  overview, 118–119
  signaling in developing brain
    morphogenesis, 121–123
    neuronal differentiation, 121–123
    pathological implications, 123–125
    proliferation, 121
Cannabis sativa, 118, 128
C-C double bond, 94
CD34-positive cells, 47, 48
Cellular birth-dating technique, 93
Central nervous system (CNS), 23, 45, 51, 67, 70, 71, 118, 119, 121
  lysophosphatidic acid, 173–174
Ceramide kinase (CERK), 138, 139, 145
Ceramide transport protein (CERT), 159
Ceramide-1-phosphate (C1P), 20, 138
  biological effect, 139
  EPCs, 142–144
  extracellular, 139
  HSPCs, 141–142
  induced migration in cancer cells, 140, 144–145
  intracellular, 139
  macrophages, 139–141
  MSCs, 142–144
  overview, 138
  prometastatic environment, 145
  serum, 139
  stromal cells, 142–144
  VSELs, 142–144
Ceramide-enriched compartments (CECs)
  ceramide, 12–17
    cilia, 15–16
    exosomes, 16–17
Ceramides, 12–17, 155, 162
  C1P, 20
  CECs (*see* Ceramide-enriched compartments (CECs))
  derivatives, 12–24
  glycosphingolipids, 21–22
  S1P, 17–20
Cerebrospinal fluid (CSF), 67
α-Chemokine stromal-derived factor 1 (SDF-1), 141, 143
Cholesterol, 97, 98
Cilia, 14–16
Ciliogenesis, 15
c-jun N-terminal kinases (JNK), 6
CNVs, 76
Corticogenesis, 73
Cytoplasmic FAs, 94, 96–97

D
Dentate gyrus (DG), 89–93, 98–100, 127
Derivatisation approach, 199
Desorption electrospray ionization (DESI), 107
Diacylglycerides/Diacylglycerols (DAG), 195, 200
Diacylglycerol lipase (DAGL), 120, 123
Docosahexaenoic acid (DHA), 100, 196
Docosapentaenoic acid (DPA), 100

E
ECBS, 120
Ecto-nucleotide pyrophosphatase/phosphodiesterase 2 (ENPP2), 41, 42
Edg family, 178
Eicosapentaenoic acid (EPA), 100
Embryo development, 12
Embryonic cortical development, 74
Embryonic development, 71, 72
Embryonic stem cells (ESCs), 2, 3, 5, 15, 16, 20, 46, 51, 156
Endocannabinoid (ECB) system, 123, 124
  neurodevelopmental role, 119–123
Endocannabinoid system
  expression, 120–121
Endocannabinoids (ECB), 127, 128, 130
Endothelial differentiation gene (EDG), 66, 173
Endothelial Progenitor Cells (EPCs), 142–144
Ependymal cells, 90
Epidermal growth factor (EGF), 74, 92, 179
Epidermal growth factor receptor (EGFR), 175–177
Epidermal growth factor receptor phosphodiesterase family member 2 (Enpp2), 68
Epileptogenesis, 124–125
Excitation/inhibition (E/I), 124
Exocytosis, 193–196
  lysophospholipids, 196–197
  role of phospholipases, 194–195
Exosomes, 16, 17, 23
Extracellular signal-regulated kinase (ERK), 156
Extracellular vesicles (EVs), 16, 17

F

FAAH, 120
Fatty acid amide hydrolase (FAAH), 120, 128
Fatty acid binding proteins (FABPs), 98
Fatty acid synthase (FASN), 99
Fatty acids (FAs), 94, 99
  classes on NSCs and neurogenesis, 100–102
  MUFAs, 101–102
  proliferation of NSCs, 99–100
  PUFAs, 100
  SFAs, 101
Fetal corticogenesis, 71, 72
Fetal intracranial hemorrhage (ICH), 75
Fibroblast growth factor (FGF), 120
Fingolimod (FTY720), 13, 23
18F-fluoro-deoxyglucose (FDG), 108
Fluoro-6-thio-heptadecanoic acid (FTHA), 108
Fragile X model, 125
Free fatty acids (FFAs), 194, 197, 200
  detection, 199
  polyunsaturated fatty acids, 196
FTY720, 160, 162, 163
Fumonisin B1 (FB1), 13
Fusarium, 22

G

G protein-coupled receptors (GPCRs), 66, 70, 71, 73
GABA, 120
GABAergic, 124
Galactosulfatide, 21
Galactosylceramide, 21
Gangliosides, 21, 157
Gas chromatography/mass spectrometry (GC/MS), 199
Gli shuttle, 15
Glioblastoma multiforme (GBM), 49, 173–175
Glioblastoma stem cells, 161, 162
Glioma stem cell theory, 172
Glioma stem cells, 158, 161–163
Gliomagenesis, 173, 174
Glucose-regulated protein 94 (Grp94), 19, 20
Glucosylceramide synthase (GCS), 155–159, 162, 163
Glycerophospholipids, 195
Glycogen synthase kinase 3β (GSK3), 15, 16
Glycosphingolipids (GSLs), 21, 22, 157
Golgi apparatus, 14, 138
G-protein coupled receptors (GPCRs), 193
G_α proteins, 69, 70
$G_{\alpha i}$ protein, 141, 144
$G\alpha_s$ proteins, 71

H

Haematopoietic stem and progenitor cells (HSPC), 160, 161
Haematopoietic stem cells (HSCs), 142, 154
Hair follicle stem cells, 47, 48
Heat shock protein 90 (HSP90), 19, 20
Hematopoietic stem cells, 47
Hematopoietic stem/progenitor cells (HSPCs), 141–142
Hemorrhagic injury, 75–76
Hippocampal neurogenesis, 127, 128, 130
Histone deacetylase (HDAC), 19, 20
Human ES cell (hESC), 6, 18
Human Lysophosphatidic acid GPCR genes, 69
HUVECs, 144
Hydrocephalus, 75–76
Hydrophilic interaction liquid chromatography (HILIC), 200
Hypoxic injury, 74

I

Imaging mass spectrometry (I-MS), 103, 107
Interkinetic nuclear migration (INM), 74
Intracerebroventricularly (ICV), 101, 102

L

Leukaemia stem cells, 153, 154, 162
L-glutamate, 72
Lipase H, 48
Lipid droplet, 95, 97
Lipid dyes, 105
Lipid homeostasis, 2
Lipidome, 193
Lipidomic experiment, 106
Lipidomics, 2, 106, 197–199
Lipids, 94, 106
  bioactive, 2, 3, 143, 144
  diversity, 192
  glycosphingolipids, 21–22
  lipid homeostasis, 2
  lysophospholipids, 3
  membrane, 193
  modulate function of proteins, 193
  morphogenetic, 12
  overview, 2
  sphingolipids, 13, 14, 21–24, 139
  uses, 2

Liquid chromatography mass spectrometry (LC/MS), 199, 200
LPA acyltransferase (LPAAT), 69
LPA receptor (LPAR), 173–175, 178
Lysophosphatidic acid (LPA), 20, 41–43, 45–51, 70–71, 73, 74, 193
  catabolism, 69
  CNS, 173–174
  distribution, 67
  EGFR/PI3K signalling, 175–179
  fetal corticogenesis, 72
  glioblastoma multiforme (GBM), 174–175
  metabolism, 67–69
  mitogenesis and neurogenesis, 73
  neural progenitor cell
    migration, 74
    survival, 73
  neurodevelopmental diseases, 74–76
  overview, 66
  potent neuromodulator, 72
  programmed cell death, 73
  receptors
    LPA_1, 70
    LPA_2, 70
    LPA_3, 70–71
    LPA_5, 71
    LPA_6, 71
  structure, 67
Lysophosphatidylcholine (LPC), 75, 196, 197
Lysophospholipase D (lysoPLD), 41–44, 52
Lysophospholipids (LPLs)
  detection, 199–200
  exocytosis, 196–197
  LPA, 3–4
  S1P, 3
Lysphosphatidic acid (LPA)
  overview, 3–4
  pluripotent stem cells, 5–7
  signaling, 4–5
  synthesis and degradation, 3–4

M

Macrophage chemoattractant protein-1 (MCP-1), 140
Macrophages, 139–141
Magnetic resonance spectroscopy (MRS), 106, 108
Mass signature, 197
Mass spectrometry (MS), 105–108, 197, 198
Matrix metalloprotease (MMP), 178
Matrix-assisted laser desorption ionization (MALDI), 107
Membrane type 1 matrix metalloprotease (MT1-MMP), 161
Mesenchymal stem cells (MSC), 47, 142–144, 157
Mitogen-activated protein kinase (MAPK), 177
Mitogenesis, 73
Mixed lineage leukaemia (MLL), 154
Modulator of oligodendrocyte differentiation and focal adhesion organization (MORFO), 43–45, 51–53
Monoacylglycerol (MAG), 69
Monoacylglycerol lipase (MAGL), 120, 123, 130
Monounsaturated fatty acids (MUFAs), 101, 102
Morphogenetic lipid, 12
Mouse (murine) ES cell (mESC), 5, 18, 20
Mouse models, 154, 162, 174
Multicellular organism, 45
Multiple sclerosis (MS), 23
Multiplex approach, 199

N

N-acyl phosphatidylethanolamine phospholipase D (NAPE-PLD), 120
Nerve growth factor (NGF), 70
Neural progenitor cell (NPC), 71–75
Neural progenitors (NPs), 120, 121
Neural stem cells (NSC), 50–51, 90–91, 173
  and adult neurogenesis, 88–94
  effects of FA classes, 100–102
  fatty acid metabolism, 99–100
  form and function
    DG, 91
    SVZ, 90–91
  heterogeneity within, 91–92
  human neurogenesis, 92–94
  neutral lipid, 98, 99
Neurodevelopment, 49, 50, 118
Neurodevelopmental diseases
  hemorrhagic injury, 75–76
  hydrocephalus, 75–76
  hypoxic injury, 74–75
  schizophrenia, 75
Neurogenesis, 50–51, 73, 99
  adult, 88–94
  effects of fatty acid classes on NSCs, 100–102
  human, 92–94
  neutral lipid, 98–99
Neurogenic niches, 126
Neuronal hyperexcitability, 124–125
Neuropsychiatric disorders, 125
Neurulation, 49
Neutral lipid, 94, 95, 97–102

adult brain
Apo, 97–98
FAs, 94
lipid droplet, 97
TAGs, 95
brain disease, 102–105
measurement techniques, 105–108
NSC maintenance and neurogenesis, 98–99
ApoE, 98–99
FABPs, 98
fatty acid, 100–102
Neutral sphingomyelinase (nSMase), 15
Niches, 46, 48, 120, 126
Normal stem cells, 157
Nucleotide pyrophosphatases/ phosphodiesterases (NPPs), 42

O
O1 epitope, 21
Oleic acid (OA), 100, 101, 103
Olfactory bulbs (OB), 90, 93
Oligodendrocyte precursor cells (OPCs), 15, 21
Oligodendrocytes, 44, 45, 178
Oligodendrogenesis, 51–53
Oncogenes, 155

P
Pancreatic cancer, 145
Peroxisome proliferator-activated receptor (PPAR), 3
Phosphatase tensin homologue (PTEN), 175
Phosphatidic acids (PA), 48, 141, 195, 197, 200
Phosphatidyl inositol phosphate (PIP), 24
Phosphatidylcholine (PC), 196
Phosphodiesterase Iα (PD-Iα), 42
Phosphoinositide 3-kinase (PI3K), 175–178
Phospholipase D (PLD), 138, 141
Phospholipases (PL), 194–195
Phospholipids, 192
Phytocannabinoids, 119, 125
Platelet-derived growth factor (PDGF), 6, 156
Pluripotency, 2, 3, 6, 15, 17–19
Pluripotent stem cells, 2, 3, 5, 6, 51
ESCs, 2
iPSCs, 2
role of LPA and S1P, 5–7
Polyunsaturated fatty acids (PUFAs), 100, 196
Positron emission topography (PET), 106, 108
Post-hemorrhagic hydrocephalus (PHH), 75
Programmed cell death (PCD), 76
Prohibitin 2 (PHB2), 20
Proliferator-activated receptors, 118
Protein kinase C (PKC), 195
Protein phosphatase 2A (PP2A), 16
Psychiatric disorders, 127
Pulmonary artery hypertension (PAH), 144

Q
Quiescent neural stem cells (qNSCs), 99, 100

R
Radial glia, 51
Reactive oxygen species (ROS), 154
Regenerative medicine, 22–24

S
S1P receptor 1 ($S1P_1$), 160
Saturated fatty acids (SFAs), 101
Schizophrenia, 75
Secondary ion mass spectrometry (SIMS), 107
Serotonin (5-HT), 75
Signaling lipids, 2
Snake presynaptic PLA_2 neurotoxins (SPANs), 197
Solid tumours, 172
Soluble N-ethylmaleimide-sensitive factor activating protein receptor (SNARE), 194, 195
Sphingolipid-induced protein scaffolds (SLIPs), 13
Sphingolipids, 13, 14, 21–24, 138, 139
CSC biology, 160–163
drug resistance, 159–160
maintenance of stemness, 156–159
overview, 155–156
self-renewal, 156–159
Sphingomyelin, 155, 156, 159, 160
Sphingomyelinase D (SMase D), 138
Sphingosine 1-phosphate (S1P), 155, 156, 158, 159, 161, 163, 174
Sphingosine kinase (SK), 155, 156, 162
Sphingosine kinase 1 (SphK1), 17
Sphingosine kinase 2 (SK2), 158
Sphingosine-1-phosphate (S1P), 66, 138, 193
ceramide, 17, 19, 20
extracellular, 18
intracellular, 19
overview, 3
pluripotent stem cells, 5, 6
signaling, 4
synthesis and degradation, 4
Stearoyl-CoA-desaturase (SCD-1), 99, 101

Stem cells, 2
  adult, 2
  biology, 45–49
  C1P-induced migration, 140
  cancer, 48–49
  differentiation, 12, 15–17
  embryonic, 46
  glioma, 158, 161
  neural, 50–51
  normal, 157
  pluripotent (*see* Pluripotent stem cells)
  therapy, 22–24
  tissue-specific, 46–48
Stemness, 156–159
Stochastic models, 154, 155
Stromal cells, 47, 142–144
Subgranular zone (SGZ), 126, 129
Subventricular zone (SVZ), 51, 52, 89–94, 99, 101, 104, 126, 129, 172, 175

T
Tamoxifen, 162
Tandem mass spectrometry, 198
Temozolomide (TMZ), 159
Δ^9-tetrahydrocannabinol (THC), 118, 119, 124, 125
Thin layer chromatography (TLC), 105, 106, 200
THP-1 cells, 140, 144
Tissue-specific stem cells, 46–48, 172
TMLHE, 104
Totipotency, 46
Transit amplifying progenitors (TAPs), 89, 90, 92
Triacylglycerols (TAGs), 95
Triple-transgenic Alzheimer's disease (3xTg-AD) mice, 103
Tumor cell, 42, 43
Tyrosine kinase receptor type 1 (TrkA), 70

V
Vascular endothelial growth factor (VEGF), 5, 68, 75
Ventricular zone (VZ), 72, 73
Very Small Embryonic-Like Stem Cells (VSELs), 142–144

X
Xenografts, 152, 157, 162

Z
Zombie cells, 23

CPSIA information can be obtained
at www.ICGtesting.com
Printed in the USA
BVOW07*2134200317
5364BVAU00002B/1/P

9 783319 493428